NATURAL ANTIOXIDANTS

Applications in Foods of Animal Origin

NATURAL ANTIOXIDANTS

Applications in Foods of Animal Origin

Edited by

Rituparna Banerjee, MVSc
Arun K. Verma, PhD
Mohammed Wasim Siddiqui, PhD

AAP | APPLE ACADEMIC PRESS

Apple Academic Press Inc.
3333 Mistwell Crescent
Oakville, ON L6L 0A2 Canada

Apple Academic Press Inc.
9 Spinnaker Way
Waretown, NJ 08758 USA

© 2017 by Apple Academic Press, Inc.
First issued in paperback 2021
Exclusive worldwide distribution by CRC Press, a member of Taylor & Francis Group
No claim to original U.S. Government works

ISBN-13: 978-1-77463-049-5 (pbk)
ISBN-13: 978-1-77188-459-4 (hbk)

Library and Archives Canada Cataloguing in Publication

Natural antioxidants : applications in foods of animal origin / edited by Rituparna Banerjee, MVSc, Arun K. Verma, PhD, Mohammed Wasim Siddiqui, PhD.
Includes bibliographical references and index.
Issued in print and electronic formats.
ISBN 978-1-77188-459-4 (hardcover).--ISBN 978-1-315-36591-6 (PDF)
1. Antioxidants. 2. Food of animal origin. 3. Food additives. I. Siddiqui, Mohammed Wasim, author, editor II. Banerjee, Rituparna, author, editor III. Verma, Arun K., author, editor

| TX553.A73N38 2017 | 613.2'86 | C2017-900924-9 | C2017-900925-7 |

Library of Congress Cataloging-in-Publication Data

Names: Banerjee, Rituparna, editor. | Verma, Arun K., editor. | Siddiqui, Mohammed Wasim, editor.
Title: Natural antioxidants : applications in foods of animal origin / editors, Rituparna Banerjee, MVSc, Arun K. Verma, PhD, Mohammed Wasim Siddiqui, PhD.
Description: Toronto : Apple Academic Press, 2017. | Includes bibliographical references and index.
Identifiers: LCCN 2017003500 (print) | LCCN 2017004858 (ebook) (print) | LCCN 2017004858 (ebook) | ISBN 9781771884594 (hardcover : alk. paper) | ISBN 9781315365916 (ebook)
Subjects: LCSH: Antioxidants. | Food--Preservation. | Oils and fats, Edible--Deterioration. | Lipids--Oxidation.
Classification: LCC TX553.A73 N386 2017 (print) | LCC TX553.A73 (ebook) | DDC 613.2/86--dc23
LC record available at https://lccn.loc.gov/2017003500

Apple Academic Press also publishes its books in a variety of electronic formats. Some content that appears in print may not be available in electronic format. For information about Apple Academic Press products, visit our website at **www.appleacademicpress.com** and the CRC Press website at **www.crcpress.com**

ABOUT THE EDITORS

Rituparna Banerjee, MVSc

Dr. Rituparna Banerjee is presently a scientist in the discipline of livestock products technology at the ICAR-National Research Centre on Meat, Hyderabad, India. Dr. Banerjee is primarily involved in research pertaining to meat science and technology. She has been awarded the prestigious Smt. Mira Mallik Gold Medal for securing the highest marks in the BVSc and AH course by the West Bengal University of Animal and Fishery Sciences. She is the only candidate to qualify on an All-India basis as a Scientist in Livestock Product Technology for Agricultural Research Service under the Indian Council of Agricultural Research in 2015. She has published 21 peer-reviewed research articles in international and national journals, three book chapters, five lead/invited papers, and 24 abstracts. She is associated with organizing training programs in microbial quality for meat food safety and value added meat products. She was trained in the fields of microbiology, chemistry, and technology of dairy functional foods and nutraceuticals and food quality and safety. Dr. Banerjee is also associated with five professional societies related to meat science and the effective dissemination of knowledge in the field of packaging of meat products. She also participated in several seminars and symposiums on meat science and food safety and has presented papers on these aspects.

Arun K. Verma, PhD

Dr. Arun K. Verma is currently working as a scientist in livestock products technology at ICAR-Central Research Institute for Research on Goats (ICAR), Makhdoom, Mathura, India. He is a graduate of Jawaharlal Nehru Krishi Vishwavidyalay, Jabalpur. He obtained his MVSc and PhD degrees from the Indian Veterinary Research Institute, Izatnagar, Bareilly, in Livestock Products Technology. He worked as an Assistant Professor (Livestock Products Technology) at Maharashtra Animal and Fishery Sciences, University, Nagpur, and later joined the Agricultural Research Service as a scientist in 2010. He has been working with functional meat products and food safety and traceability associated with goat products. He has 25 national and international peer-reviewed research papers, ten popular as well as review articles, and three book chapters to his credit. He has presented more than 25 papers at various seminars, symposia, and conferences.

Mohammed Wasim Siddiqui, PhD

Dr. Mohammed Wasim Siddiqui is an Assistant Professor and Scientist in the Department of Food Science and Post-Harvest Technology, Bihar Agricultural University, Sabour, India, and author or co-author of 31 peer-reviewed research articles, 26 book chapters, two manuals, and 18 conference papers. He has 11 edited and authored books to his credit, published by Elsevier, CRC Press, Springer, and Apple Academic Press. Dr. Siddiqui has established an international peer-reviewed journal, *Journal of Postharvest Technology*. He is Editor-in-Chief of two book series (Postharvest Biology and Technology and Innovations in Horticultural Science), published by Apple Academic Press. Dr. Siddiqui is also a Senior Acquisitions Editor in Apple Academic Press, New Jersey, USA, for horticultural science. He has been serving as an editorial board member and active reviewer of several international journals, including *LWT- Food Science and Technology* (Elsevier), *Food Science and Nutrition* (Wiley), *Acta Physiologiae Plantarum* (Springer), *Journal of Food Science and Technology* (Springer), *Indian Journal of Agricultural Science* (ICAR), etc.

Dr. Siddiqui received the Best Young Researcher Award (2015) by GRABS Educational Trust, Chennai, India, and the Young Scientist Award (2015) by Venus International Foundation, Chennai, India. He was also a recipient of the Young Achiever Award (2014) for outstanding research work by the Society for Advancement of Human and Nature (SADHNA), Nauni, Himachal Pradesh, India, where he is an honorary board member and life time author. He has been an active member of the organizing committees of several national and international seminars, conferences, and summits. He is one of key members in establishing the World Food Preservation Center (WFPC), LLC, USA, and is currently an active associate and supporter.

Dr. Siddiqui acquired a BSc (Agriculture) degree from Jawaharlal Nehru Krishi Vishwa Vidyalaya, Jabalpur, India. He received his MSc (Horticulture) and PhD (Horticulture) degrees from Bidhan Chandra Krishi Viswavidyalaya, Mohanpur, Nadia, India, with specialization in Postharvest Technology. He was awarded an Maulana Azad National Fellowship Award from the University Grants Commission, New-Delhi, India. He is a member of Core Research Group at the Bihar Agricultural University (BAU), which provides appropriate direction and assistance in prioritizing research. He has received several grants from various funding agencies to carry out his research projects. Dr. Siddiqui has been associated with postharvest technology and processing aspects of horticultural crops. He is very engaged in teaching (graduate and doctorate students) and research, and he has proved himself as an active scientist in the area of postharvest technology.

CONTENTS

LIST OF CONTRIBUTORS

Sumit Arora
Dairy Technology Division, ICAR-National Dairy Research Institute, Karnal 132001, Haryana, India

Ugochukwu Anyanwu
North Carolina Agricultural and Technical State University, Greensboro, NC, USA

Rituparna Banerjee
ICAR-National Research Centre on Meat, Chengicherla 500092, Hyderabad, India

Ajit Singh Bhatnagar
IGNOU Regional Centre, Durg, Chhattisgarh, India

A. K. Biswas
Division of Post-Harvest Technology, ICAR-Central Avian Research Institute, Izatnagar, Bareilly 243122, Uttar Pradesh, India

Sanket Borad
Dairy Technology Division, ICAR-National Dairy Research Institute, Karnal 132001, Haryana, India

Javier Carballo
Área de Tecnología de los Alimentos, Facultad de Ciencias de Ourense, Universidad de Vigo, Ourense 32004, Spain

Manat Chaijan
Food Technology and Innovation Center of Excellence, Department of Agro-Industry, School of Agricultural Technology, Walailak University, Thasala 80160, Nakhon Si Thammarat, Thailand

M. K Chatli
Department of Livestock Products Technology, GADVASU, Ludhiana 141004, Punjab, India

Ruben Domínguez
Centro Tecnológico de la Carne de Galicia, Rua Galicia No. 4, Parque Tecnológico de Galicia, San Cibrao das Viñas, Ourense 32900, Spain

Gauri Jairath
Department of Livestock Products Technology, LUVAS, Hisar 125001, Haryana, India

Tom Jones
Meat Products, Global Applications and Product Development, Kalsec®, Inc., Kalamazoo, MI 49006, USA. E-mail: tjones@kalsec.com

V. V. Kulkarni
ICAR-National Research Centre on Meat, Chengicherla 500092, Hyderabad, India

Rewa Kulshrestha
Bilaspur University, Bilaspur, Chhattisgarh, India

José M. Lorenzo
Centro Tecnológico de la Carne de Galicia, Rua Galicia No. 4, Parque Tecnológico de Galicia, San Cibrao das Viñas, Ourense 32900, Spain

Minaxi
Agricultural Structures and Environmental Control Division, ICAR- Central Institute of Post-Harvest Engineering and Technology, Ludhiana 141004, Punjab, India

B. M. Naveena
ICAR-National Research Centre on Meat, Chengicherla 500092, Hyderabad, India

Worawan Panpipat
Functional Food Research Unit, Department of Agro-Industry, School of Agricultural Technology, Walailak University, Thasala, Nakhon Si Thammarat 80160, Thailand

V. Rajkumar
Goat Products Technology Laboratory, ICAR-Central Institute for Research on Goats, Makhdoom, Farah, Mathura 281122, Uttar Pradesh, India

Mohammed Wasim Siddiqui
Department of Food Science and Postharvest Technology, Bihar Agricultural University, Sabour, Bhagalpur 813210, Bihar, India

Ashish Kumar Singh
Dairy Technology Division, ICAR-National Dairy Research Institute, Karnal 132001, Haryana, India

Reza Tahergorabi
Department of Family and Consumer Sciences, Food and Nutritional Sciences, North Carolina Agricultural and Technical State University, Greensboro, NC, USA

Neelam Upadhyay
Dairy Technology Division, ICAR-National Dairy Research Institute, Karnal 132001, Haryana, India

N. Veena
Dairy Chemistry Division, College of Dairy Science and Technology, Guru Angad Dev Veterinary and Animal Sciences University, Ludhiana 141001, Punjab, India

Arun K. Verma
Goat Products Technology Laboratory, ICAR-Central Institute for Research on Goats, Makhdoom, Farah, Mathura 281122, Uttar Pradesh, India

LIST OF ABBREVIATIONS

$^{\bullet}OH$	hydroxyl radical
$^1O_2$	singlet oxygen
4-NHE	4-hydroxynonenal
ABTS	2,2'-azino-bis3-ethylbenzothiazoline-6-sulphonic acid
ADF	antioxidant dietary fiber
ADI	acceptable daily intake
ADP	adenosine diphosphate
AF	antioxidant factor
AGEs	advanced glycation end products
AH	antioxidant
AH$^{\bullet}$	L-ascorbic acid radical
AH_2	L-Ascorbic acid
AI	atherogenic index
ALE	advanced lipid oxidation end
AOA	antioxidant activity
APE	allylic position equivalent
ARP	anti-radical power
a_w	water activity
BAPE	bis-allylic position equivalent
BCAAs	branched chain amino acids
BHA	butylated hydroxy anisole
BHT	butylated hydroxy toluene
BPR	bael pulp residue
C of V	coefficient of variance
CA	citric acid
Ca	calcium
CB	control biscuits
CF	commercially available fat
CFR	code of federal regulations
CHD	coronary heart diseases
CIE	International Commission on Illumination
CLA	conjugated linoleic acid
CO_2	carbon dioxide
CP	cauliflower powder

CPFB	*Citrus paradisi* fruit barks
CPP	caseinophosphopeptides
DABCO	diazabicyclooctane
DF	dietary fiber
DFA	desirable fatty acids
DG	dodecyl gallate
DHA	docosahexaenoic acid
DPP	dried plum puree
DPPH	2,2-diphenyl-1-picrylhydrazyl
DW	dry weight
EDTA	ethylene diaminetetraacetic acid
EFSA	European Food Safety Authority
EGCG	epigallocatechin gallate
EKWE	ethanolic kiam wood extract
EM	exxenterol
EOs	essential oils
EPA	eicosapentaenoic acid
ESR	electron spin resonance
FA	ferulic acid
FDA	Food and Drug Administration
Fe	iron
FFA	free fatty acid
FIR	far infrared
FPP	fresh plum puree
FRAP	ferric ion reducing antioxidant power
FRH	FIR treated rice hull
GADF	grape antioxidant dietary fiber
GCMS	gas chromatography mass spectroscopy
GL	glycolipids
GOMPS	good oxidation management practices
GPC	grape pomace concentrate
GRAS	generally regarded as safe
GSE	grape seed extract
H_2O_2	hydrogen peroxide
Hb	hemoglobin
HDA	hydrogen donating antioxidants
HDL	high density lipoproteins
HHE	trans-4-hydroxy-2-hexanal
HO•	hydroxyl radical
HOO˙,	hydroperoxyl radical

HOX	high oxygen
HP	high pressure
HPLC	high performance liquid chromatography
HPOs	hydroperoxides
ICA	iron chelating antioxidants
IDF	insoluble dietary fiber
IT	index of thromogenicity
IV	iodine value
K	potassium
Keq	equilibrium constant
KRP	kinnow rind powder
L*a*b*	defined color space for reflectance colorimetry
LAOX	linoleic acid accelerated by azo-initiators
LM	liposterine
LMW	low-molecular-weight
LOOH	lipid hydroperoxides
LOQ	limits of quantification
LOX	lipoxygenase
LS	lasalacid sodium salt
MAP	modified atmosphere packaging
Mbg	myoglobin
MDA	malonaldehyde
ME	mint extract
MEFS	methanol extracts of fermented soybeans
MEs	methanol extracts
MFGM	milk fat globule membrane
MFM	minced fish muscle
MHO	menhaden oil
MMbg	met-myoglobin
MST	mechanically separated turkey
MUFA	monounsaturated fatty acids
NADH	nicotinamide adenine dinucleotide
NBT	nitro blue tetrazolium
NDGA	nordihydroguaretic acid
NL	neutral lipids
NO•	oxide radical
NO_2•.	nitrogen dioxide radical
NOMb	nitrosomyoglobin
NPN	non-protein nitrogen
O_2•−	superoxide radical

O_2•	superoxide anion radical
OE	oregano extract
OEO	oregano essential oil
OFA	undesirable fatty acids
OFB	oryzanol fortified biscuits
OFF	oryzanol fortified fat
OG	octyl gallate
OMbg	oxymyoglobin
OP	olive pomace
OPE	orange peel extract
ORAC	oxygen radical absorbance capacity
ORP	oxidation reduction potential
OS	oxidative stability
Oz	oryzanol
p- AV	para- ansidine values
PCL	polycaprolactone
PDCAAS	protein digestibility corrected amino acid score
PE	polyethylene
PET	Polyethylene terephthalate
PG	propyl gallate
PHB	polydroxybutyrate
PhIP	2-amino-1-methyl-6-phenylimidazo [4, 5-*b*] pyridine
PL	phospholipids
PLA	poly lactic acid
PMS	phenazonium methosulfate
POH	phenolic compounds
POPs	phytosterol oxidation products
PP	pomegranate peels
PP	polypropylene
PPE	potato peel extract
PPO	polyphenoloxidase
PRP	pomegranate rind powder
PS	polystyrene
PSP	pomegranate seed powder
PUFA	polyunsaturated fatty acids
PV	peroxide value
PVC	polyvinyl chloride
R	radicals
RDI	recommended daily intake
RE	rosemary extract

RGE	red ginseng extract
RNS	reactive nitrogen species
RO•	alkoxyl radicals
ROO•	peroxyl radicals
ROS	reactive oxygen species
RSA	radical scavenging activity
RWGP	red wine grape pomace
SAT	saturated fatty acids
SDF	soluble dietary fiber
Se	selenium
SI	selectivity indexes
SOSG	singlet oxygen sensor green
TA	total anthocyanin
TAG	triacylglycerides
TBA	thiobarbituric acid
TBARS	thiobarbituric acid reactive substances
TBHQ	tertiary butyl hydroquinone
TC	tea catechins
TDF	total dietary fiber
TEAC	trolox equivalent antioxidant capacity
TF	total flavonol
TM	tocopherol
TOSC	total oxyradical scavenging activity
TOTOX	total oxidation
TPC	total phenolic content
UF	ultra-filtered
UMAE	ultrasonic/microwave assisted extraction
UV	ultraviolet
WGDF	white grape antioxidant dietary fiber
WHC	water holding capacity
WOF	warmed-over-flavor
WWGP	white wine grape pomace

PREFACE

Many food preservation strategies can be used for the control of oxidation in foods; however, these quality problems are not yet controlled adequately. Although synthetic antioxidant agents are approved in many countries, their excessive use has increased pressure on food manufacturers to either completely remove these agents or to adopt natural alternatives for the maintenance or extension of a product's shelf life. Therefore, the use of natural safe and effective preservatives is a demand of food consumers and producers.

Foods of animal origin are one of the key components of our diet supplying several vital nutrients like protein, fats, vitamins, and minerals. Fat is one of the most important nutrients in foods of the animal origin and is composed of various fatty acids such as saturated, monounsaturated, and polyunsaturated fatty acids. Depending upon the nature and origin of foods, proportion of these fatty acids varies. As the amount of unsaturated fatty acids increases, the animal foods become more vulnerable to oxidation. Oxidation of fat damages the nutritional and sensory characteristics, particularly flavor of food products, and thus affects their storage stability. Proteins in animal foods are also susceptible to the oxidation thus affecting the quality of foods. Moreover, during processing and storage, food products undergo changes in their physicochemical characteristics leading to development of oxygenated free radicals which initiate the oxidation of polyunsaturated fatty acids while destruction of the endogenous antioxidant system. Various approaches are being applied to minimize the oxidation of these foods such as use of antioxidants and anaerobic and active packaging.

The search of new safe substances for food preservation is being performed around the world. Many naturally occurring bioactive compounds can be considered as good alternatives to synthetic antioxidant food additives. Natural antioxidants could be extracted from various plant and animal sources. These are extracted using different approaches and solvents depending upon feasibility and yield. In foods of animal origin such as meat and meat products, dairy products, fish and fish products, and poultry products, natural antioxidants are applied in various forms and ways. As such, there is no compiled literature on the oxidation of animal products and approaches for its reduction. In this book, attempts are made to address the aforesaid issues and appraise the potential use of antioxidants from natural sources to ameliorate the oxidative stress in animal products, to prevent lipid–protein oxidation, and to improve oxidative stability considering demand for safety and nutrition.

This book, *Natural Antioxidants: Applications in Foods of Animal Origin,* will be a standard reference work describing the potential of natural antioxidants in the animal food industries. It will also contemplate the sensorial and toxicological shortcomings of using natural products. This proposed book looks forward to promoting the use of natural extracts and fulfilling consumer demands for healthier foods.

CHAPTER 1

MECHANISM OF OXIDATION IN FOODS OF ANIMAL ORIGIN

MANAT CHAIJAN* and WORAWAN PANPIPAT

Food Technology and Innovation Center of Excellence, Department of Agro-Industry, School of Agricultural Technology, Walailak University, Thasala 80160, Nakhon Si Thammarat, Thailand

Corresponding author. E-mail: cmanat@wu.ac.th

CONTENTS

ABSTRACT

Lipid and myoglobin oxidations significantly impair the quality of foods of animal origin because these reactions deteriorate flavor and color, induce the loss of nutritional value and cause technological problems during processing. Lipid and myoglobin oxidations are coupled and such reactions can occur via non-enzymatic and enzymatic routes. Several factors have been reported to enhance the oxidation of lipid in muscle foods including species, muscle type, fatty acid composition, endogenous antioxidants (AH), temperature, metal ions, sodium chloride, muscle pH, and processing parameters. It is most likely that the prooxidant effect of heme proteins, especially myoglobin, is a prime factor influencing the lipid oxidation in muscle foods. On the other hand, lipid oxidation results in a wide range of aldehyde products, which can cause the oxidation of myoglobin. The interaction between myoglobin and aldehydic lipid oxidation products can alter myoglobin redox stability and finally results in muscle discoloration. As a consequence, the oxidation of both lipid and myoglobin directly affect the quality and acceptability of muscle foods and those reactions seems to promote each other.

1.1 INTRODUCTION

The problems associated with oxidation in foods of animal origin, particularly meat and muscle foods, have gained much interest as they relate to flavor deterioration, discoloration, loss of nutritional value and safety, biological damage, ageing, and functional property changes. Meat and other muscle foods are complex foods with highly structured nutritional compositions (Rodriguez-Estrada et al., 1997). Muscles are composed of water, proteins, lipids, carbohydrates, vitamins, and minerals in variable amounts depending on several factors such as breeds, muscle types, dietary, and growth performance (Wattanachant et al., 2005). Oxidation is a major cause of quality deterioration for a variety of raw and processed muscle foods during handling, processing, and storage. Lipid, protein, pigment, and vitamin in muscle tissue are susceptible to oxidative reactions. These changes resulted from reactions of active oxygen species, free radicals, enzymes, and prooxidants with unsaturated fatty acids in lipids, amino acids in proteins, heme groups in pigments and the chains in vitamins with conjugated double bonds. However, lipid oxidation and the oxidation of heme proteins, particularly myoglobin, in muscle foods are major deteriorative reactions which occur in a concurrent manner and each process appears to enhance the other. During oxidation of oxymyoglobin, both superoxide anion and hydrogen peroxide are produced and further react with iron to produce hydroxyl radical. The hydroxyl radical has the ability to penetrate into the hydrophobic lipid region and hence facilitates lipid oxidation. The prooxidant effect of heme proteins on lipid oxidation is concentration-dependent. At equimolar concentrations, oxymyoglobin shows higher prooxidative activity toward lipid than metmyoglobin. However, the catalytic activity of metmyoglobin is promoted by hydrogen peroxide. The reaction between hydrogen peroxide and metmyoglobin results in the formation of two active hypervalent species, perferrylmyoglobin and ferrylmyoglobin, which are responsible for lipid oxidation. Additionally, lipid oxidation results in a wide range of aldehyde products, which are reported to induce the oxidation of oxymyoglobin. Studies in muscle foods have been focused mainly the interaction between myoglobin and aldehydic lipid oxidation products. Metmyoglobin formation is generally greater in the presence of unsaturated aldehydes than their saturated counterparts of equivalent carbon chain length. In addition, increasing chain length of aldehydes, from hexenal through nonenal, results in the increased metmyoglobin formation. Moreover, aldehydes alter myoglobin redox stability by increasing oxymyoglobin oxidation, decreasing the metmyoglobin reduction via enzymatic process, and enhance the prooxidant activity of metmyoglobin (Chaijan, 2008). Therefore, the oxidation of both lipid and heme proteins directly affect the quality and acceptability of muscle foods and the lowering of such a phenomenon can enhance the shelf-life stability of those foods. To design strategies to inhibit the progression of oxidative reactions in foods and biological systems, it is important to understand the nature of these reactions and how they are influenced by both intrinsic and extrinsic factors. This goal may be achieved through a better understanding of the reaction kinetics including the rate at which the reaction takes place, the effective factors (mainly temperature, concentration of reactants and products, and presence of catalysts), and how these two are related. This chapter deals with the mechanism of oxidation, especially lipid and myoglobin, in foods of animal origin emphasizing on meat and muscle foods. The interaction between lipid and myoglobin oxidations and the effect of various food-processing applications on lipid and myoglobin oxidations are also discussed.

1.2 IMPACT OF LIPID OXIDATION IN FOODS OF ANIMAL ORIGIN

Lipid oxidation is one of the important reactions in food and biological systems because it has deleterious effects on polyunsaturated fatty acids

(PUFA) and other lipid substrates, causing significant losses in food quality, health, and well-being (Chaijan et al., 2006; Chaijan, 2008). Lipid oxidation in food systems is a detrimental process. It is difficult to find a food component that would not be capable of affecting lipid oxidation because lipids are only a part of a food product (Kolakowska, 2002). Generally, lipid oxidation deteriorates the sensory quality and nutritive value of a product, poses a health hazard, and presents a number of analytical problems (Kolakowska, 2002). Lipid oxidation is also one of the main factors limiting the quality and acceptability of meats and other muscle foods, especially following refrigerated and frozen storages (Zamora & Hidago, 2001; Renerre, 2000; Morrissey et al., 1998). Oxidation of lipids is accentuated in the immediate post-slaughter period, during handling, processing, storage, and cooking. This process leads to discoloration, drip losses, off-odor and off-flavor development, texture defects, and the production of potentially toxic compounds in meat products (Richards et al., 2002; Morrissey et al., 1998).

Hydroperoxide is a primary oxidation product during storage of foods which is readily decomposed to a variety of volatile compounds including aldehydes, ketones, and alcohols (Frankel et al., 1984). The formation of the secondary lipid oxidation products is one of the main causes of the development of undesirable odors in muscle foods. Human olfactory receptors usually have remarkably low organoleptic thresholds to most of these volatile compounds (Ke et al., 1975; McGill et al., 1977). The effect of lipid oxidation and off-odor development in postmortem fish has been reported. Lipid oxidation is mainly associated with the rejection by consumer due to the off-odor and off-flavor. Flavor is a very complex attribute of meat palatability. Rancid or fishy odor has been identified as a common off-flavor associated with fish flesh and directly related with the formation of the secondary lipid oxidation products (Ke et al., 1975; McGill et al., 1977; Sohn et al., 2005; Thiansilakul et al., 2010). Varlet et al. (2006) reported that carbonyl compounds, such as heptanal or (*E,Z*)-2,6-nonadienal, show a high detection frequency and odorant intensity in salmon (*Salmo salar*), giving the flesh its typical fishy odor. The fishy volatiles identified in the boiled sardine were dimethyl sulfide, acetaldehyde, propionaldehyde, butyraldehyde, 2-ethyl-furan, valeraldehyde, 2,3-pentanedione, hexanal, and 1-penten-3-ol (Kasahara & Osawa, 1998).

Lipid oxidation usually causes a decrease in consumer acceptance. However, in some cases, lipid oxidation leads to enhancement of product quality such as the enzymatic production of fresh fish aromas and the cured meat flavor derived from lipid oxidation during ripening (Ladikos & Lougovois, 1990). A notable exception is observed in dry cured country hams and some fermented sausages and the desirable flavor of which does not occur until hydrolysis of some of the fat and a certain degree of oxidation has taken place during ripening (Pearson et al., 1977). On the other hand, lipid oxidation during cooking may be a source of intermediate which react with other components to give important constituents of the desirable flavor of normal cooked meat (Enser, 1987). The types of flavor developed from the volatile lipid oxidation compounds depend on a multitude of complex interactions, concentration ranges, and the medium in which they are tasted (Frankel, 1984). Many of the reactions involved in the formation of volatile aroma compounds from lipid, follow the same basic pathways for both thermal and rancid oxidation and similar volatile products are formed. However, subtle differences in the precise mechanisms of oxidation under storage conditions and under thermal processing lead to mixtures of volatiles exhibiting both qualitative and quantitative differences (Mottram, 1987).

1.3 BASIC MECHANISM OF LIPID OXIDATION IN FOODS OF ANIMAL ORIGIN

The lipid oxidation in foods of animal origin is assumed to proceed along a free radical route (autoxidation), photooxidation route and enzymatic route. The oxidation mechanism is basically explained by invoking free-radical reactions, while the photooxidation and lipoxygenase (LOX) routes differ from it at the initiation stage only. For this reason, they can be treated as different forms of free radical reaction initiation.

1.3.1 FREE RADICAL OXIDATION

The two major components involved in lipid oxidation are unsaturated fatty acids and oxygen. In this process, atmospheric oxygen is added to some fatty acids, producing unstable intermediates that finally breakdown to form unpleasant flavor and aroma compounds (Erickson, 2003). Although enzymatic and photogenic oxidation may play a role, the most common and important process by which unsaturated fatty acids and oxygen interact is a free radical mechanism (Erickson, 2003). A free radical reaction or autoxidation is the main reaction involved in oxidative deterioration of food lipids, including foods of animal origin (Hoac et al., 2006). It is a chain reaction that consists of initiation, propagation, and termination reactions, and involves the production of free radicals (Gunstone & Norris, 1983; Nawar, 1996;

Renerre, 2000; Fig. 1.1). Oxidation is initiated by radicals present in living organisms (e.g., hydroperoxide, hydroxide, peroxide, alcoxy, and alkyl) or by thermal or photochemical homolytic cleavage of an R-H bond. The oxidation activation energy and reaction rate at this stage depend on the type of initiator and the number of unsaturated bonds in the substrate. The dissociation energy of the C-H bonds in saturated fatty acid depend on the length of the fatty acid carbon chain and is similar in fatty acid, their esters and in triacylglycerols (Litwinienko et al., 1999). In unsaturated acids, the weakest C-H bond is found in the bis-allylic position. The activation energies are 75, 88, and 100 kcal/mol for bis-allylic, allylic, and methylene hydrogens, respectively (Simic et al., 1992). A three-step simplified free-radical scheme has been postulated as follows:

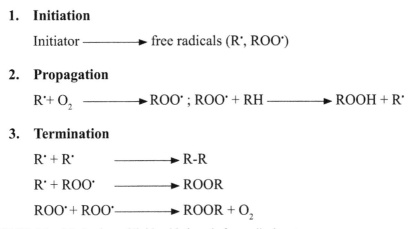

1. **Initiation**

   Initiator ⟶ free radicals (R$^{\bullet}$, ROO$^{\bullet}$)

2. **Propagation**

   R$^{\bullet}$ + O$_2$ ⟶ ROO$^{\bullet}$; ROO$^{\bullet}$ + RH ⟶ ROOH + R$^{\bullet}$

3. **Termination**

   R$^{\bullet}$ + R$^{\bullet}$ ⟶ R-R

   R$^{\bullet}$ + ROO$^{\bullet}$ ⟶ ROOR

   ROO$^{\bullet}$ + ROO$^{\bullet}$ ⟶ ROOR + O$_2$

FIGURE 1.1 Mechanism of lipid oxidation via free radical route.

Initiation occurs as hydrogen is abstracted from an unsaturated fatty acid, resulting in a lipid-free radical, which, in turn, reacts with molecular oxygen to form a lipid peroxyl radical. Initiation is frequently attributed in most foods, including muscle foods, to react ion of the fatty acids with active oxygen species (Erickson, 2003).

The propagation phase of oxidation occurs by lipid-lipid interactions, whereby the lipid peroxyl radical abstracts hydrogen from an adjacent molecule, resulting in a lipid hydroperoxide and a new lipid-free radical. This propagation continues until one of the radicals is removed by reaction with another radical or with an antioxidant (AH) whose resulting radical (A$^{\bullet}$) is much less reactive. Interactions of this type continue 10–100 times before two free radicals combine to terminate the process. The lipid radical reacts very quickly with atmospheric oxygen making a peroxyl radical which again may abstract hydrogen from another acyl chain, resulting in the formation of a lipid hydroperoxide and a new radical. By themselves, lipid hydroperoxides are not considered harmful to food quality; however, they are further degraded into compounds, especially aldehydes that are responsible for off-flavors (Erickson, 2003).

To break the repeating sequence of propagating steps, two types of termination reactions are encountered: radical–radical coupling and radical–radical disproportionation, a process in which two stable products are formed from free radicals by an atom or group transfer process. In both cases, non-radical products are formed. However, the termination reactions are not always efficient. When coupling gives rise to tertiary tetroxides, they decompose to peroxyl radicals at temperatures above −80 °C and to alkoxyl radicals at temperatures above −30 °C. On the other hand, secondary and primary peroxyl radicals, terminate efficiently by a mechanism in which the tetroxide decomposes to give molecular oxygen, an alcohol, and a carbonyl compound (Erickson, 2003). On the other word, termination of free radical oxidative reaction occurs when two radical species (peroxyl, alcoxyl, or alkyl) react with each other to form a non-radical adduct. Free radical reactions can also be terminated when one of the lipid radicals reacts with an AH proper, because hydrogen abstraction by a peroxide radical from the AH molecule produces an inert AH radical (Kolakowska, 2002).

1.3.2 PHOTOOXIDATION

Photooxidation involves the formation of hydroperoxides in a direct reaction of singlet oxygen ($^1O_2$) addition to unsaturated lipids, without radical formation. The $^1O_2$ emerges during a reaction of sensitizers (e.g., chlorophyll, hemoglobin, myoglobin, and riboflavin) with atmospheric oxygen (triplet oxygen, $^3O_2$) (Kolakowska, 2002). Photosensitizations also can occur *in vivo* (Halliwell et al., 1995). The $^1O_2$ is 1450 times more reactive than molecular oxygen. It is inserted at the end carbon of a double bond, which is shifted to an allylic position in the *trans* configuration. The resulting hydroperoxides have an allylic *trans* double bond, which renders them different from hydroperoxides formed during autoxidation. Hydroperoxides formed during photooxidation are more easily cyclized than hydroperoxy epidioxides (Frankel, 1998). In addition, light, particularly ultraviolet light may be involved in initiation of the classical free radical oxidation of lipids and catalyze other stages of the process. In the presence of light energy-activated riboflavin, which is a sensitizer, a lipid radical can form, while oxygen gives rise to superoxide radical anion ($O_2 \cdot^-$). During UV irradiation of muscle lipids, the quantity of hydroperoxides and the ratio of their formation differ, depending on the origin of lipids. For instance, the photooxidation ratio (slope of hydroperoxide accumulation over time of UV exposure of lipids) of the muscle lipids of fish varied from about one to more than 20, both within and between the species (Kolakowska, 2002). No correlations between the photooxidation ratio and monounsaturated fatty acids (MUFA), PUFA, eicosapentaenoic acid (EPA), or docosahexaenoic acid (DHA) content in fish lipids were reported (Kolakowska, 2002).

Light induced oxidation is one of the main factors limiting shelf life of milk. Exposure to visible light leads to off-flavors related to oxidation of proteins and lipids due to excitation of photosensitizers among which riboflavin has been recognized to play a major role (Bradley et al., 2003). Beta-carotene absorbs light in the same region as riboflavin, and it has therefore been suggested to protect against photooxidation since less light then reaches riboflavin (Airado-Rodríguez et al., 2011).

An et al. (2011) studied the effects of sensitizers and pH on the oil oxidation of acidic O/W emulsions under light by measuring hydroperoxide content and headspace oxygen consumption in the mixed canola and tuna oil emulsions. The emulsions consisted of canola and tuna oil (2:1 w/w, 32%), diluted acetic acid (64%), egg yolk powder (4%), chlorophyll b or erythrosine (5µM), and/or diazabicyclooctane (DABCO) or sodium azide (0.5M). From the result, chlorophyll increased oil oxidation in the emulsion under light via $^1O_2$ production while erythrosine did not. In contrast, DABCO significantly decreased photooxidation of the oil containing chlorophyll, suggesting $^1O_2$ involvement. However, sodium azide increased photooxidation of the oil containing chlorophyll possibly via azide radical production under acidic conditions. The oil photooxidation was higher in the emulsion containing chlorophyll at pH 6.27 than at pH 2.67 or 3.68, primarily by $^1O_2$ and secondarily by free radicals produced from hydroperoxide decomposition.

1.3.3 ENZYMATIC LIPID OXIDATION

LOX-catalyzed lipid oxidation differs from the free radical reaction by the formation of hydroperoxides at a certain position of the chain of a free fatty acid. LOX use molecular oxygen to catalyze the stereo and regiospecific oxygenation of PUFA with 1-*cis*, 4-*cis*-pentadiene moieties (Kolakowska, 2002). LOX react enzymatically with more than one methylene carbon on the substrate molecule to yield double oxygenation sites (German et al., 1992). The newly formed fatty acid peroxy free radical removes hydrogen from another unsaturated fatty acid molecule to form a conjugated hydroperoxy diene. LOX forms a high-energy (radical) intermediate complex with the substrate; this complex is capable of initiating the oxidation of lipids and other compounds (e.g., carotenoids, chlorophyll, tocopherols, thiol compounds, and protein), which can themselves interact with the enzyme substrate complex as well (Hammer, 1993; Hultin, 1994).

Kolakowska (2002) reported that mammalian LOX are categorized according to the positional specificity of oxygen insertion into arachidonic acid. Four isoform positions of arachidonate LOX have been identified: 5-LOX (E.C. 1.13.11.34), 8-LOX, 12-LOX (E.C. 1.13.11.31), and 15-LOX (E.C. 1.13.11.33). The LOX that catalyzes oxidation of linoleate (E.C.1.13.11.12) attacks linoleic acid, both at position 9 and position 13. In chicken meat arachidonate, 15-LOX was found to be active during 12-month storage at −20 °C (Grossman et al., 1988). In frozen-stored fish, LOX contributes to oxidative lipid deterioration. However, LOX in fish is also responsible for the formation of desirable fresh fish flavor, the seaweed flavor (Lindsay, 1994). Some species show a higher activity of 12-LOX, while 15-LOX is more active in others; for this reason, the fresh fish flavor spectrum is species dependent. The half-lives of 12- and 15-LOX at 0 °C were less than 3 h and more than 10 h, respectively (German et al., 1992). LOX was observed to be active in cold-stored fish after 48 h of storage (Medina et al., 1999). The storage of herring, three weeks at −20 °C, resulted in an increase in LOX activity. During prolonged frozen storage of herring, a decrease in LOX activity was observed (Samson & Stodolnik, 2001). Sae-leaw et al. (2013) reported that the development of fishy odor in Nile tilapia skin during iced storage was mostly governed by lipid oxidation via autoxidation or induced by LOX. Although the participation of LOX in the *post mortem* animal lipid oxidation is acknowledged, the role of LOX in lipid oxidation is much more important in plant than in animal food products. LOXs are responsible for the off flavor in frozen vegetables (Ganthavorn et al., 1991), lipid oxidation in cereal products, rapeseed, pea, avocado, and muscle foods, and for the beany and bitter flavor (Frankel, 1998).

1.4   EFFECT OF INTRINSIC FACTORS AND PROCESSING PARAMETERS IN LIPID OXIDATION IN FOODS OF ANIMAL ORIGIN

Ladikos and Lougovois (1990) reported that the nature and relative proportions of the compounds formed from lipid oxidation depend at least in part on the composition of the fat of the animal from which they are derived, which may reflect a variety of factors including the nature of diet (Pearson et al., 1977). Additionally, factors such as processing and storage conditions, type of ingredients and concentration of pro- or antioxidants, are very important in determining the rate of development and the possible deteriorative effects of lipid oxidation. Therefore, the extent of lipid oxidation can be influenced by both intrinsic and extrinsic factors such as the content and activity of pro- and antioxidants, endogenous ferrous iron, myoglobin, enzymes, pH, temperature, ionic strength, irradiation, oxygen consumption reaction, surface area in contact with oxygen, water activity (a_w), and the fatty acid composition of the meat (Andreo et al., 2003; Undeland, 2001; Harris & Tall, 1994; Renerre & Labas, 1987; Castell et al., 1965; Nawar, 1996; Slabyj & Hultin, 1984; Undeland et al., 2003). Erickson (2003) summarized typical responses exhibited by muscle tissue during storage following various treatments. Processing treatments like bleeding, curing, smoking, glazing, edible coating, freezing and packaging inhibit the lipid oxidation whereas mincing, salting, rinsing with oxidizing agent, cooking, deep fat frying, and radiation promote the lipid oxidation. However, responses by muscle foods to some processing treatments such as washing and skinning are varied among studies.

Slaughter of animals or fish is a necessary first step in converting the living organism to food. Slaughter methods and the accompanying bleeding step; however, may affect lipid oxidation through alteration in the removal of hemoglobin catalysts. For example, bleeding was a potential means in retarding lipid oxidation, fishy odor development, and microbial growth of Asian seabass slices during storage in ice (Maqsood & Benjakul, 2011a, 2011b). Electrostatic interactions between hemoglobin and muscle components may be an initial step of hemoglobin-mediated lipid oxidation (Sannaveerappa et al., 2014).

1.4.1   SPECIES, MUSCLE TYPE, AND FATTY ACID COMPOSITION

A major cause of muscle food quality deterioration is lipid oxidation and the changes associated with it. Lipid oxidation is a complex process whereby unsaturated fatty acids primarily reacting with molecular oxygen via a free radical chain mechanism (Gray, 1978). The primary autooxidation is followed by a series of secondary reactions which lead to the degradation of the lipid and the development of oxidative rancidity. The main unsaturated fatty acids comprising the lipids of animal tissues are oleic, linoleic, linolenic, and arachidonic acids. Their autooxidation gives rise to a number of different hydroperoxides which, in conjunction with the many different decomposition pathways involved, lead to a large number of volatile compounds (Mottram, 1987). Oxidation of lipids also occurs during postmortem storage of muscle tissue. Meats such as fish and poultry contain a high concentration of PUFA and are therefore more susceptible to oxidation (Pacheco-Aguilar et al., 2000; Apgar & Hultin, 1982). Thus, fish lipids undergo more rapid oxidation after capture, even at low temperature storage (Foegeding et al., 1996; Pacheco-Aguilar et al., 2000). Pacheco-Aguilar et al. (2000) reported that the shelf life of oily Monterey sardine was limited by lipid oxidation, as shown by the increase of peroxide value (PV) during storage at 0 °C up to 15 days. Sohn et al. (2005) reported that the total lipid hydroperoxide content of Pacific saury (*Cololabis saira*), Japanese Spanish mackerel (*Scomberomorus niphonius*), and chub mackerel (*Scomber japonicus*) tended to increase in both dark and ordinary muscle throughout four days of iced storage. Fatty fish such as sardine and mackerel underwent rapid lipid oxidation during iced storage due to the high content of PUFA (Chaijan et al., 2006; Chaijan et al., 2013; Pacheco-Aguilar et al., 2000). Lipid oxidation seems to be a distinct problem in surimi made from some dark-fleshed fish and particularly surimi from mammalian and avian muscle (Lanier, 2000). Lynch et al. (2001) demonstrated that lipid oxidation occurred progressively in stored ground beef at 4 °C and produced a variety of aldehydes.

1.4.2 ENDOGENOUS ANTIOXIDANTS

The normal resistance of meat to the development of rancidity depends on the balance between the presence of AH in the animal tissues and the level of unsaturation and the concentration of the fatty acids present (Enser, 1987). The living cells possess several protection mechanisms directed against lipid oxidation products (Sies, 1997). Glutathione peroxidase reduces hydroperoxides in the cellular membranes to the corresponding hydroxy-compounds. This reaction demands supply of reduced glutathione and will therefore cease post mortem when the cell is depleted of that substance. The membranes also contain the phenolic compound α-tocopherol (Vitamin

E) which is considered as the most important natural AH. Tocopherol can donate a hydrogen atom to the radicals L˙ or LOO˙ functioning as the molecule AH. It is generally assumed that the resulting tocopheryl radical reacts with ascorbic acid (Vitamin C) at the lipid/water interface, regenerating the tocopherol molecule. Muscle-based foods that contain relatively high concentrations of α-tocopherol demonstrate greater lipid and oxymyoglobin stability (Faustman et al., 1998). Poultry meat is composed of relatively high levels of unsaturated fatty acids and low levels of natural tocopherols and thus poultry products are very susceptible to the development of off-flavors due to oxidative rancidity (Dawson & Gartner, 1983). According to Wilson et al. (1976), turkey meat containing lower levels of natural tocopherol is most susceptible to warmed-over-flavor (WOF) development, followed closely by chicken, then by pork, beef, and mutton. The use of mechanically deboned poultry meat enhances the tendency of poultry products to oxidize (Moerck & Ball, 1974). However, the use of mechanically deboned beef in beef meat products did not result in flavor deterioration during storage, compared to control samples made of hand-boned beef, suggesting that lipid oxidation is not a problem as with chicken and fish; this was attributed to differences in the degree of unsaturation of fatty acids (Allen & Foegeding, 1981).

Other compounds, for example the carotenoids and phenols, have been known to function as AH (Huss, 1995). Vareltzis et al. (2008) reported that adding fish press juice to washed cod mince could inhibit hemoglobin-mediated lipid oxidation of protein isolates obtained from cod muscle (*Gadus morhua*). Press juice obtained from chicken breast muscle also showed a potent inhibitor of hemoglobin-mediated lipid oxidation in washed cod muscle (Li et al., 2005). The aqueous phase of chicken breast muscle includes low-molecular-weight (LMW) components such as ascorbate, urate, glutathione, bilirubin, and histidine-containing dipeptides (Chan et al., 1994). High-molecular-weight (HMW) components include glutathione peroxidase, superoxide dismutase, catalase, transferrin, haptoglobin, albumin, ceruloplasmin, and hemopexin (Decker, 1998). Erickson et al. (1990) reported that both the LMW (<10 kDa) and HMW (>6–8 kDa) cytosol fractions from flounder tissue inhibited iron-mediated lipid oxidation in flounder sarcoplasmic reticulum and the effect was believed to be due to the binding of iron. Furthermore, Slabyj and Hultin (1984) reported that both LMW and HMW components in herring cytosol could inhibit iron-mediated lipid oxidation in microsomes. Han and Liston (1989) reported the ability of rainbow trout cytosol to inhibit iron-mediated lipid oxidation in fish muscle microsomes.

1.4.3   TEMPERATURE

Like most chemical reactions, lipid oxidation rates increase with increasing temperature and time (Hultin, 1992). Saeed and Howell (2002) reported that the rate of lipid oxidation in frozen Atlantic mackerel increased with increasing storage time and storage temperature. Furthermore, freezing can facilitate lipid oxidation, partly because of concentration effects (Foegeding et al., 1996). The influence of long-term frozen storage, temperature (-25 and $-45\ °C$) and type of packaging materials (low and medium oxygen barriers) on the lipid oxidation of frozen Atlantic herring fillets (*Clupea harengus*) was studied by Tolstorebrov et al. (2014). The lowest lipid oxidation in term of PV and thiobarbituric acid reactive substances (TBARS) was detected in frozen Atlantic herring fillets kept at $-45\ °C$ and the packaging material with a medium oxygen barrier. From the result, the oxygen concentration in the package was considered to be the dominating factor for the herring's oxidation during frozen storing.

Cooked meats held in a refrigerator develop rancid odors and flavors which usually become apparent within 48 h at 4 °C. These flavors are particularly noticeable after reheating the meat and are referred to as WOF (Tims & Watts, 1958). The rapid development of oxidized flavor in refrigerated cooked meats is in marked contrast to the slow onset of rancidity commonly encountered in raw meats, fatty tissues, rendered fat, or lard, which is normally not apparent until they have been stored for weeks or months (Pearson et al., 1977).

Heating results in the release of heme-bound iron and in forming other polymers with proteins; those polymers enhance the catalytic effect of iron. This is also true with respect to the thermal inactivation of enzymes that contain metals acting as prosthetic groups (e.g., LOX and peroxidases). These enzymes, even after denaturation, are capable of catalyzing oxidation. On the other hand, heating does not release iron from ferritin, but does enhance its reduction (Kanner, 1992). The rate of oxidation in the presence of metals is higher at lower pH than at neutral pH for Fe^{3+} and Fe^{2+} (Richards & Hultin, 2000).

The extent of lipid oxidation in cooked meat appears to be related to the intensity of heat treatment. Pearson et al. (1977) reported that meat heated at 70 °C for 1 h developed rancidity rapidly. However, thiobarbituric acid (TBA) values decreased when the cooking temperature was raised above 80 °C. According to Huang and Greene (1978), meat subjected to high temperatures and/or long periods of heating developed lower TBA values, than did samples subjected to lower temperature for a shorter period of time.

This phenomenon was postulated to have resulted from AH substances produced from browning reactions during the heating of meat. It has been reported that Maillard reaction products possessed AH activity which can be applied as natural AH in food products (Limsuwanmanee et al., 2014).

The effect of drying method on oxidative stability of microencapsulated fish oil with ratio of 33/22, EPA: DHA which emulsified with four combinations of matrices was studied by Anwar and Kunz (2011). The emulsions were dried by spray granulation, spray drying, and freeze drying to produce 25% oil powders. A combination of 10% soybean soluble polysaccharide and 65% octenyl succinic anhydride dried by spray granulation was the best procedure for fish oil encapsulation due to its having a very low propanal content. The microcapsules produced by spray granulation might be then covered by successive layers resulting in multiple encapsulations which provide maximum protection to the oil droplets. They also suggested that combination of matrices, drying temperature, microcapsule morphology, and processing time are among the most critical factors governing oxidative stability of fish oil.

1.4.4 METAL IONS

Most biological and food studies of lipid peroxidation involve transition metal ions (Fe^{n+}, Cu^{n+}, etc.), and it is generally accepted that iron is pivotal in catalyzing oxidative changes in tissues (Gutteridge & Halliwell, 1990; Kanner, 1994). It was found that iron concentration in cod muscle is very low, with an average value of 6 ppm (Vareltzis et al., 2008). "Free" catalytic iron in the muscle might be generated by the destruction of the heme and release of iron. However, Undeland et al. (2003) suggested that the "free" iron has a negligible effect on the oxidation of washed cod muscle system. LMW iron added at a concentration of 23.2 µM did not induce oxidation of washed minced cod lipids while hemoglobin at this concentration was very pro-oxidative (Richards & Hultin, 2000).

Ladikos and Lougovois (1990) suggested that lipid oxidation is enhanced by metals such as iron, cobalt, and copper which facilitate the transfer of electrons leading to increased rates of free radical formation. The most common way that metal ions enter food is via the water used and in some instances via salt and spices (Taylor, 1987). The form of the metal is as important as the amount of metal present (Taylor, 1987). Ferrous iron has been shown to have greater pro-oxidant activity than ferric iron in cooked uncured meats (Pearson et al., 1977). Low levels of ascorbic acid may increase the efficiency of iron as a catalyst for lipid oxidation, presumably by regenerating the active ferrous form (Sato & Hegarty, 1971).

Kolakowska (2002) reported the effect of transition metals in lipid oxidation. Hydroperoxides formed at the propagation stage of the free radical oxidation, as well as those produced by photooxidation and enzyme-catalyzed oxidation, can disintegrate and yield alcoxy, alkyl, and peroxyl radicals, which reinitiate the oxidation of unsaturated fatty acid. Hydroperoxide decomposition may be triggered by temperature and/or light, but most important in this respect is the activity of transition metals, mainly iron and copper.

The Fe^{2+} ions are more reactive than Fe^{3+} ions and decompose hydrogen peroxide over 100 times faster (Girotti, 1998). Iron occurs in human and animal bodies, in up to 90% in a bound form in: hemoglobin, myoglobin, cytochromes, the storage protein ferritin and hemosiderin, the iron transport proteins, transferrins, and as prosthetic groups of enzymes. A small amount of iron occurs in a "free" form, that is, primarily as LMW iron. It complexes with organic phosphates, inorganic phosphates, amino acids (histidine, glycine, and cysteine), and organic acids (citric acid) (Decker & Hultin, 1992). LMW iron contributes between 2.5 and 3.8% to the total iron content in muscle tissue of lamb, pork, and chicken. Dark muscles of chicken, turkey, and mackerel contain twice as much LMW iron and more ferritin than light muscles (Kanner, 1992). LMW iron acts as a catalyst. Protein-bound Fe and Cu are minimally catalytic in oxidation. Ascorbate, NAD(P)H, thiol compounds, reduced glutathione, cysteine, and protein thiol groups release iron, which can catalyze the Fenton reaction. This occurs *post mortem* during, for example, the storage of fish or turkey, but the amount of reductants is then also decreased (Kanner, 1992; Hultin, 1994).

1.4.5 SODIUM CHLORIDE

Sodium chloride is able to catalyze lipid oxidation in muscle tissue (Nambudiry, 1980; Love & Pearson, 1971). Alternatively, the Na^+ may replace iron from a cellular complex via an ion exchange reaction (Kanner & Kinsella, 1983). The displaced iron may then participate in the initiation of lipid oxidation. It is most likely that meat or meat products containing salt such as surimi and cured meat are susceptible to lipid oxidation (Chaijan, 2008).

Ladikos and Lougovois (1990) reported that sodium chloride induces rancidity in freezer-stored, cooked, cured meat, in cured pork and in raw and cooked beef, both during cooking and subsequent storage. Chaijan (2011) reported that salting caused an increase in lipid oxidation in tilapia muscle. The effect of sodium chloride on fat oxidation depends on the level of free moisture in the system (Pearson et al., 1977). According to Love and Pearson (1971), the oxidative effect of sodium chloride may be attributed to the action of the reactive chloride ion on lipids, or to a modification of the heme proteins catalyzing lipid oxidation.

1.4.6 MUSCLE pH

Several studies have shown that lipid oxidation in muscle foods increases with a decrease in pH (Chen & Waimaleongora, 1981; Ogden et al., 1995; Tichivangana & Morrissey, 1985). Acid injection and marination have been used as a practice to improve the water-holding capacity and tenderness of muscle foods (Ke et al., 2009). In general, the lower the pH values, the stronger the prooxidant effect. Rapid formation of the oxidized forms of heme proteins (methemoglobin and metmyoglobin) might contribute to the rapid oxidation of muscle at acidic pH values (Chaijan et al., 2007). An increase in iron solubility with decreasing pH could also increase oxidation rates. Vareltzis and Hutin (2007) observed that exposing hemoglobin and microsomes to low pH affected lipid oxidation rates. When isolated membranes alone were exposed to low pH, they were less susceptible to hemoglobin-promoted lipid oxidation. Exposure of cod hemoglobin to pH 3 decreased its prooxidant activity compared to untreated cod hemoglobin (Vareltzis & Hultin, 2007). However, if cod hemoglobin and isolated cod microsomes were exposed to low pH together, oxidation was promoted (Vareltzis & Hultin, 2007).

Hemoglobin-mediated lipid oxidation can be accelerated by reduction in pH and could be due to enhanced autoxidation of hemoglobin at reduced pH. Hemoglobin from different fish is known to promote lipid oxidation in fish muscle differently (Maqsood et al., 2012). Maqsood and Benjakul (2011a) monitored the lipid oxidation of washed Asian seabass mince added with hemoglobin from various fish at different pH (6, 6.5, and 7) during 10 days of iced storage. Hemoglobins accelerated lipid oxidation more effectively at pH 6, compared with pH 6.5 and 7 as indicated by the higher PV and TBARS.

1.4.7 PARTICLE SIZE REDUCTION AND TUMBLING

Ladikos and Lougovois (1990) postulated that any process causing disruption of the muscle membrane system, such as grinding, cooking, and deboning, results in exposure of the labile lipid components to oxygen, and thus accelerate development of oxidative rancidity. Destruction of the extremely well organized structure of living animal cells will bring together lipids, oxidation catalysts, and enzymes responsible for lipid oxidation. Pearson et al. (1977) suggested that chopping and emulsification are at least as likely to cause WOF as grinding or mincing of samples. Dawson and Gartner (1983) attributed the high oxidative potential of mechanically deboned poultry to the extreme stress and aeration during the process and the compositional nature (bone marrow, heme, and lipids) of the product; TBA values increase most rapidly with decreasing particle sizes, as the latter are related to greater cell disruption. On the other hand, comminuted beef has a storage life similar to that of intact pork, despite the differences in fatty acid composition (Enser, 1987).

The mechanical force of tumbling can break the structure of the cell and organelle membranes which could lead to an exposure of phospholipids to cellular prooxidants (e.g., iron and heme proteins) or free radicals. The tumbling process has also been found to promote lipid oxidation in beef bottom round (Cheng & Ockerman, 2003). However, inhibition of lipid oxidation by the citric acid marination could be due the removal of prooxidants such as heme proteins by the marination/tumbling procedure. Alternately, inhibition of lipid oxidation could be due to the presence of citric acid since these molecules are strong metal chelators (Ke et al., 2009).

1.4.8 HIGH PRESSURE PROCESSING

High pressure (HP) processing has been shown to initiate lipid oxidation in fresh meat, especially a threshold pressure around 500 MPa seems to exist, which can lead to reduced quality and shelf life (Bolumar et al., 2014). Two mechanisms have been proposed to explain the pressure-induced lipid oxidation: (a) increased accessibility of iron from hemoproteins and (b) membrane disruption. Several studies have observed that the addition of ethylenediaminetetraacetic acid (EDTA), which can chelate metal ions like iron, can be correlated with a reduction of the lipid oxidation in meat processed by HP, which suggests that transition metal ion catalysis is the major factor underlying the increased lipid oxidation (Beltran et al., 2004; Ma et al., 2007).

However, iron release was not observed after HP treatment of chicken breast (Orlien et al., 2000). In the same study, it was also concluded that the catalytic activity of metmyoglobin did not increase during HP-treatment, indicating that pressure-induced changes of the metmyoglobin conformation that facilitates the access to the catalytic heme group did not take place (Orlien et al., 2000). So far, the role of iron in the induction of lipid oxidation of meats treated by HP is not well established. Membrane disruption facilitates contact between unsaturated lipids from the membrane and enzymes and catalysts like heme, nonheme iron as well as other metal cations and, thus, may contribute to the initiation of lipid oxidation. Recently, the formation of free radicals during HP has been proposed as a possible mechanism behind the induction of lipid oxidation in HP processed meats (Bolumar et al., 2011). Radical formation in the aqueous and lipid phases from HP-treated meat was first reported by Mariutti et al. (2008), and further studied by Bolumar et al. (2012), who characterized the kinetics of the formation of radicals in chicken meat during the application of different HP treatments. It was found that there is a threshold for the formation of radicals under HP conditions at 400 MPa at 25 °C and 500 MPa at 5 °C. The chemical mechanism which leads to the formation of radicals in meats by HP was intensively described by Bolumar et al. (2014) using electron spin resonance detection. The higher level of spin adducts was observed in the beef loin than in the chicken breast with radicals forming in the sarcoplasmic and myofibrillar fractions as well as in the non-soluble protein fraction due to the HP treatment, indicating that other radicals than iron-derived radicals were formed, and most likely protein derived radicals. The addition of EDTA reduced the radical formation suggesting iron-species (protein-bound or free) catalyzes the formation of radicals when meat systems are submitted to HP.

1.5   MYOGLOBIN OXIDATION IN FOODS OF ANIMAL ORIGIN

Myoglobin is a globular heme protein found in the muscle of meat-producing animals (Faustman & Phillips, 2001). It has been known to be a major contributor to the color of muscle, depending upon its redox state and concentration. Myoglobin concentration is affected by both genetics and environment (Giddings, 1974; Livingston & Brown, 1981; Faustman et al., 1996). The content of myoglobin in skeletal muscle will vary depending on the metabolic profile of the muscle, animal species, and age of the animal (Chaijan, 2008). Chaijan et al. (2004) reported that lipid and myoglobin contents were higher in dark muscle than in ordinary muscle of both sardine and mackerel, and higher contents of both constituents were found in sardine muscle than mackerel muscle.

Myoglobin is made up of a single polypeptide chain, globin, consisting of 153 amino acids and a prosthetic heme group, an iron (II) protoporphyrin-IX complex (Hayashi et al., 1998; Pegg & Shahidi, 1997). This heme group gives myoglobin and its derivatives their distinctive color (Dunn et al., 1999; Pegg & Shahidi, 1997). The structure and chemistry of the iron atom have an impact on the reactions and color changes that myoglobin undergoes (Livingston & Brown, 1981). The oxidation of ferrous-oxymyoglobin (Fe^{2+}) to ferric-metmyoglobin (Fe^{3+}) is responsible for discoloration of meat during storage. Ferrous iron (Fe^{2+}) can react with molecular oxygen to produce superoxide anion ($O_2^{\cdot-}$) with concomitant oxidation to ferric iron (Fe^{3+}). Hydrogen peroxide (H_2O_2), which may be produced by dismutation of $O_2^{\cdot-}$, can react with Fe^{2+} to produce hydroxyl radical ($OH^{\cdot}$) (Hultin, 1992). This reaction termed as Fenton reaction is the principal mechanism for myoglobin oxidation (Fig. 1.2).

$$Fe^{2+} + O_2 \longrightarrow Fe^{3+} + O_2^{\cdot-}$$

$$2O_2^{\cdot-} + 2H^+ \longrightarrow H_2O_2 + O_2$$

$$Fe^{2+} + H_2O_2 \longrightarrow Fe^{3+} + OH^- + OH^{\cdot}$$

FIGURE 1.2   Reactive oxygen species generated by the Fenton reaction.

In general, fish myoglobins are more readily oxidized than the mammalian counterpart (Haard, 1992). Discoloration of tuna meat during frozen storage is associated with the formation of metmyoglobin (Haard, 1992). This phenomenon can be influenced by many factors such as pH, temperature, ionic strength, and oxygen consumption reaction (Renerre & Labas, 1987). Metmyoglobin formation is positively correlated with lipid oxidation (Chan et al., 1997a, 1997b; Lee et al., 2003a, 2003b). Benjakul and Bauer (2001) suggested that the freeze-thaw process caused damage of cell and heme-proteins, resulting in the release of prooxidants. Haard (1992) also reported that fish myoglobins are at least 2.5 times more sensitive to autoxidation than mammalian myoglobins. Autoxidation of myoglobin becomes greater as temperature increased and pH decreased (Livingston et al., 1981; Chaijan et al., 2007). Chaijan et al. (2007) demonstrated that sardine myoglobin was prone to oxidation and denaturation at temperature above 40 °C and at very acidic or alkaline pHs as evidenced by the formation of metmyoglobin, the changes in tryptophan fluorescence intensity as well as the disappearance of Soret absorption. Furthermore, the rate of myoglobin autoxidation was related to oxygen concentration (Brown & Mebine, 1969). Atmospheres enriched in carbon dioxide (CO_2) are effective in delaying spoilage of meat; however, one problem is that CO_2 can promote the oxidation of oxymyoglobin to metmyoglobin, thereby causing the discoloration (Haard, 1992). Post-harvest discoloration of fish muscle has been reviewed by Chaijan and Panpipat (2009).

It has been suggested that myoglobin has a close relationship with lipid oxidation (O'Grady et al., 2001; Ohshima et al., 1988). Besides the unpleasant color, the oxidation of myoglobin is the main cause in the development of the undesirable odor during ice storage of fish muscle. Lee et al. (2003a) reported that surface metmyoglobin accumulation and lipid oxidation of refrigerated tuna (*Thunnus albacares*) steaks increased during six days of storage, leading to discoloration and lowered odor acceptability. The total lipid hydroperoxide content and TBARS of the yellowtail (*Seriola quinqueradiata*) dark muscle were higher than those of the ordinary muscle during two days of ice storage. Those changes were accompanied with the increasing intensity of fishy, spoiled, and rancid off-odor smells as well as increasing metmyoglobin formation. However, no correlation was found between the content of total lipid hydroperoxide and the odor intensities in ordinary muscle (Sohn et al., 2005). It is believed that the formation of metmyoglobin by the oxidation of myoglobin predominantly in dark muscle accelerates lipid oxidation and leads to the generation of greater amounts of hydroperoxide. Thus, the lipid oxidation associated with metmyoglobin formation may have caused the development of the rancid off-odor and fishy smell in dark muscle. For ordinary muscle of yellowtail which contained a low level of metmyoglobin, the influence of myoglobin oxidation on the development of rancid off-odor appeared to be insignificant (Sohn et al., 2005). The suppression of myoglobin oxidation will in turn decrease lipid oxidation and off-odor development of muscle foods.

1.6 INTERRELATIONSHIP BETWEEN LIPID OXIDATION AND MYOGLOBIN OXIDATION IN FOODS OF ANIMAL ORIGIN

The heme proteins including hemoglobin and myoglobin are effective promoters of lipid oxidation (Love, 1983; Han et al., 1994). Myoglobin consists of a globin portion plus a porphyrin heme, the latter containing an iron atom coordinated inside the heme ring (Grunwald & Richards, 2006a,

2006b). Heme is the nomenclature used to describe the porphyrin ring containing ferrous (Fe^{2+}) iron, while hemin describes the porphyrin ring containing ferric (Fe^{3+}) iron. Ferrous myoglobin are typically either liganded with O_2 or no ligand is present (e.g., deoxymyoglobin). The problem in understanding the pathway by which heme proteins promote lipid oxidation is that heme protein autoxidation, ferryl radical formation, heme dissociation, heme destruction, and iron release can all occur in a very short time sequence and simultaneously so that the most relevant step related to lipid oxidation is obscured ring (Grunwald & Richards, 2006a, 2006b). Heme-initiated lipid oxidation, especially myoglobin, has been extensively reported in meats (Ledward, 1987; Love & Pearson, 1974; Richards & Hultin, 2002). The interrelationship between lipid and myoglobin oxidations in muscle foods has been reported by Chaijan (2008). Ohshima et al. (1988) proposed that the lipid oxidation in fish muscle was promoted by autoxidation of myoglobin. Moreover, O'Grady et al. (2001) reported a relationship between oxymyoglobin oxidation and lipid oxidation in bovine muscle. There are numerous potential mechanisms by which myoglobin can promote lipid oxidation in muscle foods. The process by which ferrous myoglobin (or hemoglobin) is converted to ferric metmyoglobin is called autoxidation. Superoxide anion radical (O_2^{-}) or •OOH is liberated in this process depending on whether deoxy or oxy heme protein undergoes autoxidation (Brantley et al., 1993). O_2^{-} and •OOH can readily be converted to hydrogen peroxide (H_2O_2), which enhances the ability of heme proteins to promote lipid oxidation. Moreover, metmyoglobin can react with H_2O_2 or lipid hydroperoxides to generate ferryl heme protein radicals, which can abstract hydrogen from PUFA and hence initiate lipid oxidation. Alternatively, displaced hemin or released iron can stimulate lipid oxidation. Metmyoglobin is an effective prooxidant at acidic pH and in the presence of hydroperoxides. Morey et al. (1973) found that H_2O_2 acting as an oxidizing agent caused changes in the oxidation state of the iron in myoglobin. The reaction between H_2O_2 and metmyoglobin results in the formation of red pigment, ferrylmyoglobin (MbFe(IV)=O). Under physiological conditions (pH 7.4), ferrylmyoglobin is a strong prooxidant, which is able to abstract a hydrogen atom from fatty acids with subsequent stereospecific addition of oxygen (Rao et al., 1994). The prooxidative activity of ferrylmyoglobin is independent of pH and of lipid concentration (Chan et al., 1997a, 1997b; Kanner et al., 1987).

The concentration of ferrous iron and its ability to be active in the lipid oxidation reaction will be a key factor causing differences in lipid oxidation among species. In general, dark meats tend to have more reactive iron. Chaijan et al. (2004) reported that lipid and myoglobin contents were higher in dark muscle than in ordinary muscle of both sardine and mackerel. Saturation of red color in meat was directly related to myoglobin concentration (Faustman et al., 1992). Other constituents of meat including enzymatic and non-enzymatic reducing systems can accelerate oxidation by converting iron from the inactive ferric form to the active ferrous state (Foegeding et al., 1996). Changes of PV, conjugated diene (CD) and TBARS in sardine muscle indicated that lipid oxidation occurred throughout 15 days of iced storage. Apart from a plenty of unsaturated fatty acids, heme protein as well as reactive iron in the muscle might contribute to the accelerated oxidation (Chaijan et al., 2006).

Under fluctuating oxygen supply and pH decrease of post mortem system, the heme pigments like hemoglobin and myoglobin become catalytic in lipid peroxidation by mechanisms involving both one- and two-electron transfer processes, which are different from mechanisms for lipid oxidation by the non-heme-iron LOX (Carlsen et al., 2005). Reeder and Wilson (1998) suggested that myoglobin plays a role of photosensitizer which may be responsible for the initial formation of lipid hydroperoxides and increase the rate of oxygen uptake of fish oil via a photosensitized oxidation.

1.6.1 ROLE OF DEOXYMYOGLOBIN IN LIPID OXIDATION

The physiologically active myoglobin species are the purple high-spin iron (II) myoglobin (deoxymyoglobin), which has the sixth coordination site of the heme iron vacant, and the bright cherry-red low-spin oxy-iron (II) myoglobin (oxymyoglobin), which bind a molecule of oxygen at the sixth coordination of the heme iron, due to their high affinity for oxygen (Baron & Andersen, 2002; Faustman et al., 1999; Gorelik & Kanner, 2001). Disturbance of the globin structure can result in binding of the unusual ligands (e.g., the distal histidine in the heme cavity, exogenous amino acids as histidine and methionine, or a hydroxyl group) at the sixth coordination of the heme iron and induce the formation of a low-spin iron (II) species, known as hemochromes. Hemochromes in its oxidation state II can be found either reversible (binding to the imidazole group of the distal histidine or hydroxyl ion) or irreversible (binding to the imidazole group of free histidine) (Baron & Andersen, 2002).

The study regarding prooxidative activity of deoxymyoglobin in biological system including muscle foods is scarce (Baron & Andersen, 2002). This is mainly due to the fact that deoxymyoglobin initiated lipid oxidation demands strictly anaerobic condition; to exclude oxymyoglobin initiated lipid oxidation and the subsequent propagation of lipid oxidation. However, Richards and Dettmann (2003) postulated that perch and trout deoxyhemoglobin could stimulate lipid oxidation in washed cod muscle during storage at 4 °C as evidenced by the formation of lipid peroxides and TBARS. A more rapid formation of methemoglobin from deoxygenated molecules, deoxyhemoglobin, likely increases the lipid oxidation (Richards et al., 2002).

1.6.2 ROLE OF OXYMYOGLOBIN IN LIPID OXIDATION

Oxymyoglobin oxidation and lipid oxidation are coupled (Yin & Faustman, 1993). A high correlation between oxymyoglobin oxidation and lipid oxidation both in microsomes and liposomes was reported by Yin and Faustman (1993, 1994) and O'Grady et al. (2001). Lipid oxidation in fresh meat is influenced by the oxidation of oxymyoglobin since the oxymyoglobin oxidation results in production of two species known as the prooxidants, namely metmyoglobin and H_2O_2 (Chan et al., 1997a, 1997b). It has been proposed that $O_2^{\cdot-}$ and H_2O_2 are produced during oxidation of oxymyoglobin to metmyoglobin (Gotoh & Shikama, 1976). $O_2^{\cdot-}$ can further react with H_2O_2 and Fe^{3+} via the Fenton reaction to produce $OH^{\cdot}$ and facilitate lipid oxidation (Chan et al., 1997a, 1997b). The prooxidative effect of oxymyoglobin toward lipid oxidation was concentration-dependent (Chan et al., 1997a, 1997b). Additionally, H_2O_2 can react with metmyoglobin to form a prooxidative ferrylmyoglobin radical (Decker & Hultin, 1992). Kanner and Harel (1985) reported that H_2O_2-activated metmyoglobin caused the rapid oxidation of poultry skeletal muscle microsomes. Moreover, reactive oxygen species including superoxide, hydroperoxyl radical ($HOO^{\cdot}$), and H_2O_2, originated by the autoxidation of oxymyoglobin (Kruger-Ohlsen & Skibsted, 1997), can cause damage to muscle lipids via oxidation reaction (Skulachev, 1996; Hultin & Kelleher, 2000).

Prooxidative activity of oxymyoglobin in a myoglobin-liposome system was concentration-dependent and showed the higher activity than did metmyoglobin. The added sperm whale myoglobin was found to promote lipid oxidation in washed cod by which lipid oxidation occurred more rapidly at pH 5.7 compared to pH 6.3 (Grunwald & Richards, 2006b). Prooxidative activity of oxymyoglobin is difficult to assess because of continuous autoxidation of oxymyoglobin to metmyoglobin. Chan et al. (1997a) and Yin and Faustman (1993) reported a high correlation between oxymyoglobin oxidation and lipid oxidation both in microsomes and liposomes system. The role of oxymyoglobin oxidation in lipid oxidation was reported by Chan et al.

(1997b). Hogg et al. (1994) showed that oxymyoglobin can promote oxidative modification of low-density lipoprotein. Galaris et al. (1990) showed visible absorption spectral change of oxymyoglobin upon incubation with linoleic acid at physiological pH. This could be attributed to the formation of the noncatalytic low-spin myoglobin derivative, hemochrome (Akhrem et al., 1989).

1.6.3 ROLE OF METMYOGLOBIN IN LIPID OXIDATION

High-spin iron (III) myoglobin, commonly known as metmyoglobin, binds a molecule of water at the sixth coordination site of the heme iron (Pegg & Shahidi, 1997). Like hemochromes, the low-spin iron (III) myoglobin species known as hemichromes can be formed by disturbance of the globin structure. Hemichrome formation is either reversible or irreversible depending on the type of ligand at the sixth coordination site of the iron and the extent of globin denaturation. Hemichrome formation from iron (III) myoglobin is the intermediate step in the heat denaturation of myoglobin in muscle foods (Baron & Andersen, 2002). Post mortem process, especially the pH fall, continuously inactivate the reductive enzyme systems and stimulate acid-catalyzed autoxidation of the iron (II) states to the iron (III) state of myoglobin, resulting in the accumulation of metmyoglobin in meats (George & Stratmann, 1954; Gotoh & Shikama, 1976).

Formation of metmyoglobin is highly correlated with lipid oxidation in muscle foods (Andersen & Skibsted, 1991). Baron et al. (1997) found that metmyoglobin is an effective prooxidant at acidic pH and in the presence of hydroperoxides. In contrast, at physiological pH and in the presence of lipids, metmyoglobin can undergo a rapid neutralization due to formation of the noncatalytic heme pigment. However, further denaturation of the heme proteins due to a high lipophilic environment may result in heme release or further exposure of the heme group to the surrounding lipids, thereby inducing lipid peroxidation (Baron & Andersen, 2002). Metmyoglobin acts as a prooxidant in raw fish more effectively than in raw turkey, chicken, pork, beef, and lamb (Livingston et al., 1981).

In addition, the lipid to heme protein ratio is an important factor affecting the prooxidative activity of heme proteins (Kendrick & Watts, 1969). At lower linoleate/heme protein ratios, heme proteins become ineffective initiators of lipid oxidation (Nakamura & Nishida, 1971). The mechanism responsible for the inhibition of lipid oxidation at low linoleate/heme protein ratios has been proposed. The fatty acid anions bind reversibly to metmyoglobin, resulting in a spin transition, to yield the low-spin metmyoglobin derivative which is not prooxidative (Baron et al., 1998). At high linoleate-to-heme ratios, metmyoglobin immediately denatures and results in exposure or release of the heme group to the environment that instantly initiates hematin-induced lipid peroxidation in the system (Baron & Andersen, 2002). The result of Grunwald and Richards (2006a) also confirmed that sperm whale metmyoglobin caused a more rapid formation of lipid peroxides and TBARS in washed cod muscle as compared to ferrous myoglobin during 2 °C storage.

1.6.4 ROLE OF FERRYLMYOGLOBIN IN LIPID OXIDATION

The reaction between H_2O_2 and metmyoglobin results in the formation of a red pigment, ferrylmyoglobin (Baron & Andersen, 2002). During this interaction, the production of free radicals occurs in the globin part of the heme protein. H_2O_2 activation of metmyoglobin is a necessary step in the conversion of metmyoglobin to a prooxidant (Kanner & Harel, 1985). Interaction between metmyoglobin and H_2O_2 is a complex mechanism, resulting in the generation of two distinct hypervalent myoglobin species, perferrylmyoglobin ($^{\cdot}MbFe(IV)=O$) and ferrylmyoglobin ($MbFe(IV)=O$) (Davies, 1990, 1991) as follows:

$$\text{Metmyoglobin} + H_2O_2 \longrightarrow {^{\cdot}MbFe(IV)=O} \longrightarrow MbFe(IV)=O$$

Perferrylmyoglobin is a transient species with a very short half-life and autoreduces rapidly to the more stable ferrylmyoglobin (Baron & Andersen, 2002). Ferrylmyoglobin is a relatively stable species, which is slowly reduced back to metmyoglobin at physiological pH but with an increasing rate at decreasing pH due to an acid-catalyzed process (Mikkelsen & Skibsted, 1995). Perferrylmyoglobin can effectively transfer its radical to other proteins and subsequently induces lipid oxidation (Baron & Andersen, 2002). However, the ability of perferrylmyoglobin to initiate lipid oxidation by abstracting a hydrogen atom from fatty acids (LH) was suggested by Kanner and Harel (1985) as shown in the following reaction:

$$^{\cdot}MbFe(IV)=O + LH \longrightarrow MbFe(IV)=O + L^{\cdot} + H^{+}$$

Ferrylmyoglobin is responsible for the oxidation of a variety of substrates (Baron & Andersen, 2002). Under conditions similar to those found in muscle foods, ferrylmyoglobin is able to initiate lipid oxidation (Hogg et al., 1994). However, under the conditions found in fresh meat (pH 5.5–5.8), ferrylmyoglobin autoreduces rapidly to metmyoglobin. Nevertheless, under physiological conditions (pH 7.4), ferrylmyoglobin is a strong prooxidant, which is able to abstract a hydrogen atom from fatty acids with subsequent stereospecific addition of oxygen (Rao et al., 1994). The prooxidative activity of ferrylmyoglobin is independent of pH and of lipid concentration (Baron & Andersen, 2002). Therefore, ferrylmyoglobin is expected to be an effective prooxidant under the conditions found in muscle food, as well as under physiological conditions. However, ferrylmyoglobin formation in muscle tissues is determined by hydrogen peroxide and lipid hydroperoxide production. Its potential to oxidize lipids depends on the concentration of reducing agents and their compartmentalization in the muscle cells (Baron & Andersen, 2002).

The interrelationship between myoglobin oxidation, lipid oxidation, and discoloration in oxeye scad fish during ice storage has been reported by Wongwichian et al. (2015). The myoglobin autoxidation rate, hydrogen peroxide, and ferrylmyoglobin concentrations increased with increasing storage time. The CD and PV of oxeye scad lipids tended to stabilize during the initial phase of storage, increased in the differentiation phase and had declined at the end of storage. However, TBARS increased markedly. Overall, lipid and myoglobin oxidations in oxeye scad occurred in a concurrent manner and each process appeared to enhance the other.

Heme, hematin and hemin are normally used interchangeably to describe the existence of non-protein bound heme-iron (or "free heme iron"). Heme in solution is mainly found as hematin (ferriprotoporphyrin hydroxide). Hemin is ferriprotoporphyrin chloride which readily converts to hematin in aqueous solution and accordingly the term hematin should be used for non-protein bound heme-iron (Carlsen et al., 2005). Grunwald and Richards (2006b) suggested that sperm whale myoglobin having a more rapid hemin loss rate possessed a more effective prooxidative activity than did modified myoglobin with high hemin affinity. It was found that hemin concentrations in mackerel light muscle increased around 3-fold during ice storage (Decker & Hultin, 1990). Following release of hemin from the globin, hemin is proposed to intercalate within phospholipids membranes due to hydrophobic attractions. Also the propionate groups of hemin can bind with phospholipid headgroup amines by electrostatic interactions (Cannon et al., 1984). Hemin can react with lipid hydroperoxide to form alkoxyl radical and ferryl-hydroxo complex (Dix & Marnett, 1985). Ferryl-hydroxo complex can react with another lipid hydroperoxide to form a peroxyl radical and regenerate hemin:

$$\text{hemin}(3+) + \text{LOOH} \longrightarrow \text{LO}^\cdot + \text{hemin}(4+)\text{-OH}$$

$$\text{hemin}(4+)\text{-OH} + \text{LOOH} \longrightarrow \text{LOO}^\cdot + \text{hemin}(3+) + H_2O$$

Alkoxyl and peroxyl radicals are capable of abstracting a hydrogen atom from a PUFA which will stimulate the lipid oxidation processes (Grunwald & Richards, 2006b).

Recently, a new heme protein determination method for fish muscle overcoming such extractability problems faced by previous reported methods was developed by Chaijan and Undeland (2015). The principle was to homogenize and heat samples in an SDS-containing phosphate buffer to dissolve major muscle components and convert ferrous/ferric heme proteins to hemichromes with a unique absorption peak at 535 nm.

1.7   INTERACTION BETWEEN LIPID OXIDATION PRODUCTS AND MYOGLOBIN

Lipid oxidation generates a wide range of secondary aldehyde products, which are predominantly *n*-alkanals, *trans*-2-alkenals, 4-hydroxy-*trans*-2-alkenals, and malondialdehyde (Lynch & Faustman, 2000). Lynch et al. (2001) demonstrated that propional, pentenal, hexanal, and 4-hydroxynonenal (4-HNE) were the primary aldehydes formed during lipid oxidation in stored ground beef at 4 °C. The aldehyde products are more stable than free radical species and readily diffuse into the cellular media, where they may exert toxicological effects by reacting with critical biomolecules *in vivo* (Esterbauer et al., 1991). Aldehydes produced during lipid oxidation can form adducts with proteins and this may have an impact on protein stability and functionality as well as the color stability of meat. Aldehyde products can alter myoglobin stability (Lynch & Faustman, 2000). Covalent modification of equine, bovine, porcine and tuna myoglobin by 4-hydroxynonenal (4-HNE) has been demonstrated (Faustman et al., 1999; Phillips et al., 2001a, 2001b; Lee et al., 2003a, 2003b). Lynch and Faustman (2000) also determined the effect of aldehydic lipid oxidation products on oxymyoglobin oxidation, metmyoglobin reduction and the catalytic activity of metmyoglobin as a lipid prooxidant *in vitro*. Metmyoglobin formation was greater in the presence of α,β-unsaturated aldehydes than their saturated counterparts of equivalent carbon chain length (Faustman et al., 1999). The covalent attachment of aldehydes to oxymyoglobin rendered oxymyoglobin more susceptible to oxidation (Faustman et al., 1998). Alderton et al. (2003) studied the effect of 4-HNE on bovine metmyoglobin formation. Increased metmyoglobin formation was found in the presence of 4-HNE. Similar results were observed by Faustman et al. (1999) during the incubation of 4-HNE with equine oxymyoglobin and by Lee et al. (2003a, 2003b) with porcine and tuna oxymyoglobin. The covalent binding of α,β-unsaturated aldehydes to oxymyoglobin at key amino acid residues may subsequently lead to alter tertiary structure of the protein and increases susceptibility to oxidation. This would result in a loss of physiological activity and the brown discoloration in fresh meat (Alderton et al., 2003). Alderton et al. (2003) and Lee et al. (2003a) demonstrated that 4-HNE covalently attached to bovine and porcine myoglobin, respectively. The liquid chromatography-mass spectrometry (LC-MS) spectra revealed the covalent binding of up to three molecules of 4-HNE to bovine myoglobin and a di-adduct formed under the reaction of 4-HNE with porcine myoglobin.

1.8 CONCLUSION

Chemical constituents in foods of animal origin including lipid, protein, pigment, and vitamin in muscle tissue are susceptible to oxidative reactions resulting in the loss of quality, including discoloration, development of off-flavors, loss of nutrients, textural changes, and progression of spoilage and/or pathogenicity. Among such compositions, lipid and heme proteins especially myoglobin are prone to oxidation. Lipid oxidation and myoglobin oxidation in meat are coupled and both reactions appear capable of influencing each other. The oxidation of oxymyoglobin results in the production of metmyoglobin and H_2O_2 necessary to induce lipid oxidation. On the other hand, aldehyde lipid oxidation products alter myoglobin redox stability, resulting in the promoted oxidation of oxymyoglobin and the formation of adduct with myoglobin through covalent modification. Thus, studies of the relationship between lipid oxidation and myoglobin oxidation processes in muscle foods are important in understanding reactions and mechanisms that may affect the quality and acceptability and could be useful in minimizing lipid oxidation of meats and meat products during handling, processing, and storage.

KEYWORDS

- **mechanism**
- **lipid oxidation**
- **myoglobin**
- **muscle foods**
- **quality**

REFERENCES

Airado-Rodríguez, D.; Intawiwat, N.; Skaret, J.; Wold, J. P. The Effect of Naturally Occurring Tetrapyrroles on Photooxidation in Cow Milk. *J. Agric. Food Chem.* **2011,** *59,* 3905–3914.

Akhrem, A. A.; Andreyuk, G. M.; Kisel, M. A.; Kiselev, P. A. Hemoglobin Conversion to Hemichrome Under the Influence of Fatty Acids. *Biochim. Biophys. Acta.* **1989,** *992,* 191–194.

Alderton, A. L.; Faustman, C.; Liebler D. C.; Hill, D. W. Induction of Redox Instability of Bovine Myoglobin by Adduction with 4-Hydroxy-2-Nonenal. *Biochemistry.* **2003,** *42,* 4398–4405.

Allen, C. E; Foegeding, E. A. Some Lipid Characteristics and Interactions in Muscle Foods Overview. *Food Technol.* **1981,** *35,* 253–257.

An, S.; Edwald Lee, E.; Choe, E. Effects of Solubility Characteristics of Sensitiser and pH on the Photooxidation of Oil in Tuna Oil-Added Acidic O/W Emulsions. *Food Chem.* **2011,** *128,* 358–363.

Andersen, H. J.; Skibsted, L. H. Oxidative Stability of Frozen Pork Patties: Effect of Light and Added Salt. *J. Food Sci.* **1991,** *56,* 1182–1184.

Andreo, A. I.; Doval, M. M.; Romero, A. M.; Judis, M. A. Influence of Heating Time and Oxygen Availability on Lipid Oxidation in Meat Emulsions. *Eur. J. Lipid Sci. Technol.* **2003,** *105,* 207–213.

Anwar, S. H.; Kunz, B. The Influence of Drying Methods on the Stabilization of Fish Oil Microcapsules: Comparison of Spray Granulation, Spray Drying, and Freeze Drying. *J. Food Eng.* **2011,** *105,* 367–378.

Apgar, M. E.; Hultin, H. O. Lipid Peroxidation in Fish Muscle Microsomes in the Frozen State. *Cryobiology.* **1982,** *19,* 154–162.

Baron, C. P.; Andersen, H. J. Myoglobin-Induced Lipid Oxidation. A Review. *J. Agric. Food Chem.* **2002,** *50,* 3887–3897.

Baron, C. P.; Skebsted, L. H.; Andersen, H. J. Peroxidation of Linoleate at Physiological pH: Hemichrome Formation by Substrate Binding Protects against Metmyoglobin Activation by Hydrogen Peroxides. *Free Rad. Biol. Med.* **1998,** *28,* 549–558.

Baron, C. P.; Skebsted, L. H.; Andersen, H. J. Prooxidative Activity of Myoglobin Species in Linoleic Acid Emulsions. *J. Agric. Food Chem.* **1997,** *45,* 1704–1710.

Beltran, E.; Pla, R.; Yuste, J.; Mor-Mur, M. Use of Antioxidants to Minimize Rancidity in Pressurized and Cooked Chicken Slurries. *Meat Sci.* **2004,** *66,* 719–725.

Benjakul, S.; Bauer, F. Biochemical and Physicochemical Changes in Catfish (*Silurus glanis* Linne) Muscle as Influenced by Different Freeze-Thaw Cycles. *Food Chem.* **2001,** *72,* 207–217.

Bolumar, T.; Andersen, M. L.; Orlien, V. Antioxidant Active Packaging for Chicken Meat Processed by High Pressure Treatment. *Food Chem.* **2011,** *129,* 1406–1412.

Bolumar, T.; Andersen, M. L.; Orlien, V. Mechanisms of Radical Formation in Beef and Chicken Meat During High Pressure Processing Evaluated by Electron Spin Resonance Detection and the Addition of Antioxidants. *Food Chem.* **2014,** *150,* 422–428.

Bolumar, T.; Skibsted, L. H.; Orlien, V. Kinetics of the Formation of Radicals in Chicken Breast During High Pressure Processing. *Food Chem.* **2012,** *134,* 2114–2120.

Bradley, D. G.; Lee, H. O.; Min, D. B. Singlet Oxygen Detection in Skim Milk by Electron Spin Resonance Spectroscopy. *J. Food Sci.* **2003,** *68,* 491–494.

Brantley, R. E.; Smerdon, S. J.; Wilkinson, A. J.; Singleton, E. W.; Olson, J. S. The Mechanism of Autooxidation of Myoglobin. *J. Biol. Chem.* **1993,** *268,* 6995–7010.

Brown, W. D.; Mebine, L. B. Autoxidation of Oxymyoglobins. *J. Biol. Chem.* **1969,** *244,* 6696–6701.

Cannon, J. B.; Kuo, F. S.; Pasternack, R. F.; Wong, N. M.; Muller-Eberhard, U. Kinetics of the Interaction of Hemin Liposomes with Heme Binding Proteins. *Biochemistry.* **1984,** *23,* 3715–3721.

Carlsen, C. U.; Meller, J. K. S.; Skibsted, L. H. Heme-Iron in Lipid Oxidation. *Coord. Chem. Rev.* **2005,** *249,* 485–498.

Castell, C. H.; MacLean, J.; Moore, B. Rancidity in Lean Fish Muscle. IV. Effect of Sodium Chloride and Other Salts. *J. Fish. Res. Board Can.* **1965,** *22,* 929–944.

Chaijan, M. Physicochemical Changes of Tilapia (*Oreochromis niloticus*) Muscle During Salting. *Food Chem.* **2011,***129,* 1201–1210.

Chaijan, M. Review: Lipid and Myoglobin Oxidations in Muscle Foods. *Songklanakarin J. Sci. Tech.* **2008,** *30,* 47–53.

Chaijan, M.; Benjakul, S.; Visessanguan, W.; Faustman, C. Changes of Lipids in Sardine (*Sardinella gibbosa*) Muscle during Iced Storage. *Food Chem.* **2006,** *99,* 83–91.

Chaijan, M.; Benjakul, S.; Visessanguan, W.; Faustman, C. Characteristics and Gel Properties of Muscles from Sardine (*Sardinella gibbosa*) and Mackerel (*Rastrelliger kanagurta*) Caught in Thailand. *Food Res. Int.* **2004,** *37,* 1021–1030.

Chaijan, M.; Benjakul, S.; Visessanguan, W.; Faustman, C. Characterization of Myoglobin from Sardine (*Sardinella gibbosa*) Dark Muscle. *Food Chem.* **2007,** *100,* 156–164.

Chaijan, M.; Klomklao, S.; Benjakul, S. Characterisation of Muscles from Frigate Mackerel (*Auxis thazard*) and Catfish (*Clarias macrocephalus*). *Food Chem.* **2013,** *139,* 414–419.

Chaijan, M.; Panpipat, W. Post Harvest Discoloration of Dark-Fleshed Fish Muscle: A Review. *Walailak J. Sci. Tech.* **2009,** *6,* 149–166.

Chaijan, M.; Undeland, I. Development of a New Method for Determination of Total Haem Protein in Fish Muscle. *Food Chem.* **2015,** *173,* 1133–1141.

Chan, W. K. M.; Decker, E. A.; Chow, C. K.; Boissonneault, G. A. Effect of Dietary Carnosine on Plasma and Tissue Antioxidant Concentrations and on Lipid Oxidation in Rat Skeletal Muscle. *Lipids.* **1994,** *29,* 461–466.

Chan, W. K. M.; Faustman, C.; Yin, M.; Decker, E. A. Lipid Oxidation Induced by Oxymyoglobin and Metmyoglobin with Involvement of H_2O_2 and Superoxide Anion. *Meat Sci.* **1997a,** *46,* 181–190.

Chan, W. K. M.; Faustman, C.; Renerre, M. Model Systems for Studying Pigment and Lipid Oxidation Relevant to Muscle Based Foods. In *Natural Antioxidants. Chemistry, Health Effects, and Applications;* AOCS Press: Champaign, IL, 1997b; pp 319–330.

Chen, T. C.; Waimaleongora, E. K. C. Effect of pH on TBA Values of Ground Raw Poultry Meat. *J. Food Sci.* **1981,** *46,* 1946–1947.

Cheng, J.; Ockerman, H. W. Effect of Phosphate with Tumbling on Lipid Oxidation of Precooked Roast Beef. *Meat Sci.* **2003,** *65,* 1353–1359.

Davies, M. Detection of Myoglobin-Derived Radicals on Reaction of Metmyoglobin with Hydrogen Peroxide and other Peroxide Compounds. *Free Rad. Res. Com.* **1990,** *10,* 361–370.

Davies, M. Identification of a Globin Free Radical in Equine Myoglobin Treated with Peroxides. *Biochim. Biophys. Acta.* **1991,** *1077,* 86–90.

Dawson, L. E.; Gartner, R. Lipid Oxidation in Mechanically Deboned Poultry. *Food Technol.* **1983,** *37,* 112–116.

Decker, E. A. Antioxidant Mechanisms. In *Food Lipids;* Akoh, C. C., Min, D. B., Eds.; Marcel Dekker: New York, 1998; p 397.

Decker, E. A.; Hultin, H. O. Factors Influencing Catalysis of Lipid Oxidation by the Soluble Fraction of Mackerel Muscle. *J. Food Sci.* **1990,** *55,* 947–950.

Decker, E. A.; Hultin, H. O. Lipid Oxidation in Muscle Foods Via Redox Iron. In *Lipid Oxidation in Food;* St. Angelo, A. J., Ed.; ACS Symposium Series: Washington, DC, 1992; p 500.

Dix, T. A.; Marnett, L. J. Conversion of Linoleic Acid Hydroperoxide to Hydroxy, Keto, Epoxyhydroxy, and Trihydroxy Fatty Acids by Hematin. *J. Biol. Chem.* **1985,** *260,* 5351–5357.

Dunn, C. J.; Rohlfs, R. J.; Fee, J. A.; Saltman, P. Oxidation of Deoxy Myoglobin by $[Fe(CN)_6]^{3-}$. *J. Inorg. Biochem.* **1999,** *75,* 241–244.

Enser, M. What is Lipid Oxidation. *Food Sci. Technol. Today.* **1987,** *1,* 151–153.

Erickson, M. C. Lipid Oxidation in Muscle Foods. In *Food Lipids Chemistry, Nutrition and Biotechnology;* Akoh, C. C., Min, D. B., Eds.; CRC Press: New York, 2003; p 322.

Erickson, M. C.; Hultin, H. O.; Borhan, M. Eeffect of Cytosol on Lipid Peroxidation in Flounder Sarcoplasmic Reticulum. *J. Food Biochem.* **1990,** *14,* 407–419.

Esterbauer, H.; Schaur, R. J.; Zollner, H. Chemistry and Biochemistry of 4-Hydroxynonenal, Malonaldehyde and Related Aldehydes. *Free Rad. Biol. Med.* **1991,** *11,* 81–128.

Faustman, C.; Chan, W. K. M.; Lynch, M. P.; Joo, S. T. In *Strategies for Increasing Oxidative Stability of Fresh Meat Color,* Reciprocal Meat Conference Proceedings, AMSA: Brigham Young University: Utah, 1996; p 73.

Faustman, C.; Chan, W. K. M.; Schaefer, D. M.; Havens, A. Beef Color Update: The Role for Vitamin E. *J. Anim. Sci.* **1998,** *76,* 1019–1026.

Faustman, C.; Liebler, D. C.; McClure, T. D.; Sun, Q. α,β-Unsaturated Aldehydes Accelerate Oxymyoglobin Oxidation. *J. Agric. Food Chem.* **1999,** *47,* 3140–3144.

Faustman, C.; Phillips, A. L. Measurement of Discoloration in Fresh Meat. In *Current Protocols in Food Analytical Chemistry;* Wrolstad, R. E., Ed.; Wiley & Sons, Inc: New York, 2001; p F3.3.1.

Faustman, C.; Yin, M. C.; Nadeau, D. B. Color Stability, Lipid Stability, and Nutrient Composition of Red and White Veal. *J. Food Sci.* **1992,** *57,* 302–304.

Foegeding, E. A; Lanier, T. C.; Hultin, H. O. Characteristics of Edible Muscle Tissues. In *Food Chemistry;* Fennema, O. R., Ed.; Marcel Dekker, Inc: New York, 1996; p 880.

Frankel, E. N. *Lipid Oxidation;* The Oily Press: Dundee, Scotland, 1998.

Frankel, E. N. Recent Advances in the Chemistry of the Rancidity of Fats. In *Recent Advances in the Chemistry of Meat;* Bailey, A. J., Ed.; The Royal Society of Chemistry: London, 1984; p 87.

Frankel, E. N.; Neff, W. E.; Selke, E. Analysis of Autoxidized Fats by Gas Chromatography-Mass Spectrometry. IX. Homolytic vs. Heterolytic Cleavage of Primary and Secondary Oxidation Products. *Lipids.* **1984,** *19,* 790–800.

Galaris, D.; Sevanian, A.; Cadenas, E.; Hochstein, P. Ferrylmyoglobin-Catalyzed Linoleic Acid Peroxidation. *Arch. Biochem. Biophys.* **1990,** *281,* 163–169.

Ganthavorn, C.; Nagel, C. W.; Powers, J. R. Thermal Inactivation of Asparagus Lipoxidase and Peroxidase. *J. Food Sci.* **1991,** *56,* 47–49.

George, P.; Stratmann, C. J. The Oxidation of Myoglobin to Metmyoglobin by Oxygen. *Biochem. J.* **1954,** *57,* 568–573.

German, J. B.; Zhang H.; Berger, R. Role of Lipoxygenases in Lipid Oxidation in Foods. In *Lipid Oxidation in Food;* St. Angelo, A. J., Ed.; ASC Symposium Series: Washington, DC, 1992; Vol. 500, p 74.

Giddings, G. G. Reduction of Ferrimyoglobin in Meat. *CRC Crit. Rev. Food Technol.* **1974,** *5,* 143–173.

Girotti, A.W. Lipid Hydroperoxide Generation, Turnover, and Effect or Action in Biological Systems. *J. Lipid Res.* **1998,** *39,* 1529–1542.

Gorelik, S.; Kanner, J. Oxymyoglobin Oxidation and Membranal Lipid Peroxidation Initiated by Iron Redox Cycle. *J. Agric. Food Chem.* **2001,** *49,* 5939–5944.

Gotoh, T.; Shikama, K. Generation of the Superoxide Radical during Autooxidation of Oxymyoglobin. *J. Biochem.* **1976,** *80,* 397–399.

Gray, J. I. Measurement of Lipid Oxidation. A Review. *J. Am. Oil Chem. Soc.* **1978,** *55,* 539–546.

Grossman, S.; Bergman, M.; Sklan, D. Lipoxygenase in Chicken Muscle. *J. Agric. Food Chem.* **1988,** *36,* 1268–1270.

Grunwald, E. W.; Richards, M. P. Mechanisms of Heme Protein-Mediated Lipid Oxidation Using Hemoglobin and Myoglobin Variants in Raw and Heated Washed Muscle. *J. Agr. Food Chem.* **2006a,** *54,* 8271–8280.

Grunwald, E. W.; Richards, M. P. Studies with Myoglobin Variants Indicate that Released Hemin is the Primary Promoter of Lipid Oxidation in Washed Fish Muscle. *J. Agr. Food Chem.* **2006b,** *54,* 4452–4460.

Gunstone, F. D.; Norris, F. A. Oxidation. In *Lipid in Foods: Chemistry, Biochemistry and Technology;* Gunstone, F. D., Norris, F. A., Eds.; Pergamon Press: New York, 1983; p 58.

Gutteridge, J. M.; Halliwell, B. The Measurement and Mechanism of Lipid Peroxidation in Biological Systems. *Trends Biochem. Sci.* **1990,** *15,* 129–135.

Haard, N. F. Biochemistry and Chemistry of Color and Color Change in Seafoods. In *Advance in Seafood Biochemistry;* Flick, G. J., Martin, R. E., Eds.; Technomic Publishing Co., Inc: Lancaster, PA, 1992; p 312.

Halliwell, B.; Murcia, M. A.; Chirico, S.; Auroma, O. I. Free Radicals and Antioxidants in Food and *in Vivo*: What They do and How They Work. *Crit. Rev. Food Sci. Nutr.* **1995,** *35,* 7–20.

Hammer, F. E. Oxidoreductases. In *Enzymes in Food Processing;* Nagodawithana, T., Reed, G., Eds.; Academic Press, Inc.: Salt Lake City, UT, 1993; p 221.

Han, D.; Mcmillin, K. W.; Godber, J. S. Hemoglobin, Myoglobin, and Total Pigments in Beef and Chicken Muscles: Chromatographic Determination. *J. Food Sci.* **1994,** *59,* 1279–1282.

Han, T. J.; Liston, J. Lipid Peroxidation Protection Factors in Rainbow Trout (*Salmo gaird-nerii*) Muscle Cytosol. *J. Food Sci.* **1989**, *54*, 809–813.

Harris, P.; Tall, J. Substrate Specificity of Mackerel Flesh Lipopolygenase. *J. Food Sci.* **1994**, *59*, 504–506.

Hayashi, T.; Takimura, T.; Aoyama, Y.; Hitomi, Y. Structure and Reactivity of Reconstituted Myoglobins: Interaction between Protein and Polar Side Chain of Chemically Modified Hemin. *Inorg. Chim. Acta.* **1998**, *275*, 159–167.

Hoac, T.; Daun, T; Trafikowska, U.; Zackrisson, J.; Åkesson, B. Influence of Heat Treatment on Lipid Oxidation and Glutathione Peroxidase Activity in Chicken and Duck Meat. *Inno. Food Sci. Emer. Tech.* **2006**, *7*, 88–93.

Hogg, N.; Rice-Evens, C.; Darley-Usmar, V.; Wilson, M. T.; Paganga, G.; Bourne, L. The Role of Lipid Hydroperoxides in the Myoglobin-Dependent Oxidation of LDL. *Arch. Biochem. Biophys.* **1994**, *314*, 39–44.

Huang, W. H.; Greene, B. E. Effect of Cooking Method on TBA Numbers of Stored Beef. J. *Fd. Sci.* **1978**, *43*, 1201–1203.

Hultin, H. O. Lipid Oxidation in Fish Muscle. In *Advance in Seafood Biochemistry;* Flick, G. J., Martin, R. E., Eds.; Technomic Publishing Co., Inc: Lancaster, PA, 1992; p 99.

Hultin, H. O. Oxidation of Lipids in Seafoods. In *Seafoods: Chemistry, Processing Technology and Quality;* Shahidi, F., Botta, J. R., Eds.; Blackie Academic & Professional: London, 1994; p 49.

Hultin, H. O; Kelleher, S. D. Surimi Processing from Dark Muscle Fish. In *Surimi and Surimi Seafood;* Park, J. W., Ed.; Marcel Dekker: New York, 2000; p 59.

Huss, H. H. *Quality and Quality Changes in Fresh Fish*, FAO Fishery Technical Paper 348; Food and Agriculture Organization of the United Nations: Rome, 1995; p 48.

Kanner, J. Mechanisms of Nonezymic Lipid Peroxidation in Muscle Foods. In *Lipid Oxidation in Foods;* St. Angelo, A. J., Ed.; American Chemical Society Symposium Series: Washington, DC, 1992; Vol. 500, p 55.

Kanner, J. Oxidative Processes in Meat and Meat Products: Quality Implications. *Meat Sci.* **1994**, *36*, 169–189.

Kanner, J.; German, J. B.; Kinsella, J. E. Initiation of Lipid Peroxidation in Biological Systems. *CRC Crit. Rev. Food Sci. Nutr.* **1987**, *25*, 317–364.

Kanner, J.; Harel, S. Initiation of Membranal Lipid Peroxidation by Activated Metmyoglobin and Methemoglobin. *Arch. Biochem. Biophys.* **1985**, *237*, 314–321.

Kanner, J.; Kinsella, J. E. Lipid Deterioration Initiated by Phagocytic Cells in Muscle Foods: β-Caroene Destruction by a Myeloperoxidase-Hydrogenperoxide-Halide System. *J. Agric. Food Chem.* **1983**, *31*, 370–376.

Kasahara, K.; Osawa, C. Combination Effects of Spices on Masking of Odor in Boiled Sardine. *Fish. Sci.* **1998**, *64*, 415–418.

Ke, P. J.; Ackman, R. G.; Linke, B. A. Autoxidation of Polyunsaturated Fatty Compounds in Mackerel Oil: Formation of 2, 4, 7-Decatrienals. *J. Am. Oil Chem. Soc.* **1975**, *53*, 349–353.

Ke, S.; Eric, Y. H.; Decker, A.; Hultin, H. O. Impact of Citric Acid on the Tenderness, Microstructure and Oxidative Stability of Beef Muscle. *Meat Sci.* **2009**, *82*, 113–118.

Kendrick, J.; Watts, B. M. Acceleration and Inhibition of Lipid Oxidation by Heme Compounds. *Lipids.* **1969**, *4*, 454–458.

Kolakowska, A. Lipid Oxidation in Food Systems. In *Chemical and Functional Properties of Food Lipids;* Sikorski, Z. E., Kolakowska, A., Eds.; CRC Press: Boca Raton, FL, 2002; pp 133–160.

Kruger-Ohlsen, M.; Skibsted, L. H. Kinetics and Mechanism of Reduction of Ferrylmyoglobin by Ascorbate and D-Isoascorbate. *J. Agric. Food Chem.* **1997,** *45* (3), 668–676.

Ladikos, D.; Lougovois, V. Lipid Oxidation in Muscle Foods: A Review. *Food Chem.* **1990,** *35,* 295–314.

Lanier, T. C. Surimi Gelation Chemistry. In *Surimi and Surimi Seafood;* Park, J. W. Ed.; Marcel Dekker: New York, 2000; p 237.

Ledward, D. Interaction between Myoglobin and Lipid Oxidation in Meat and Meat Products. *Food Sci. Technol. Today.* **1987,** *1,* 153–155.

Lee, S.; Joo, S. T.; Alderton, A. L.; Hill, D. W.; Faustman, C. Oxymyoglobin and Lipid Oxidation in Yellowfin Tuna (*Thunnus albacares*) Loins. *J. Food Sci.* **2003b,** *68,* 1664–1668.

Lee, S.; Phillips, A. L.; Liebler, D. C.; Faustman, C. Porcine Oxymyoglobin and Lipid Oxidation *in Vitro. Meat Sci.* **2003a,** *63,* 241–247.

Li, R.; Richards, M. P.; Undeland, I. Characterization of Aqueous Components in Chicken Breast Muscle as Inhibitors of Hemoglobin-Mediated Lipid Oxidation. *J. Agric. Food Chem.* **2005,** *53,* 767–775.

Limsuwanmanee, J.; Chaijan, M.; Manurakchinakorn, M.; Panpipat, W.; Klomklao, S.; Benjakul, S. Antioxidant Activity of Maillard Reaction Products Derived from Stingray (*Himantura signifier*) Non-Protein Nitrogenous Fraction and Sugar Model Systems. *LWT Food Sci. Tech.* **2014,** *57,* 718–724.

Lindsay, R. C. Flavour of Fish. In *Seafoods: Chemistry, Processing Technology and Quality;* Shahidi, F., Botta, J. R., Eds.; Blackie Academic & Professional: London, 1994; p 75.

Litwinienko, G.; Daniluk, A.; Kasprzycka–Guttman, T. A Differential Scanning Calorimetry Study on the Oxidation of C12-C18 Saturated Fatty Acids and Their Esters. *J. Am. Oil Chem. Soc.* **1999,** *76,* 655–657.

Livingston, D. J.; Brown, W. D. The Chemistry of Myoglobin and its Reactions. *Food Technol.* **1981,** *35,* 244–252.

Love, J. D. The Role of Heme Iron in the Oxidation of Lipids in Red Meats. *Food Technol.* **1983,** *12,* 117–120.

Love, J. D.; Pearson, A. M. Lipid Oxidation in Meat and Meat Products. A Review. *J. Am. Oil Chem. Soc.* **1971,** *48,* 547–549.

Love, J. D.; Pearson, A. M. Metmyoglobin and Nonheme Iron as Prooxidants in Cooked Meat. *J. Agr. Food Chem.* **1974,** *22,* 1032–1034.

Lynch, M. P.; Faustman, C. Effect of Aldehyde Lipid Oxidation Products on Myoglobin. *J. Agric. Food Chem.* **2000,** *48,* 600–604.

Lynch, M. P.; Faustman, C.; Silbart, L. K.; Rood, D.; Furr, H. C. Detection of Lipid-Derived Aldehydes and Aldehyde: Protein Adducts *In Vitro* and in Beef. *J. Food Sci.* **2001,** *66,* 1093–1099.

Ma, H. J.; Ledward, D. A.; Zamri, A. I.; Frazier, R. A.; Zhou, G. H. Effects of High/Pressure Thermal Treatment on Lipid Oxidation in Beef and Chicken Muscle. *Food Chem.* **2007,** *104,* 1575–1579.

Maqsood, S.; Benjakul, S. Comparative Studies on Molecular Changes and Pro-Oxidative Activity of Haemoglobin from Different Fish Species as Influenced by pH. *Food Chem.* **2011a,** *124,* 875–883.

Maqsood, S.; Benjakul, S. Effect of Bleeding on Lipid Oxidation and Quality Changes of Asian Seabass (*Lates calcarifer*) Muscle during Iced Storage. *Food Chem.* **2011b,** *124,* 459–467.

Maqsood, S.; Benjakul, S.; Kamal-Eldin, A. Haemoglobin-Mediated Lipid Oxidation in the Fish Muscle: A Review. *Trends Food Sci. Tech.* **2012,** *28,* 33–43.

Mariutti, L. R.; Orlien, V.; Bragagnolo, N.; Skibsted, L. H. Effect of Sage and Garlic on Lipid Oxidation in High-Pressure Processed Chicken Meat. *Eur. Food Res. Tech.* **2008,** *227,* 337–344.

McGill, A. S.; Hardy, R.; Gunstone, F. D. Further Analysis of the Volatile Components of Frozen Cold Stored Cod and the Influence of These on Flavour. *J. Sci. Food Agric.* **1977,** *28,* 200–205.

Medina, I.; Saeed, S.; Howell, N. Enzymatic Oxidative Activity in Sardine (*Sardina pilchardus*) and Herring (*Clupea harengus*) during Chilling and Correlation with Quality. *Eur. Food Res. Technol.* **1999,** *210,* 34–38.

Mikkelsen, A.; Skibsted, L. H. Acid-Catalysed Reduction of Ferrylmyoglobin: Product Distribution and Kinetics of Autoreduction and Reduction by NADH. *Z. Lebsensm. Unters. Forsch.* **1995,** *200,* 171–177.

Moerck, K. E.; Ball, H. R. Lipid Oxidation in Mechanically Deboned Chicken Meat. *J. Food Sci.* **1974,** *39,* 876–881.

Morey, K. S.; Hansen, S. P.; Brown, W. D. Reaction of Hydrogen Peroxide with Myoglobins. *J. Food Sci.* **1973,** *38,* 1104–1107.

Morrissey, P. A.; Sheehy, P. J. A.; Galvin, K.; Kerry, J. P.; Buckley, D. J. Lipid Stability in Meat and Meat Products. *Meat Sci.* **1998,** *49,* 73–86.

Mottram, D. S. Lipid Oxidation and Flavour in Meat and Meat Products. *Food Sci. Technol. Today.* **1987,** *1,* 159–162.

Nakamura, Y.; Nishida, T. Effect of Hemoglobin Concentration on the Oxidation of Linoleic Acid. *J. Lipids Res.* **1971,** *12,* 149–154.

Nambudiry, D. D. Lipid Oxidation in Fatty Fish: The Effect of Salt Content in Meat. *J. Food Sci. Technol.* **1980,** *17,* 176–178.

Nawar, W. W. Lipids. In *Food Chemistry;* Fennema, O. R., Ed.; Marcel Dekker: New York; 1996; pp 225–319.

O'Grady, M. N; Monahan, F. J.; Brunton, N. P. Oxymyoglobin Oxidation in Bovine Muscle-Mechanistic Studies. *J. Food Sci.* **2001,** *66,* 386–392.

Ogden, S. K.; Gurrero, I.; Taylor, A. J.; Buendia, H. E.; Gallardo, F. Changes in Odour, Color and Texture during the Storage of Acid Preserved Meat. *Lebensm. Wiss. Technol.* **1995,** *28,* 521–527.

Ohshima, T.; Wada, S.; Koizumi, C. Influences of Heme Pigment, Non-Heme Iron, and Nitrite on Lipid Oxidation in Cooked Mackerel Meat. *Nippon Suisan Gakk.* **1988,** *54,* 2165–2171.

Orlien, V.; Hansen, E.; Skibsted, L. H. Lipid Oxidation in High-Pressure Processed Chicken Breast Muscle during Chill Storage: Critical Working Pressure in Relation to Oxidation Mechanism. *Eur. Food Res.Tech.* **2000,** *211,* 99–104.

Pacheco-Aguilar, R.; Lugo-Sanchez, M. E.; Robles-Burgueno, M. R. Postmortem Biochemical Characteristic of Monterey Sardine Muscle Stored at 0°C. *J. Food Sci.* **2000,** *65,* 40–47.

Pearson, A. M.; Love, J. D.; Shorland, F. B. Warmed-Over Flavour in Meat, Poultry and Fish. *Adv. Food Res.* **1977,** *23,* 1–74.

Pegg, R. B.; Shahidi, F. Unraveling the Chemical Identity of Meat Pigments. *CRC Crit. Rev. Food Sci. Nutr.* **1997,** *37,* 561–589.

Phillips, A. L.; Faustman, C.; Lynch, M. P.; Gonovi, K. E.; Hoagland, T. A.; Zinn, S. A. Effect of Dietary α-Tocopherol Supplementation on Color and Lipid Stability in Pork. *Meat Sci.* **2001a,** *58,* 389–393.

Phillips, A. L.; Lee, S.; Silbart, L. K.; Faustman, C. In *in-Vitro Oxidation of Bovine Oxymyo-globin as Affected by 4-Hydroxy-Nonenal.* The 54th Annual Reciprocal Meat Conference,

Indianapolis, July 24–28, 2001; American Meat Science Association: Chicago, 2001b; p 378.

Rao, S. I.; Wilks, A.; Hamberg, M.; Ortiz de Montellano, P. R. The Lipoxygenase Activity of Myoglobin. Oxidation of Linoleic Acid by the Ferryl Oxygen Rather than Protein Radical. *J. Biol. Chem.* **1994,** *269,* 7210–7216.

Reeder, B. J; Wilson, M. T. Mechanism of Reaction of Myoglobin with the Lipid Hydroperoxide Hydroperoxyoctadecadienoic Acid. *Biochem. J.* **1998,** *330,* 1317–1323.

Renerre, M. Oxidative Processes and Myoglobin. In *Antioxidants in Muscle Foods, Nutritional Strategies to Improve Quality;* Decker, E., Faustman, C., Lopez-Bote, C. J., Eds.; John Wiley & Sons, Inc.: New York, 2000; p 113.

Renerre, M.; Labas, R. Biochemical Factors Influencing Metmyoglobin Formation in Beef Muscles. *Meat Sci.* **1987,** *19,* 151–165.

Richards, M. P.; Hultin, H. O. Effect of pH on Lipid Oxidation using Trout Hemolysate as a Catalyst: A Possible Role for Deoxyhe Moglobin. *J. Agric. Food Chem.* **2000,** *48,* 3141–3147.

Richards, M. P.; Modra, A. M.; Li, R. Role of Deoxyhemoglobin in Lipid Oxidation of Washed Cod Muscle Mediated by Trout, Poultry and Beef Hemoglobin. *Meat Sci.* **2002,** *62,* 157–163.

Richards, M. P.; Dettmann, M. A. Comparative Analysis of Different Hemoglobins: Autoxidation, Reaction with Peroxide, and Lipid Oxidation. *J. Agric. Food Chem.* **2003,** *51,* 3886–3891.

Rodriguez-Estrada, M. T.; Penazzi, G.; Caboni, M. F.; Bertacco, G.; Lercker, G. Effect of Different Cooking Methods on Some Lipid and Protein Components of Hamburgers. *Meat Sci.* **1997,** *45,* 365–375.

Saeed, S.; Howell, N. K. Effect of Lipid Oxidation and Frozen Storage on Muscle Proteins of Atlantic Mackerel (*Scomber scombrus*). *J. Sci. Food Agric.* **2002,** *82,* 579–586.

Sae-leaw, T.; Benjakul, S.; Gokoglu, N.; Nalinanon, S. Changes in Lipids and Fishy Odour Development in Skin from *Nile tilapia* (*Oreochromis niloticus*) Stored in Ice. *Food Chem.* **2013,** *141,* 2466–2472.

Samson, E.; Stodolnik, L. Effect of Freezing and Salting on the Activity of Lipoxygenase of the Muscle Tissue and Roe of Baltic Herring. *Acta Ichthyol. Piscat.* **2001,** *31,* 97–111.

Sannaveerappa, T.; Cai, H.; Richards, M. P.; Undeland, I. Factors Affecting the Binding of Trout Hbi and Hbiv to Washed Cod Mince Model System and their Influence on Lipid Oxidation. *Food Chem.* **2014,** *143,* 392–397.

Sato, K.; Hegarty, G. R. Warmed Over Flavour in Cooked Meats. *J. Food Sci.* **1971,** *36,* 98–102.

Sies, H. Oxidative Stress: Oxidants and Antioxidants. *Exp Physiol.* **1997,** *82,* 291–295.

Simic, G. M.; Jovanovic, S. V.; Niki, E. Mechanisms of Lipid Oxidative Processes and Their Inhibition. In *Lipid Oxidation in Food;* St. Angelo, A. J., Ed.; ACS Symposium Series, Washigton, DC, 1992; Vol. 500, p 14.

Skulachev, V. P. Role of Uncoupled and Non-Coupled Oxidations in Maintenance of Safety Low Levels of Oxygen and its One-Electron Reductants. *Quart. Rev. Biophys.* **1996,** *29,* 169–202.

Slabyj, B. M.; Hultin, H. O. Oxidation of a Lipid Emulsion by a Peroxidizing Microsomal Fraction from Herring Muscle. *J. Food Sci.* **1984,** *49,* 1392–1393.

Sohn, J. H.; Taki, Y.; Ushio, H.; Kohata, T.; Shioya, I.; Ohshima, T. Lipid Oxidations in Ordinary and Dark Muscles of Fish: Influences on Rancid Off-Odor Development and Color Darkening of Yellowtail Flesh During Ice Storage. *J. Food Sci.* **2005,** *70,* s490–s496.

Taylor, A. J. Effect of Water Quality on Lipid Oxidation. *Fd Sci. TechnoL Today.* **1987**, *1*, 158–159.

Thiansilakul, Y.; Benjakul, S.; Richards, M. P. Changes in Heme Proteins and Lipids Associated with Off-Odour of Seabass (*Lates calcarifer*) and Red Tilapia (*Oreochromis mossambicus* × *O. niloticus*) during Iced Storage. *Food Chem.* **2010**, *121*, 1109–1119.

Tichivangana, J. Z.; Morrissey, P. A. Metmyoglobin and Inorganic Metals as Pro-Oxidants in Raw and Cooked Muscle Systems. *Meat Sci.* **1985**, *15*, 107–116.

Tims, M. J.; Watts, B. M. Protection of Cooked Meats with Phosphates. *Food Technol.* **1958**, *12*, 240–243.

Tolstorebrov, I.; Eikevik, T. M.,; Indergard, E. The Influence of Long-Term Storage, Temperature and Type of Packaging Materials on the Lipid Oxidation and Flesh Color of Frozen Atlantic Herring Fillets (*Clupea harengus*). *Int. J. Refrig.* **2014**, *40*, 122–130.

Undeland, I. Lipid Oxidation in Fatty Fish during Processing and Storage. In *Farmed Fish Quality;* Kestin, S. C., Warris, P. D., Eds.; Fishing News Books, Blackwell Science: Oxford, 2001; p 261.

Undeland, I.; Hultin, H. O.; Richards, M. P. Aqueous Extracts from Some Muscles Inhibit Hemoglobin-Mediated Oxidation of Cod Muscle Membrane Lipids. *J. Agric. Food Chem.* **2003**, *51*, 3111–3119.

Vareltzis, P.; Hultin, H. O.; Autio, W. R. Hemoglobin-Mediated Lipid Oxidation of Protein Isolates Obtained from Cod and Haddock White Muscle as Affected by Citric Acid, Calcium Chloride and pH. *Food Chem.* **2008**, *108*, 64–74.

Vareltzis, P.; Hutin, H. O. Effect of Low Ph on the Susceptibility of Isolated Cod Microsomes to Lipid Oxidation. *J. Agric Food Chem.* **2007**, *55*, 9859–9867.

Varlet, V.; Knockaert, C.; Prost, C; Serot, T. Comparison of Odor-Active Volatile Compounds of Fresh and Smoked Salmon. *J. Agric. Food Chem.* **2006**, *54*, 3391–3401.

Wattanachant, S.; Benjakul, S.; Ledward, D. A. Effect of Heat Treatment on Changes in Texture, Structure and Properties of Thai Indigenous Chicken Muscle. *Food Chem.* **2005**, *93*, 337–348.

Wilson, B. R.; Pearson, A. M.; Shorland, F. B. Effect of Total Lipids and Phospholipids on Warmed-Over Flavour in Red and White Muscle from Several Species as Measured by Thiobarbituric Acid Analysis. *J. Agric. Food Chem.* **1976**, *24*, 7–11.

Wongwichian, C.; Klomklao, S.; Panpipat, W.; Benjakul, S.; Chaijan, M. Interrelationship between Myoglobin and Lipid Oxidations in Oxeye Scad (*Selar boops*) Muscle during Iced Storage. *Food Chem.* **2015**, *174*, 279–285.

Yin, M.; Faustman, C. α-Tocopherol and Ascorbate Delay Oxymyoglobin and Phospholipid Oxidation *in Vitro*. *J. Food Sci.* **1993**, *58*, 1273–1276.

Yin, M.; Faustman, C. The Influence of Microsomal and Cytosolic Components on the Oxidation of Myoglobin and Lipid *in Vitro*. *Food Chem.* **1994**, *51*, 159–164.

Zamora, R.; Hidalgo, F. J. Inhibition of Proteolysis in Oxidized Lipid-Damaged Proteins. *J. Agric. Food Chem.* **2001**, *49*, 6006–6011.

CHAPTER 2

NATURAL ANTIOXIDANTS: OCCURRENCE AND THEIR ROLE IN FOOD PRESERVATION

AJIT SINGH BHATNAGAR and REWA KULSHRESTHA*

Department of Food Processing and Technology, Bilaspur University, Bilaspur, Chhattisgarh, India

Corresponding author. E-mail: rewakumar11@gmail.com

CONTENTS

ABSTRACT

Antioxidants neutralize or inhibit free radicals by preventive and radical scavenging modes. Free radicals are unstable reactive molecules; rapidly attack the molecules in nearby cells leading to aging of body. The repair process involves scavenging free radicals by antioxidant compounds. Antioxidants have dual role: shelf-life prolongation and combating oxidative stress. Consumers' concerns regarding the bio-safety of synthetic antioxidants have pushed the food industry to seek natural alternatives such as ascorbic acid (AA), tocols, carotenoids, phenolics, oryzanol (OZ), and so forth. In contrast to animal foods, foods of vegetable origin usually contain natural antioxidants, such as tocopherols, carotenoids, or flavonoids in sufficient amounts due to higher degree of unsaturation. Natural antioxidants provide preservative action in various foods, namely cereal based breakfast foods, baked foods like bread and crackers, dried products, and processed fruit products. This chapter provides full insight on different naturally occurring antioxidants' structure, levels, and effectiveness as food preservative.

2.1 ANTIOXIDANTS: WHAT, WHY, AND HOW

Antioxidants are compounds which neutralize free radicals or inhibit free radicals. Free radicals are unstable molecules with an unpaired electron and highly reactive. To become stable, they take electron from other molecules/cells and in the process eventually damage the DNA and cause aging, degenerative diseases, and may also lead to cancer. Free radicals can be classified as reactive oxygen species (ROS) and reactive nitrogen species (RNS). Few examples of ROS are alkoxyl radicals (RO$^\bullet$), peroxyl radicals (ROO$^\bullet$), hydroxyl radical (HO$^\bullet$), and superoxide anion radical ($O_2^{\bullet-}$) while examples of RNS would be nitric oxide radical (NO$^\bullet$) and nitrogen dioxide radical ($NO_2^\bullet$). Potential sources of free radicals could be ultraviolet (UV) or ionizing radiations, metabolic processes, inflammatory reactions, air-pollution, and tobacco (smoking or chewing). ROS are formed *in vivo* both usefully and "accidentally." Their formation increases in all human disease, and sometimes makes a significant contribution to disease severity (Halliwell, 1995). Free radicals damage the cell molecules like DNA, proteins, and lipids, potentially causing a variety of disorders, including diabetes mellitus, hypertension, cancer, Alzheimer, and aging of body. Free radicals like ROS and RNS are generally produced due to the oxidative stress in the body. The oxidative stress leads to over 200 disorders including aging of body, degenerative disorders, and cancer. Oxidative stress can cause single/multi-organ disorders/diseases including brain disorders like Alzheimer, Parkinson, obsessive–compulsive disorder (OCD), attention deficit hyperactivity disorder (ADHD), autism, migraine, stroke, trauma, and cancer; lung disorders like asthma, chronic obstructive pulmonary disease (COPD), allergies, acute respiratory distress syndrome (ARDS), and cancer; eye disorders like macular or retinal degeneration and cataract; heart disorders like coronary heart diseases (CHD), cardiac fibrosis, hypertension, ischemia, and myocardial infarction; kidney disorders like chronic kidney diseases, renal graft, and nephritis; bone and joint disorders like rheumatism, osteoarthritis, and psoriasis; blood vessel disorders like restenosis, atherosclerosis, endothelial dysfunction, and hypertension; skin disorders like skin aging, sunburn, psoriasis, dermatitis, and melanoma; multi-organ disorders like diabetes, aging, and chronic fatigue; and immune system disorders like chronic inflammations, auto-immune disorders, lupus, inflammatory bowel disease (IBD), multiple sclerosis (MS), and cancer. Oxidative stress in the body leads to the generation of free radicals like ROS and RNS which cause the above-mentioned disorders/diseases. Now, the million-dollar question is how to reduce the oxidative stress in the body? The answer would be by slowing down the oxidative processes inside the body. How would it be achieved? The answer is ANTIOXIDANTS.

As we said earlier, antioxidants are compounds which neutralize free radicals or inhibit free radicals. An antioxidant can be defined as: "any substance that, when present in low concentrations compared to that of an oxidisable substrate, significantly delays or inhibits the oxidation of that substrate." This oxidisable substrate (lipid, protein, and carbohydrate) can generate free radicals if it involves transfer of unpaired single electrons. Examples of oxygen-centered free radicals, also known as ROS, are superoxide (O_2^-), hydroxyl ($HO^{\bullet}$), peroxyl ($ROO^{\bullet}$), alkoxyl ($RO^{\bullet}$), and nitric oxide ($NO^{\bullet}$). The hydroxyl (half-life of 10^{-9} s) and the alkoxyl (half-life of seconds) free radicals are very reactive and rapidly attack the molecules in nearby cells, and probably the damage caused by them is unavoidable and is dealt with by repair processes (Gülcin, 2012). The repair process involves scavenging free radicals by antioxidant compounds. Antioxidants have dual role: shelf-life prolongation and combating oxidative stress. Antioxidants are often added to foods to prevent the radical chain reactions of oxidation, and they act by inhibiting the initiation and propagation step leading to the termination of the reaction and delay the oxidation process (Shahidi & Wanasundara, 1992; Gülcin, 2006). Antioxidants may be broadly grouped according to their mechanism of action into primary or chain-breaking antioxidants and secondary or preventive antioxidants. The main difference with primary antioxidants is that the secondary antioxidants do not convert free radicals into stable molecules (Wanasundara & Shahidi, 2005). The primary antioxidants consist mainly of hindered phenols and hindered aromatic amines. They scavenge and destroy the chain propagating peroxy and alkoxy radicals before they can react with the polymer. Gum guaiac was the first antioxidant approved for the stabilization of animal fats, especially lard.

2.2 ANTIOXIDANTS VERSUS FREE RADICALS

The modes of action of an antioxidant can be broadly classified as preventive/inhibitory and radical scavenging. Some examples of preventive/inhibitory antioxidant enzymes are superoxide dismutase, catalase, and glutathione peroxidase while radical scavenging antioxidants can be broadly classified as hydrogen donating and iron-chelating antioxidants (ICA). Hydrogen donating antioxidants (HDA) donate a hydrogen to peroxy radicals of lipid or protein, which are mainly responsible for free radical chain reaction in a biological system. Few examples of synthetic HDA are butylated hydroxy anisole (BHA), butylated hydroxy toluene (BHT), propyl gallate (PG), tertiary butyl hydroquinone (TBHQ), and so forth, while natural HDA are AA (vitamin C), tocols (tocopherols and tocotrienols), carotenoids, polyphenols, phenolics, phytosterols (PS), lignans, OZ, gossypol, and so forth. ICA usually deactivate radically active trace metals by the formation of complex ion. Examples of synthetic and natural ICA are ethylene diamine tetra acetic acid (EDTA) and phytic acid, which form ions EDTA complex and phytate complex, respectively. Several mechanisms by which antioxidants provide defense mechanism against free radicals are:

1. Scavenging species that initiate peroxidation,
2. Chelating metal ions such that they are unable to generate reactive species or decompose lipid peroxides,
3. Quenching O_2^- preventing formation of peroxides,
4. Breaking the autoxidative chain reaction, and/or
5. Reducing localized O_2 concentrations (Nawar, 1996).

Several assays for quantifying antioxidant capacity are known and well established. Few of them include:

1. ORAC (oxygen radical absorbance capacity).

2. DPPH (2,2-diphenyl-1-picrylhydrazyl).
3. ABTS (2,2'-azino-bis(3-ethylbenzothiazoline-6-sulphonic acid)).
4. FRAP (ferric ion reducing antioxidant power).

It should be noted that antioxidant activity of food extracts can be determined using a variety of tests (stable free radical scavengers: galvinoxyl, diphenylpicrylhydrazyl (DPPH); lipid oxidation: peroxide oxygen, conjugated dienes, Rancimat (measurements of oxygen consumption of a linoleic acid emulsion and oxidation induction period in lard at 100 °C), oxygen radical absorbance capacity (ORAC) values), active oxygen method, iodine value (IV) (measure of the change in number of double bonds that bind I), anisidine value (reaction of acetic acid p-anisidine and aldehydes to produce a yellow color that absorbs at 350 nm), measurement of absorbance at 234 nm (conjugated dienes) and 268 nm (conjugated trienes) to assess oxidation in the early stages, and chromatographic methods; however, extraction procedures strongly influence the composition of the extracts and, therefore, also influence the antioxidant activity results (Halliwell, 1996; Schwarz et al., 2001; Trojakova et al., 2001; Brewer, 2011). In addition, the effect of the antioxidant compound in a food matrix may be significantly different than the activity of a purified extract (Brewer, 2011).

2.3   SYNTHETIC ANTIOXIDANTS VERSUS NATURAL ANTIOXIDANTS

The use of synthetic antioxidants (such as BHA, BHT, PG, and TBHQ) to preserve food products for a longer shelf life with retained quality and organoleptic attributes has become common commercially. However, the consumers' concerns regarding their bio-safety have motivated the food industry to seek natural alternatives. Synthetic phenolic antioxidants BHA, BHT, PG, and TBHQ effectively inhibit oxidation by scavenging free radicals. Chelating agents, such as EDTA, can bind radically active trace metals reducing their contribution to the process. Some natural substances like vitamins (AA, tocols, and phylloquinone), carotenoids, polyphenols/phenolics, PS, sesame lignans, OZ, and phytic acid can also perform the role of radical scavenging and metal ion chelating effectively and as efficiently as their synthetic counterparts. Natural antioxidants, mostly absorb light in the UV region (100–400 nm), can effectively scavenge free radicals, and chelate transition metals, thus stopping progressive autoxidative damage and production of off-odors and off-tastes in food products. Consumers have expressed concern about the safety of preservatives and additives in their food (Brewer et al., 1994; Brewer & Prestat, 2002; Rojas & Brewer, 2008; Brewer, 2011). Sloan (1999) reported that one of the top 10 trends for the food industry to watch included the sales of natural, organic, and vegetarian foods. There is a clear trend in consumer preference for clean labeling (Hillmann, 2010; Brewer, 2011), for food ingredients and additives that are organic/natural with names that are familiar, and that are perceived to be healthy (Joppen, 2006; Brewer, 2011). In addition, the call for sustainable sources and environment friendly production is forcing the food industry to move in that direction (Berger, 2009; Brewer, 2011). Phenolic antioxidants can inhibit free radical formation and/or interrupt propagation of autoxidation. Fat-soluble vitamin E (α-tocopherol (α-T)) and water-soluble vitamin C (L-ascorbic acid (AH_2)) are both effective in the appropriate matrix. Plant extracts, generally used for their flavoring characteristics, often have strong H-donating activity thus making them extremely effective antioxidants (Brewer, 2011).

2.4 NATURAL ANTIOXIDANTS: DYNAMICS AND MECHANISM

Chain-breaking antioxidants differ in their antioxidative effectiveness depending on their chemical characteristics and physical location within a food (proximity to membrane phospholipids (PL), emulsion interfaces, or in the aqueous phase). The chemical potency of an antioxidant and solubility in oil influence its accessibility to peroxy radicals especially in membrane, micellar and emulsion systems, and the amphiphilic character required for effectiveness in these systems (Wanatabe et al., 2010; Brewer, 2011). Antioxidant effectiveness is related to activation energy, rate constants, oxidation–reduction potential, ease with which the antioxidant is lost or destroyed (volatility and heat susceptibility), and antioxidant solubility (Nawar, 1996; Brewer, 2011). In addition, inhibitor and chain propagation reactions are both exothermic. As the A:H and R:H bond dissociation energies increase, the activation increases and the antioxidant efficiency decreases. Conversely, as these bond energies decrease, the antioxidant efficiency increases. The most effective antioxidants are those that interrupt the free radical chain reaction. Usually containing aromatic or phenolic rings, these antioxidants donate H to the free radicals formed during oxidation becoming a radical themselves. These radical intermediates are stabilized by the resonance delocalization of the electron within the aromatic ring and formation of quinone structures (Nawar, 1996; Brewer, 2011). In addition, many of the phenolics lack positions suitable for molecular oxygen attack. Both synthetic (BHA, BHT, and PG) and natural antioxidants contain phenolics (flavonoids) function in this manner. Natural extracts with antioxidant activity generally quench free radical oxygen with phenolic compounds (POH) as well. Because bivalent transition metal ions, Fe^{2+} in particular, can catalyze oxidative processes, leading to formation of hydroxyl radicals, and can decompose hydroperoxides via Fenton reactions, chelating these metals can effectively reduce oxidation (Halliwell et al., 1987; Brewer, 2011). Food materials containing significant amounts of these transition metals (red meat) can be particularly susceptible to metal-catalyzed reactions. Food tissues, because they are (or were) living, are under constant oxidative stress from free radicals, ROS, and pro-oxidants generated both exogenously (heat and light) and endogenously (H_2O_2 and transition metals). For this reason, many of these tissues have developed antioxidant systems to control free radicals, lipid oxidation catalysts, oxidation intermediates, and secondary breakdown products (Nakatani, 2003; Agati et al., 2007; Brown & Kelly, 2007; Chen, 2008; Iacopini et al., 2008; Brewer, 2011). These antioxidant compounds include flavonoids, phenolic acids, carotenoids, and tocopherols that can inhibit Fe^{3+}/AA induced oxidation, scavenge free radicals, and act as reductants (Khanduja, 2003; Ozsoy et al., 2009; Brewer, 2011). Spices and herbs, used in foods for their flavor and in medicinal mixtures for their physiological effects, often contain high concentrations of POH that have strong H-donating activity (Lugasi et al., 1995; Muchuweti et al., 2007; Brewer, 2011).

2.5   ROLE OF LIPID FRACTIONS IN RADICAL SCAVENGING AND ANTIOXIDANT ACTIVITY

Radical scavenging activity (RSA) tests are used to evaluate the health impact of many bioactive compounds found in foods. RSA tests like DPPH and galvinoxyl free radicals are generally used for seed oils and their free radical quenching ability are measured by spectrophotometric and ESR assays. Generally, it is accepted that the higher the degree of unsaturation of an oil, the more susceptible it is to oxidative deterioration. RSA of seed oil fractions like neutral lipids (NL), glycolipids (GL), and PL is also studied. The results revealed that the PL fraction had the strongest antiradical action followed by GL and NL, respectively (Ramadan et al., 2003). The radical quenching property of GL was expected to be due to reducing sugars in all GL components and the sterol moiety in steryl glucoside. Moreover, less polar POH that have been extracted with GL may be responsible for the strong antiradical action. On the other hand, four postulates have been proposed to explain the antioxidant activity of PL:

i)   synergism between PL and tocopherols;
ii) chelation of pro-oxidant metals by phosphate groups;
iii) formation of Maillard-type products between PL and oxidation products; and
iv) action as an oxygen barrier between oil/air interfaces (Ramadan, 2012).

2.6   ROLE OF FATTY ACIDS IN OXIDATIVE STABILITY

All edible oils and fats consist of triglycerides with a variety of fatty acids that differ in chain-length (number of carbon atoms in molecule), degree of saturation (number of double bond in carbon chain), position of double bond within the carbon chain, and geometry of each double bond (cis and trans isomers). Oleic acid is the most abundant monounsaturated fatty acid (MUFA) in all the common edible oils (Gunstone, 2000; Abdulkarim et al., 2007). Compared with polyunsaturated fatty acids (PUFA), oleic acid is more stable toward oxidation both at ambient storage temperatures and at the high temperatures that prevail during the cooking and frying of food. Therefore, oils with high amounts of oleic acid are slower to develop oxidative rancidity during shelf life or undergo oxidative decomposition during frying than those oils that contain high amounts of PUFA. The various strengths of hydrogen–carbon bond of fatty acids explain the differences of oxidation rates of stearic, oleic, linoleic, and linoleic acids during thermal oxidation or autoxidation. Compared with (PUFA) (ω-6 and ω-3 PUFA), oleic acid (ω-9 MUFA) and saturated fatty acids are more stable toward oxidation both at ambient storage temperatures and at the high temperatures that prevail during the cooking and frying of food (Abdulkarim et al., 2007).

The oil rich in linoleic acid is more easily polymerized during deep-fat frying than the oil rich in oleic acid. The energy required to break carbon–hydrogen bond on the carbon 11 of linoleic acid is 50 kcal/mol (Min & Boff, 2002; Choe & Min, 2007). The double bonds at carbon 9 and carbon 12 decrease the carbon–hydrogen bond at carbon 11 by withdrawing electrons. The carbon–hydrogen bond on carbon 8 or 11, which is α to the double bond of oleic acid, is about 75 kcal/mol. The carbon–hydrogen bond on the saturated carbon without any double bond next to it is ~100 kcal/mol (Min & Boff, 2002; Choe & Min, 2007). Oxidation produces hydroperoxides and then low molecular volatile compounds such as aldehydes, ketones, carboxylic acids, and short-chain alkanes and alkenes. In a study reported by Bhatnagar et al. (2009), it was found that the oxidative stability (OS) of oil blends depended upon the PUFA and MUFA content of the oil blends. The higher the PUFA and MUFA content, the lower would be the OS, while the RSA of oil blends depended upon the total tocopherols' content. The higher the total tocopherols' content the higher the DPPH scavenging activity would be. Antioxidant decreases the frying oil oxidation, but the effectiveness of antioxidant decreases with high frying temperature.

OS of stripped and crude seed oil was studied during 21 days under accelerated oxidative conditions at 60 °C (Ramadan & Moersel, 2004). Peroxide value (PV) and UV absorptivity were determined to monitor lipid oxidation during the experiment. The crude oil had a much lower PV than that of stripped oil over the entire storage period. PV in crude oil remained increased at a low level over 21 days, whereas the peroxides accumulated in the stripped oils to high levels. Absorption at 232 nm and 270 nm, due to the formation of primary and secondary compounds of oxidation, showed a pattern similar to that of the PV. The high content of conjugated oxidative products is attributed to high PUFA content which is readily decomposed to form conjugated hydroperoxides. It was concluded that the low OS of PUFA rich oil could be partly explained by the fact that it has a high proportion of PUFA. Aside from the fatty acid profile, factors such as oxygen concentration, metal contaminants, lipid hydroxy compounds, enzymes, and light may also influence the OS of the oil (Ramadan, 2012).

2.7   VITAMIN E OR TOCOLS (TOCOPHEROLS AND TOCOTRIENOLS)

Tocols (tocopherols and tocotrienols) constitute a series of benzopyranols (or methyl tocols) that occur in plant tissues and vegetable oils and are powerful lipid-soluble antioxidants. In the tocopherols, the C16 side chain is saturated, and in the tocotrienols it contains three *trans* double bonds. Together, these two groups are termed the tocochromanols. In essence, the tocopherols have a 20-carbon phytyl tail (including the pyranol ring), and the tocotrienols a 20-carbon geranyl tail with double bonds at the 3', 7', and 11' positions, attached to the benzene ring. The side-chain methyl groups have *R,R,R* stereochemistry. The four main constituents of the two classes are termed—*alpha* (5,7,8-trimethyl), *beta* (5,8-dimethyl), *gamma* (7,8-dimethyl), and *delta* (8-methyl) (Christie, 2013). The plant chloroplast is site for tocol biosynthesis and the aromatic amino acid tyrosine is considered to be its precursor. The mechanism of biosynthesis of tocols involves coupling of phytyl diphosphate with homogenestic acid (2,5-dihydroxyphenylacetic acid), followed by cyclization and methylation reactions. Vitamin E compounds include the tocopherols and tocotrienols. Tocotrienols have a conjugated triene double bond system in the phytyl side chain, while tocopherols do not. Methyl substitution affects the bioactivity of vitamin E, as well as its *in vitro* antioxidant activity. Tocopherols or vitamin E have eight known homologues, that is, α-, β-, γ-, δ- tocopherols and α-, β-, γ-, δ- tocotrienols. Tocopherols are fat-soluble antioxidants that function as scavengers of lipid peroxyl radicals. Tocopherols' content is found to be related to RSA and antioxidant activity of oils. Tocopherols' content decreases during processing of oils. There have been many reports on the protective effect of tocopherols as food antioxidants (Dougherty, 1988). Tocopherols protect PL and cholesterol against oxidation (Faustman et al., 1989; Li et al., 1996). Total tocopherol content and major tocopherol homologues differ from one oil to another (Table 2.1 and Fig. 2.1). The level of tocopherols decreases with time of storage and heating of oils (Li et al., 1996).

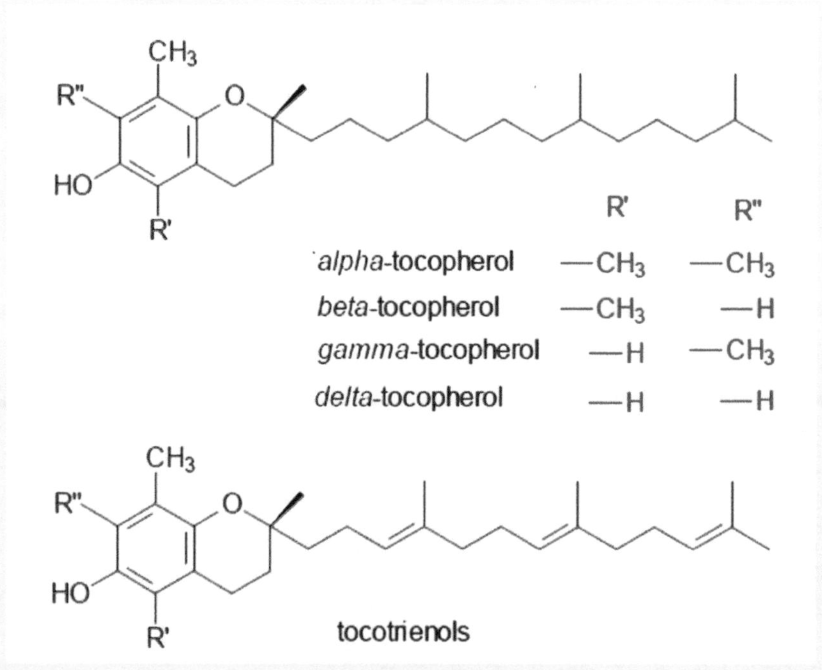

	R'	R"
alpha-tocopherol	—CH₃	—CH₃
beta-tocopherol	—CH₃	—H
gamma-tocopherol	—H	—CH₃
delta-tocopherol	—H	—H

FIGURE 2.1   Different tocopherol homologues of vegetable oils.

TABLE 2.1 Levels of Tocopherols and Tocotrienols in Some Crude Vegetable Oils (mg/kg) (Kamal-Eldin & Andersson, 1997; Bhatnagar et al., 2009; Codex Stan 210, 2011).

Tocopherols	Coconut oil	Groundnut oil	Mustard oil	Olive oil	Palm oil	Rice bran oil	Safflower seed oil	Sesame oil	Soyabean oil	Sunflower seed oil
α-tocopherol	ND-17.0	49.0–373.0	268.0–380.0	4.0–280.0	4.0–193.0	49.0–583.0	234.0–660.0	ND-3.3	9.0–352.0	403.0–935.0
β-tocopherol	ND-11.0	ND-41.0	ND	1.0–10.0	ND-234.0	ND-47.0	ND-17.0	ND	ND-36.0	ND-45.0
γ-tocopherol	ND-14.0	88.0–389.0	426.0–550.0	1.0–10.0	ND-526.0	ND-212.0	ND-12.0	521.0–983.0	89.0–2307.0	ND-34.0
δ-tocopherol	ND	ND-22.0	97.0–150.0	ND	ND-123.0	ND-31.0	ND	4.0–21.0	154.0–932.0	ND-7.0
α-tocotrienol	ND-44.0	ND	ND	ND	4.0–336.0	ND-627.0	ND	ND	ND-69.0	ND
γ-tocotrienol	ND-1.0	ND	ND	ND	14.0–710.0	142.0–790.0	ND-12.0	ND-20.0	ND-103.0	ND
δ-tocotrienol	ND	ND	ND	ND	ND-377.0	ND-59.0	ND	ND	ND	ND
Total tocopherols	ND-50.0	170.0–850.0	790.0–1050.0	5.0–300.0	150.0–1500.0	191.0–2349.0	240.0–670.0	330.0–1010.0	600.0–3370.0	440.0–1020.0

ND: not detected.

2.7.1 ANTIOXIDATIVE MECHANISM OF TOCOPHEROLS

Vitamin E or tocopherols which are benzopyranols or methylated tocols are natural antioxidants and integral bioactive molecules of an oil or fat. The primary task of tocopherols is to act as antioxidants to prevent free radical damage to unsaturated lipids or other membrane constituents of the tissues. Tocopherols are powerful antioxidants *in vitro* and *in vivo*. They are certainly extremely useful as antioxidants in non-biological systems, including foods, cosmetics, pharmaceutical preparations, and so forth (Christie, 2013). Because of their lipophillic character, tocopherols are located in the membranes or with storage lipids where that are immediately available to interact with lipid hydroperoxides. They react rapidly in a non-enzymic manner to scavenge lipid peroxyl radicals, that is, the chain-carrying species that propagate lipid peroxidation. In model systems *in vitro*, all the tocopherols ($\alpha > \gamma > \beta > \delta$) and tocotrienols are good antioxidants (Christie, 2013). In general, the oxidation of lipids is known to proceed by a chain process mediated by a free radical, in which the lipid peroxyl radical serves as a chain carrier. In the initial step of chain propagation, a hydrogen atom is abstracted from the target lipid by the peroxyl radical as shown:

$$LOO^{\cdot} + LH \rightarrow LOOH + L^{\cdot} \tag{2.1}$$

$$L^{\cdot} + O_2 \rightarrow LOO^{\cdot} \tag{2.2}$$

where LH is a lipid; LOO$^{\cdot}$ is the lipid peroxyl radical, and LOOH is the lipid hydroperoxide. The main function of α-T is to scavenge the lipid peroxyl radical before it is able to react with the lipid substrate as shown:

$$LOO^{\cdot} + TOH \rightarrow LOOH + TOO^{\cdot} \tag{2.3}$$

where TOH is tocopherol and TOO$^{\cdot}$ is the tocopheroxyl radical. As shown in eq 2.3 tocopherols thus prevent propagation of the chain reaction. The potency of an antioxidant is determined by the relative rates of reactions eqs 2.1 and 2.2. Studies of the relative rates of chain propagation to chain inhibition by α-T in model systems have demonstrated that α-T is able to scavenge peroxyl radicals much more rapidly than the peroxyl radical can react with a lipid substrate (Christie, 2013). In biological systems, oxidant radicals can spring from a number of sources, including singlet oxygen, alkoxyl radicals, superoxide, peroxynitrite, nitrogen dioxide, and ozone. α-T is most efficient at providing protection against peroxyl radicals in a membrane environment

(Table 2.2). When a tocopheroxyl radical is formed, it is stabilized by delocalization of the unpaired electron about the fully substituted chromanol ring system rendering it relatively unreactive. This also explains the high first order rate constant for hydrogen transfer from α-T to peroxyl radicals. Reaction of the tocopheroxyl radical with a lipid peroxyl radical yields 8α-substituted tocopherones, which are readily hydrolysed to 8α-hydroxy tocopherones that rearrange spontaneously to form α-T quinones (Fig. 2.2).

TABLE 2.2 Approximate Biological Activity Relationships of Vitamin-E Compounds (Akoh & Min, 2002).

Compound	Activity of D-α-tocopherol (%)
D-α-tocopherol	100
L-α-tocopherol	29
DL-α-tocopherol	74
DL-α-tocopheryl acetate	68
D-β-tocopherol	8
D-γ-tocopherol	3
D-δ-tocopherol	–
D-α-tocotrienol	22
D-β-tocotrienol	3
D-γ-tocotrienol	–
D-δ-tocotrienol	–

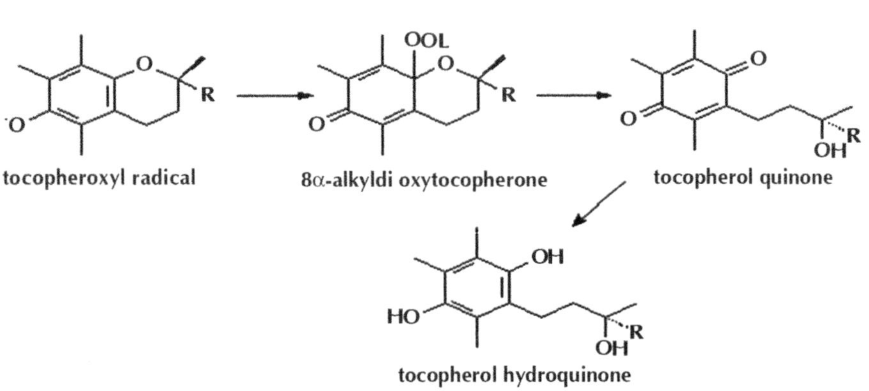

tocopheroxyl radical   8α-alkyldi oxytocopherone   tocopherol quinone

tocopherol hydroquinone

FIGURE 2.2   Mechanism for radical quenching action of α-tocopherol.

2.7.2 TOCOPHEROLS IN FOOD PRESERVATION

The antioxidant function of tocopherols in various foods has been reported by many researchers in various types of food. Instant ramen, a Japanese dried noodle product, is fried in lard to produce the characteristic flavor. Kuwahara et al. (1971) reported that natural vitamin E at a level of 0.03% prevented the lard oxidation in the noodles better than the synthetic antioxidants such as BHA and BHT (Christine, 2010). Kanematsu et al. (1972) studied the effect of tocopherols, such as synthetic α-T, mixed natural tocopherol concentrate, and BHA, on OS of margarines. As a control, margarine without antioxidants was used. Margarines were stored at 5 or 25 °C for six months. Natural mixed tocopherols and BHA were found to have nearly the same antioxidant effect (Christine, 2010). King (1986) determined the effects of antioxidants and modified atmospheres (vacuum and N_2 gas) on the stability of pecan kernels stored at 85 °C for 15 weeks. Chopped pecan kernels were treated with various antioxidants such as α-T at 0.05%, γ-tocopherol and mixed tocopherols at 0.02 and 0.05%, BHA, BHT, and TBHQ at 0.02% sealed in cans with or without headspace modification by vacuum or N_2 flush. α-T, γ-tocopherol, and mixed tocopherols at 0.05% had significantly protected the color and reduced flavor changes. γ-tocopherol and mixed tocopherols also reduced the headspace pentane production (Christine, 2010). Ochi et al. (1988) studied the effects of α- and δ-tocopherols on OS of cookies. Both α- and δ-tocopherols were decreased by 20% during baking. During storage between 25 and 60 °C, the degradation of α-T was faster than the degradation of δ-tocopherol. But the loss of α- and δ-tocopherols decreased with an increase in added amounts of whole milk powder and egg because they protected the fats from oxidative degradation. Cookies with added α-T (50 mg/100 g dough) had relatively low PV of their lipid fraction after baking (Christine, 2010). Inagaki (1968) investigated the antioxidative components in unshu-orange flavedo (the white part in the peel of unshu orange), which inhibited autoxidation of limonene. The antioxidants identified by thin layer and gas–liquid partition chromatography were: 100–160 µg/g α-T, and 60–70 µg/g γ-tocopherol on fresh weight basis (Christine, 2010). Elez-Martínez et al. (2007) studied the ability of α-T to extend the shelf life of an avocado puree by determining the PV and IV of the stored puree with and without added tocopherol (100 ppm) and sorbic acid, an antimicrobial agent. They found that the α-T was very effective in extending the shelf life but sorbic acid had the effect of enhancing oxidation in the avocado puree. Therefore, the best quality product was obtained with the addition of α-T and using low oxygen packaging (Christine, 2010). It is proposed that understanding the interfacial phenomena is a key to understand the actions of tocopherols in heterogeneous food systems (Frankel, 1996).

2.8 VITAMIN K$_1$ (PHYLLOQUINONE)

Vitamin K is a fat-soluble vitamin that functions as a co-enzyme and is involved in the synthesis of a number of proteins participating in blood clotting and bone metabolism (Damon et al., 2005). Vitamin K also plays a role as a co-factor for blood coagulation and coagulation inhibitors in the liver, as well as a variety of extra hepatic proteins such as the bone protein osteocalcin (Shearer, 1992). The importance of vitamin K as a blood-clotting agent is well known. Moreover, it is demonstrated that vitamin K may play a variety of health-promoting roles. Vitamin K reduces the risk of heart disease, kills cancer cells, and enhances skin health and have antioxidant properties (Otles & Cagindi, 2007). A recent study concluded also that high phylloquinone intakes are markers of a dietary and lifestyle pattern that is associated with lower CHD risk in men (Erkkilä et al., 2007). Vitamin K$_1$ (phylloquinone) is a polycyclic aromatic ketone which contains a functional naphthoquinone ring and a phytyl side chain, that is, 2-methyl-1,4-naphthoquinone, with a 3-phytyl substituent (Fig. 2.3). Its molecular formula is C$_{31}$H$_{46}$O$_2$ and molar mass is 450.70 g/mol.

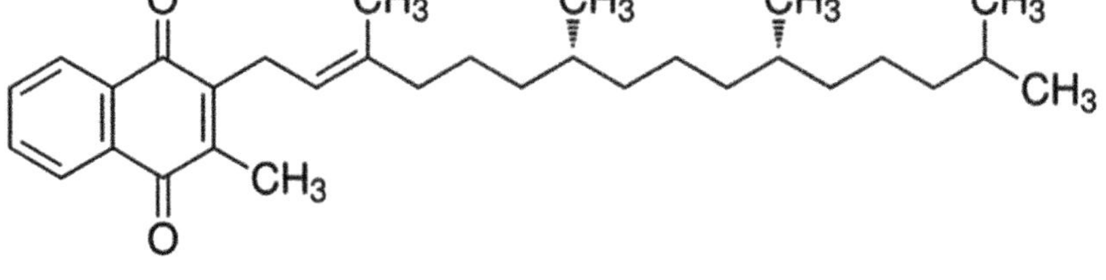

FIGURE 2.3 Structure of vitamin K$_1$ (phylloquinone).

2.8.1 FOOD APPLICATIONS OF VITAMIN K$_1$

The vitamin K$_1$ (phylloquinone) level is very low in most foods (<10 mg/100 g), and the majority of the vitamin is obtained from a few green and leafy vegetables (e.g., spinach and broccoli). Many studies have shown that some vegetable oils (especially soybean, cottonseed, and rapeseed oils) are important dietary sources of phylloquinone (Gao & Ackman, 1995; Piironen et al., 1997; Booth & Suttie, 1998; Koivu et al., 1999; Jakob & Elmadfa, 2000). Niger seed oil has been characterized by extremely high level of vitamin K_1 (0.2%) (Ramadan & Moersel, 2002). Among edible oils, the best sources of phylloquinone are niger seed oil (*ca* 2.0 mg/g), rapeseed oil (*ca* 1.5 ug/g) and soybean oil (*ca* 1.3 ug/g). Sunflower oil is the poorest source (*ca* 0.10 ug/g) of phylloquinone (Piironen et al., 1997). Its levels are also moderate in olive oil (Jakob & Elmadfa, 2000; Shearer et al., 1996; Piironen et al., 1997). A study reported that in olive oil, the mean content of phylloquinone ranged from 12.7 to 18.9 µg/100 g while in human plasma, phylloquinone content varied between 0.22 and 0.56 ng/mL (Otles & Cagindi, 2007). The addition of phylloquinone-rich oils in the processing and cooking of foods that are otherwise poor sources of vitamin K (e.g., peanut and corn oils) makes them potentially important dietary sources of the vitamin. The significance of dietary vitamin K has recently increased. Blending of niger seed oil with other vegetable oils would enrich them with vitamin K_1 (phylloquinone) (Bhatnagar & Gopala Krishna, 2015).

2.9   VITAMIN C (ASCORBIC ACID)

AA (vitamin C) is considered to be one of the most powerful, least toxic natural antioxidants. It is a water-soluble vitamin and is found in high concentrations in many foods or plants (Table 2.3). The varied roles of AH_2 related to its antioxidant property in foods, browning reaction, and its anaerobic loss are discussed here. AH_2 is the trivial name for *L-threo-2-hexenono-l, 4-* lactone, the molecule responsible for preventing scurvy (Fig. 2.4). AH_2 or vitamin C is ubiquitous and has multiple functions in all metabolically active plant and animal cells. One of the principal biochemical reactions of AH_2 is to destroy toxic free radicals (hydroxyl and perhydroxyl) resulting from the metabolic products of oxygen. In this role, the mixture of AH_2 and its oxidation product dehydroascorbic acid (A) is thought of as a "redox buffer" (Sapper et al., 1982; Ming-Long & Paul, 1988). When terminating free radicals, AH_2 is converted to A, which is then recycled to AH_2 by reductase enzymes and co-factors. Besides the redox functions in cells, other physiological actions of AH_2 (Loewus & Loewus, 1987; Ming & Paul, 1988) may be related to the compound's chelation with metals and complexing with protein (Gorman & Clydesdale, 1983; Fleming & Bensch, 1983; Ming-Long & Paul, 1988). The uses of AH_2, including those in food, continue to increase because of the compound's vitamin C activity, useful properties, and low toxicity. AA reacts with superoxide radical (O_2^-), perhydroxyl radical (HO_2), hydroxyl radical ($HO^•$), and singlet oxygen (Fessenden & Verma, 1978; Nanni et al., 1980; Cabelli & Bielski, 1983; Ming-Long & Paul, 1988). Those reactions by AH_2 retard lipid autoxidation. Ascorbate radical is the initial oxidation product of two enzyme reactions that occur in plants (Loewus & Loewus, 1987; Ming & Paul, 1988). Ascorbate oxidase is the enzyme that oxidizes AH_2 in the presence of oxygen as shown by the overall reaction below:

$$2 \text{ Ascorbate} + O_2 \rightarrow 2 \text{ Dehydroascorbic acid} + 2H_2O$$

Ascorbate peroxidase is another plant enzyme that oxidizes AH_2, but it uses hydrogen peroxide instead of oxygen as the electron acceptor.

$$\text{Ascorbate} + H_2O_2 \rightarrow \text{Dehydroascorbic acid} + 2H_2O$$

TABLE 2.3 Vitamin C in Selected Foods (Steinberg & Rucker, 2013).

Sources of vitamin C	mg of ascorbic acid per 100 g of wet weight or edible portion
Animal products	
Cow's milk	0.5–2
Human milk	3–6
Oysters (raw)	30
Beef, pork, veal	2–10
Beef, pork, Liver	20–30
Fruits	
Apple	3–30
Banana	8–16
Blackberry	8–10
Cherry	15–30
Currant, red	20–50
Currant, black	150–200
Grapefruit	30–70
Kiwi fruit	80–90
Lemon, orange	40–50
Lime	30–45
Melon	9–60

TABLE 2.3 *(Continued)*

Sources of vitamin C	mg of ascorbic acid per 100 g of wet weight or edible portion
Strawberry	59–70
Pineapple	15–25
Rose hips	250–800
Vegetables	
Beans, various	10–15
Broccoli	70–90
Brussels sprouts	100–120
Cabbage	30–70
Carrot	5–10
Cucumber	6–8
Cauliflower	50–70
Eggplant	15–20
Kale	70–100
Onion	10–15
Parsley	90–130
Peas	8–12
Potato	4–30
Pumpkin	15
Radish	25
Spinach	35–40
Tomato	15–20
Condiments	
Chicory	30–40
Coriander (spice)	90
Garlic	15–25
Horseradish	50
Lettuce, various	10–30
Leek	15
Parsley	200–300
Pepper, various	150–200

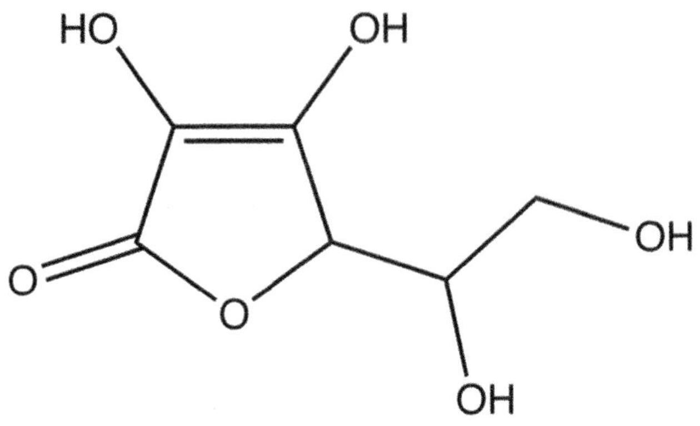

FIGURE 2.4   Ascorbic acid.

2.9.1   ROLE OF VITAMIN C IN FOOD PRESERVATION

AA (E-300) generally regarded as safe (GRAS) substance prevents oxidative browning in heat-processed foods, enzyme-catalyzed oxidation in frozen fruits, rusting and rancidity in frozen fish, discolorations and rancidity in meat products, and oxidized flavor in dairy and beverage products (Bauern-feind, 1953). It acts synergistically with other antioxidants (known to regenerate α-Ts) in edible fats and also acts as flour and dough improver. It is important to add the AA as late as possible during processing or preservation to maintain highest levels during the shelf life of the food commodity (Wiley, 1994). The beneficial use of AA has been established for the stabilization of beer (Wales, 1956) and other food applications where it can serve to reduce the oxygen from the headspace of a closed system (Cort, 1974). AH_2 may preserve or promote the reduced oxidation state of a metal ion in food (Hay et al., 1967). The oxidation state is an important variable in mineral nutrition (Solomons & Viteri, 1982; Keypour et al., 1986). AA has strong singlet oxygen and superoxide anion quenching ability and has been shown to protect riboflavin loss in milk (Lee et al., 1998).

The acid-catalyzed degradation of AH_2 is thought to be responsible for anaerobic loss of vitamin C in foods, such as canned grapefruit and orange juices, which have a pH of ~3–5 (Kefford et al., 1959; Smoot & Nagy, 1980; Ming-Long & Paul, 1988). At 50 °C, the juices lose 70–95% of AH_2 in 12 weeks; the degradation reaction is zero-order with respect to AH_2. The anaerobic loss of AH_2 is often one-tenth the rate of loss under aerobic conditions. Categories of reactions AA undergoes are mentioned in Table 2.4 and the level of permitted AA derivatives in different foods is presented in Table 2.5.

TABLE 2.4 Categories of Ascorbic Acid Reactions.

S. no.	Type of reaction	Substrate involved
1	Redox	Glutathione/glutathione disulfide and ascorbic acid/dehydroascorbic acid
2	Oxidation	Ascorbic acid and dehydroascorbic acid
3	Reduction	o-quinone back to phenolic compound
4	Anaerobic degradation	L-ascorbic acid decarboxylate and dehydrate to give almost quantitative yields of furfural and CO_2

TABLE 2.5 Antioxidants Permitted in Foodstuffs for Infants and Young Children (Miková, 2003).

E number	Name	Foodstuff	Maximum level
E 300	L-ascorbic acid	Fruit and vegetable based drinks, juices, and baby foods	0.3 g/kg
E 301	Sodium L-ascorbate		
E 302	Calcium L-ascorbate	Fat-containing cereal-based foods including biscuits	0.2 g/kg
E 304	L-ascorbyl palmitate	Fat-containing cereals, biscuits, rusks, and baby foods	*100 mg/kg individually or in combination
E 306	Tocopherol rich extract		
E 307	α-tocopherol		
E 308	γ-tocopherol		
E 309	β-tocopherol		

*10 mg/kg for follow-on formulae for infants in good health.

AH_2 retards enzymic browning by at least two mechanisms (Golan-Goldhirsh et al., 1984; Ming & Paul, 1988). AH_2 chemically reduces benzoquinone intermediates to colorless o-dihydroxyphenols, and it also irreversibly denatures polyphenoloxidase (PPO). Evidence (Golan-Goldhirsh et al., 1987; Ming-Long & Paul, 1988) suggests that PPO is denatured mainly by Cu^{+2}-catalyzed oxidative cleavage of imidazole groups of the histidine residues on the enzyme to aspartic acid and urea. The oxidation is mediated through a quaternary complex thought to contain the imidazole group, cupric ion, AH_2 radical ($AH^{\cdot}$), and O_2. The anti-browning effects of AA have been widely demonstrated in several fruit fresh-cut products under a wide range of conditions (Soliva-Fortuny et al., 2001; Senesi et al., 1999). Ascorbate reacts with nitrite forming NO, NO_2, and N_2, thereby inhibits carcinogenic nitrosamine formation.

2.10 PHYTOSTEROLS

PS are steroid alcohols and resemble cholesterol in structure, the predominant sterol found in animals both in their biological function and chemical structure. PS are fat-soluble nutrients and are biosynthetically derived from squalene and belong to the group of triterpenes. They are made of tetracyclic cyclopentaphenanthrene ring and long flexible side chain at the C-17 carbon atom (Clifton, 2002; Moreau et al., 2002). The 3-hydroxyl group of free sterols may be esterified by a fatty acid or a phenolic acid or β-linked to a carbohydrate. Plant sterols are primarily present in plasma membrane, the mitochondria, and endoplasmic reticulum, and to a large extent determine the properties of the membranes (Jonker et al., 1985). Plant sterols are white powder and solid at room temperature and the melting point of sitosterol, campesterol, and stigmasterol are 140 °C, 157–158 °C, and 170 °C (Nes, 1987). Plant sterols are divided into three different kinds based on structures namely, 4-desmethyl sterols, 4-methyl sterols, and 4,4-dimethyl sterols. The 4-desmethyl sterols family includes three different kinds of PS which accounts for most of the total PS mass. They are β-sitosterol (include an extra methyl group at C-24 position), campesterol (includes an additional ethyl group at C-24 poisition), and stigmasterol (includes an additional ethyl group at C-24 position and a double bond at C-22 position). All these PS account for about 65, 30, and 3% of the total dietary PS intake (Weihrauch & Gardner, 1978; Moreau et al., 2002; Ostlund, 2002). The chemical difference among 4-desmethyl sterols resides in number of carbon atoms in carbon-17 branch chain and in presence or absence of a double bond at position 22. The 4-methyl sterols and 4,4-dimethyl sterols are minor components in plant sources. PS are structurally similar to cholesterol (steroid nucleus and a hydroxyl group at C-3 position) and are differentiated by their degree of saturation and side chain configuration at C-24 position (Clifton, 2002; Moreau et al., 2002). Analysis of sterols provides a powerful tool for quality control of vegetable oils, and for the detection of oil as well as blends not recognized by the fatty acids' profile (Ramadan, 2012). Structures of some 4-desmethylsterols of vegetable oils are provided in Figure 2.5. The PS levels and composition of some vegetable oils are provided in Table 2.6.

2.10.1 FOOD APPLICATIONS OF PHYTOSTEROLS

A dark chocolate containing PS esters was developed to reduce cholesterol in individuals. However, oxidative instability during chocolate processing

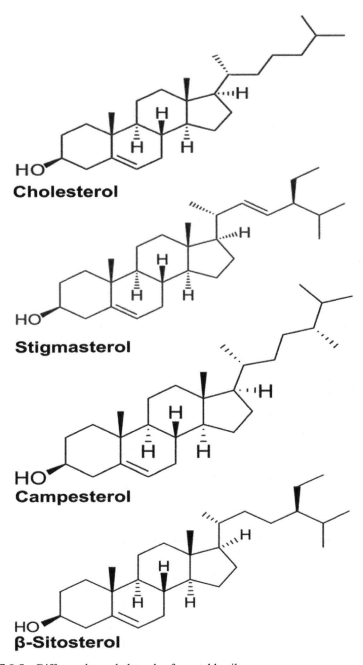

FIGURE 2.5 Different desmethylsterols of vegetable oils.

TABLE 2.6 Levels of Total Phytosterols (mg/kg) and Desmethylsterols (wt % of Total Sterols) in Some Crude Vegetable Oils (Kamal-Eldin & Appelqvist, 1994; Gunstone, 2002; Codex Stan 210, 2011).

Sterols	Coconut oil	Groundnut oil	Mustard oil	Olive oil	Palm oil	Rice bran oil	Safflower seed oil	Sesame oil	Soyabean oil	Sunflower seed oil
Total sterols	400–1200	900–2900	8000–8800	800–1000	300–700	10500–31000	2100–4600	4500–19000	1800–4500	2400–5000
Cholesterol	ND-3.0	ND-3.8	ND-0.4	ND-0.5	2.6–6.7	ND-0.5	ND-0.7	0.1–0.3	0.2–1.4	ND-0.7
Brassicasterol	ND-0.3	ND-0.2	10.0–13.2	ND-0.1	ND	ND	ND-0.4	0.1–0.2	ND-0.3	ND-0.2
Campesterol	6.0–11.2	12.0–19.8	30.0–34.4	3.5–4.0	18.7–27.5	11.0–35.0	9.2–13.3	10.3–20.5	15.8–24.2	6.5–13.0
Stigmasterol	11.4–15.6	5.4–13.2	ND-0.3	2.5–3.0	8.5–13.9	6.0–40.0	4.5–9.6	4.4–14.0	14.9–19.1	6.0–13.0
β-Sitosterol	32.6–50.7	47.4–69.0	40.0–47.9	93.0	50.2–62.1	25.0–67.0	40.2–50.6	57.7–61.9	47.0–60.0	50.0–70.0
δ-5-Avenasterol	20–40.7	5.0–18.8	1.0–2.1	ND-0.5	ND-2.8	ND-9.9	0.8–4.8	6.2–7.8	1.5–3.7	ND-6.9
δ-7-Stigmastenol	ND-3.0	ND-5.1	0.8–1.6	ND-0.5	0.2–2.4	ND-14.1	13.7–24.6	0.5–7.6	1.4–5.2	6.5–24.0
δ-7-Avenasterol	ND-3.0	ND-5.5	1.0–2.1	ND-0.5	ND-5.1	ND-4.4	2.2–6.3	1.2–5.6	1.0–4.6	3.0–7.5
Others	ND-3.6	ND-1.4	ND-0.5	ND-0.5	ND	ND	0.5–6.4	0.7–9.2	ND-1.8	ND-5.3

ND: not detected.

and storage could reduce the PS bioactivity. Chocolate bars were prepared containing palm oil (CONT) or 2.2 g of PS (PHYT). All samples were stored at 20 and 30 °C during five months. A peak of hydroperoxides formation was observed after 60 days at 20 °C and after 30 days at 30 °C. PS-enriched samples presented higher values of hydroperoxides than control samples, which could be attributed to the higher level of α-linolenic acid present in the PHYT samples. All chocolate bars became lighter and softer after 90 days of storage. However, these physical changes did not reduce their sensory acceptability. In addition, PS bioactivity was kept during the storage, since no significant alterations in the PS esters were observed up to five months. However, some PS oxidation occurred in the PHYT bars, being sitostane-triol, 6-ketositosterol, 6β-hydroxycampesterol, and 7-ketocampesterol the major phytosterol oxidation products (POPs). The POPs/PS ratio was low (0.001). Therefore, the dark chocolate bars developed in this study kept their potential functionality after five months of storage at room temperature, representing an option as a functional food (Botelho et al., 2014).

2.11   PHENOLIC COMPOUNDS

POH have aromatic hydrocarbon ring with one or more hydroxyl group. POH contribute to the quality of edible oils by enhancing the shelf life and flavor stability of oil (Bendini et al., 2006). Several POH have been identified that inhibit oxidation of fats and oils by interrupting the free radical mechanism of oxidation. Structural groups influencing the antioxidant activity of POH include position and number of hydroxyl groups, polarity, solubility, and stability of POH during processing (Soobrattee et al., 2005). Phenolics present in fats and oils are potent antioxidants and have anticarcinogenic properties. The antioxidant property of phenolics depends on the structure and type of phenolics present in oil. Hydroxybenzoic acid derivatives, that is, gallic acid, vanillic acid, and vanillin show an UV-absorption maxima at around 280 nm while hydroxycinnamic acid derivatives, that is, caffeic, cinnamic, p-coumaric, and ferulic acids show an UV-absorption maxima at around 320 nm.

Virgin olive oil has a unique place among vegetable oils because of its polyphenols and their beneficial role in human health (Visioli, 2000; Tricho-poulou & Vasilopoulou, 2000). The polyphenols are an important class of minor constituents linked both to the flavor of virgin olive oil and to its keeping ability. POH present in olive oil are conventionally character-ized as "polyphenols," though not all of them are polyhydroxy aromatic compounds. Compounds which often appear in lists of olive oil polyphenols are 4-acetoxy-ethyl-1,2-dihydroxybenzene, 1-acetoxy-pinoresinol, apigenin, caffeic acid, cinnamic acid, *o*- and *p*-coumaric acids, elenolic acid, ferulic acid, gallic acid, homovanillic acid, *p*-hydroxybenzoic acid, *p*-hydroxyphenylacetic acid, hydroxytyrosol, luteolin, oleuropein, pinoresinol, protocatechuic acid, sinapic acid, syringic acid, tyrosol, vanillic acid, and vanillin (Morales & Tsimidou, 2000; Garcia et al., 2001; Mateos et al., 2001; Boskou, 2006). Structures of some phenolics of plant origin are provided in Figure 2.6. Phenolic acid content of some vegetable oils is provided in Table 2.7.

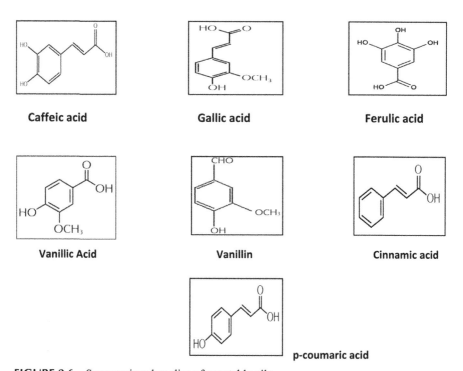

FIGURE 2.6   Some major phenolics of vegetable oils.

2.11.1 ANTIOXIDATIVE MECHANISM OF PHENOLICS

POH or "phenolics" constitute a class of chemical compounds consisting of one or several hydroxyl group (–OH) covalently bonded to one or several aromatic rings. It is difficult to say how many different phenolics exist on earth. In the sole example of flavonoids, there are more than 10,000 known formulas. Phenolics may be subdivided in various classes such as flavonoids

TABLE 2.7 Phenolic Acid Content (µg/100 g oil) of Some Vegetable Oils (Siger et al., 2008).

Vegetable oils	p-hydroxybenzoic acid	Vanillic acid	Caffeic acid	p-coumaric acid	Ferulic acid	Sinapic acid	Total
Soyabean	0.77–0.83	1.04–1.16	0.73–0.87	1.44–1.56	1.12–1.28	0.87–0.93	5.97–6.63
Sunflower	1.45–1.55	6.75–7.05	4.8–5.0	1.74–1.86	1.22–1.38	1.37–1.43	17.33–18.27
Rapeseed	1.55–1.65	ND	0.25–0.35	12.98–13.22	5.5–5.7	235.5–236.5	255.87–257.33
Corn	1.68–1.72	ND	ND	1.82–1.98	5.7–5.9	0.57–0.63	9.77–10.23
Grape seed	ND	0.75–0.85	ND	ND	ND	0.12–0.28	0.87–1.13
Hemp	5.94–6.06	1.90–2.10	ND	1.85–2.15	0.92–1.08	2.95–3.05	13.56–14.44
Linseed	3.03–3.17	0.85–1.15	ND	ND	0.95–1.05	ND	4.83–5.37
Rice bran	ND	ND	ND	ND	0.37–0.43	ND	0.37–0.43
Pumpkin seed oil	2.95–3.25	11.3–11.5	ND	3.74–3.86	3.74–3.86	ND	21.73–22.47

ND: not detected.

(quercetin, catechin, and condensed tannins), phenolic acids (caffeic acid) and esters (chlorogenic acid, gallotannins), stilbenes (resveratrol), phenolic alcohols (hydroxytyrosol), and so forth (Laguerre et al., 2014). POH are natural antioxidants and integral bioactive molecules of an oil or fat. POH act as free radical scavengers and metal chelators (Dai & Mumper, 2010). POH act as free radical acceptors and chain breakers. They interfere with the oxidation of lipids and other molecules by rapid donation of a hydrogen atom to radicals (R) ,that is, R + POH → RH + PO•. The phenoxy radical intermediates (PO•) are relatively stable due to resonance and therefore a new chain reaction is not easily initiated. Moreover, the phenoxy radical intermediates also act as terminators of propagation route by reacting with other free radicals, that is, PO• + R• → POR. POH possess ideal structure chemistry for free RSAs because they have, (a) phenolic hydroxyl groups that are prone to donate a hydrogen atom or an electron to a free radical and (b) extended conjugated aromatic system to delocalize an unpaired electron. Several relationships between structure and reduction potential of phenolics have been established. For phenolic acids and their esters, the reduction activity depends on the number of free hydroxyl groups in the molecule. Hydroxycinnamic acids were found to be more effective than their hydroxybenzoic acid counterparts, possibly due to the aryloxy-radical stabilizing effect of the —CH=CH—COOH linked to the phenyl ring by resonance (Dai & Mumper, 2010).

2.12 CAROTENOIDS

Carotenoids are a class of hydrocarbons consisting of eight isoprenoid units joined in such a manner that the arrangement of isoprenoid units is reversed at the center of the molecule so that the two central methyl groups are in a 1,6-positional relationship and the remaining non-terminal methyl groups are in a 1,5-positional relationship. Carotenoids are defined by their chemical structure. The majority of carotenoids are derived from a 40-carbon polyene chain, which could be considered the backbone of the molecule. This chain may be terminated by cyclic end-groups (rings) and may be complemented with oxygen-containing functional groups (Zeb & Mehmood, 2004). These hydrocarbons are commonly known as carotenes, while oxygenated derivatives of these hydrocarbons are known as xanthophylls. β-carotene, the principal carotenoid in carrots, is a familiar carotene, while lutein, the major yellow pigment of marigold petals, is a common xanthophyll. The structure of a carotenoid ultimately determines what potential biological function that pigment may have. The characteristic pattern of alternating single and double bonds in the polyene backbone of carotenoids allows them to absorb excess energy from other molecules, while the nature of the specific end groups on carotenoids may influence their polarity. The former may account for the antioxidant properties of biological carotenoids, while the latter may explain the differences in the ways that individual carotenoids interact with biological membranes (Britton, 1995).

The most important carotenoids are α-carotene, β-carotene, β-cryptoxanthin, lutein, violaxanthin, neoxanthin, and lycopene. β-carotene, α-carotene, and β-cryptoxanthin are carotenes that are converted into vitamin A or retinol in the body. β-carotene is the most widely studied carotenoid. Lutein and zeaxanthin are both stored in the retina of the eye; however, neither converts to vitamin A. Both are powerful antioxidants and may be very important for healthy eyes. Carotenoids are singlet oxygen quenchers and protect the oil from photo-oxidation (Psomiadou & Tsimidou, 1998). The structure of some carotenoids of plant origin is provided in Figure 2.7. Among the vegetable oils, palm oil is the richest source of carotenoids, especially β- and α- carotenes. In comparison, other vegetable oils contain little amounts of carotenoids, especially coconut, palmkernel, sesame, and groundnut oils have very low content of carotenoids. The dark red-orange color of oil palm fruit is due to the high concentration of carotenoids and anthocynanins. Crude palm oil, extracted commercially by pressing, contains 400–1000 ppm of carotenoids, the variation being due to process conditions, species of oil palm, and level of oxidation. Carotenoids in palm oil are α-carotene, β-carotene, phytoene, phytofluene, *cis* β-carotene, *cis* α-carotene, δ-carotene, γ-carotene, ζ-carotene, neurosporene, β-zeacarotene, α-zeacarotene, and lycopene (Table 2.8) (Yap et al., 1991; Jalani et al., 1997). Carotenoids are fat-soluble nutrients and categorized as either xanthophylls or carotenes according to their chemical composition. Carotenoids are singlet oxygen quenchers and protect the oil from photo-oxidation (Psomiadou & Tsimidou, 1998).

TABLE 2.8 Composition of Carotenoids in Palm Oil, Given as % of Total Carotenoids (Yap et al., 1991; Jalani et al., 1997).

Carotenoids	*Elaeis guineesis* variety crude palm oil	*Elaeis oleifera* variety crude palm oil
Total (ppm)	500–700	4300–4600
Phytoene	1.27	1.12
Cis β-carotene	0.68	0.48
Phytofluene	0.06	Trace

TABLE 2.8   *(Continued)*

Carotenoids	*Elaeis guineesis* variety crude palm oil	*Elaeis oleifera* variety crude palm oil
β-Carotene	56.02	54.08
α-Carotene	35.06	40.38
Cis α-carotene	2.49	2.30
ζ-Carotene	0.69	0.36
γ-Carotene	0.33	0.08
δ-Carotene	0.83	0.09
Neurosporene	0.29	0.04
β-Zeacarotene	0.74	0.57
α-Zeacarotene	0.23	0.43
Lycopene	1.30	0.07

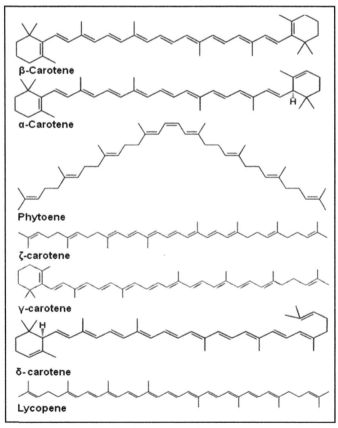

FIGURE 2.7   Different carotenoids of vegetable oils.

2.13   PHYTIC ACID

Phytic acid is one of the bioactive compounds that are being intensively studied to evaluate their effects on health. It has been shown to have potential as anticancer agent which only affects malignant cells and does not affect normal cells and tissues (Vucenik & Shamsuddin, 2003). Phytic acid is a simple ranged carbohydrate with six phosphate groups attached to each carbon (Shamsuddin, 2002). It serves as the major phosphorus storage compound in plant in the seed, as well as being a natural antioxidant by its chelating properties and reduction of the catalytic activities of many divalent transition metals (Verghese et al., 2006). The chelation ability of phytic acid with minerals has been suggested to have beneficial effects toward lowering serum cholesterol and triglycerides and suppression of iron-mediated oxidation (Lee et al., 2005). A variety of benefits of phytic acid on human health has also been reported including its potential as an anti-cancer property in soft tissue, colon, prostate, metastatic, and mammary cancers. It may also act as an inhibitor for renal stone development (Dost & Tokul, 2006). In whole grain cereals such as corn, wheat, and rice, the ranges of phytic acid is from 1.5 to 6.4% while defatted and dehulled oilseed meals such as soy, peanut, and sesame contain 1.5% or more of the compound (Grases et al., 2004). Phytic acid is primarily found in the outer layers (bran) of unpolished rice.

Phytin is a white amorphous powder, odorless and tasteless, almost insoluble in water, soluble in dilute mineral acids and in some organic acids. One part phytin dissolves in 10 parts of 1 N hydrochloric acid and forms a clear solution. According to some authors, phytin contains 36% organically bound phosphoric acid. Upon heating with dilute acids, alkali, and water, phytin hydrolyzes to give o-phosphoric acid and the cyclitol myo-inositol as end products. These are obtained together with some other products of semi-degradation. Phytic acid (myo-inositol 1,2,3,4,5,6 hexakisphosphate) is the most abundant form of phosphorus in rice and is virtually indigestible by humans or non-ruminant livestock and hampers the nutritional value of rice and its milling by-product rice bran (Larson et al., 2000). Studies have demonstrated that the antinutrient effect of phytic acid can be manifested only when large quantities of phytic acid are consumed in combination with a diet poor in oligoelements (Shamsuddin & Vucenik, 2005). IP6 (phytic acid, phytin) is a 6-phosphate ester of inositol (Saad et al., 2011). It exists in almost all plants as its mixed calcium and magnesium salts (phytin) and, especially, seeds and grains contain it in a lot of amount.

It is considered as the storage of organic phosphates of plants with 60~90% of entire phosphorus quantity being in the form of phytin (Saad et al., 2011). For rice bran, its concentration is particularly high and 9.5~14.5% of rice bran is occupied by phytin (Saad et al., 2011), therefore, rice bran is likely to be appropriate material for IP6. The most outstanding feature of phytic acid is its strong metal chelate function (Saad et al., 2011), allowing metal ions such as ferrum which often adversely affect the production or storage of food in various forms to be removed or deactivated. Compared to other chelate agents, it is distinctively effective in wider pH range (Saad et al., 2011). Besides this function, it is known to have strong pH buffer action (Saad et al., 2011), their derived effects of preventing the change of properties or colors and antioxidation effect (Saad et al., 2011). In recent years, various physiologically active functions of phytic acid within living bodies have been reported including the prevention of urinary and nephritic calculi (Shamsuddin, 2002; Vucenik & Shamsuddin, 2003; Shamsuddin & Vucenik, 2005), prevention of colic cancer (Shamsuddin, 2002; Vucenik & Shamsuddin, 2003; Shamsuddin & Vucenik, 2005), and suppression of bacterial plaque formation (Shamsuddin, 2002; Vucenik & Shamsuddin, 2003; Shamsuddin & Vucenik, 2005). In addition, other carcinostatic effects have also been suggested (Shamsuddin, 2002; Vucenik & Shamsuddin, 2003; Shamsuddin & Vucenik, 2005). Furthermore, as notable functions of phytic acid, the deodorant effect of body odor, bad breath or uraroma (Shamsuddin, 2002; Vucenik & Shamsuddin, 2003; Shamsuddin & Vucenik, 2005), prevention of acute alcoholism (Saad et al., 2011), and enrichment of meat or fish taste (Saad et al., 2011) are popular. These effects of phytic acid provide food products with practical added values.

2.13.1 PHYTIC ACID IN FOOD PRESERVATION

Phytic acid (known as inositol hexakisphosphate, IP6, or phytate when in salt form) is an organic acid extracted from rice bran (Fig. 2.8). Phytic acid is used as an acidulant for pH adjustment. Phytic acid binds to metals strongly because of strong chelating effect. Moreover, phytic acid shows antioxidant action and prevention of color degradation. Phytic acid has been approved as GRAS by the Food and Drug Administration (FDA) in the United States. The most outstanding feature of phytic acid is its strong metal chelate function, allowing metal ions such as ferrum (Fe) which often adversely affect the production or storage of food in various forms to be removed or deactivated. Phytic acid is the best acidulant because it lowers pH level the most at the same concentration. In conclusion, phytic acid has a mild, not a strong or sharp acid taste relatively. Phytic acid does not affect the original taste of food or beverage and only low concentration of it is required to achieve the desired pH level. Phytic acid has the potential to prevent color degradation in food or beverage including anthocyanin. Phytic acid is the most potent natural iron chelator and has strong bacteriostatic and antioxidant action (Graf et al., 1987; Graf & Eaton, 1990). Phytic acid is found to have similar iron-chelating properties as desferrioxamine, a drug commonly used to kill germs, tumor cells, or to remove undesirable minerals from the body (Hawkins et al., 1993). The chelating stability constants of magnesium ion and calcium ion of phytic acid are compared favorably with that of EDTA. In fruits and vegetables, phytic acid helps to prevent oxidative browning by inhibiting polyphenol oxidase. Phytic acid may be used as a safe preservative and antioxidant in food products (Graf et al., 1987). Prevention of browning of cut lotus root by phytic acid was investigated by immersing cut lotus root in 0.5% phytic acid, 1.0% phytic acid, and distilled water with no additives (control) then removed after 1 h. The cut lotus root with phytic acid showed significant prevention of browning. The chelate action of phytic acid compared to synthetic chelating agent, sodium metaphosphate was studied. Sodium metaphosphate, an effective metal ion chelator has the greatest salt forming activity among phosphates, particularly with calcium salts. Iron chelate ability of phytic acid was superior to sodium metaphosphate at pH 5.0. Phytic acid sequesters metal ions promoted oxidation, discoloration, and loss of flavor. Iron may cause discoloration in wine or fruit juice. Hence, phytic acid can be added to chelate polyvalent iron cations to prevent or treat these problems and make a wine more stable and commercially acceptable. Phytic acid is a natural antioxidant. Phytic acid forms a chelate with iron, thereby preventing the radical formation and oxidative damage. It blocks the formation of hydroxyl radicals and suppresses lipid peroxidation. In fruits and vegetables, phytic acid helps to prevent oxidative browning by inhibiting polyphenol oxidase. Phytic acid may be used as a safe preservative and antioxidant in food products (Graf et al., 1987). Graf et al. (1987) reported the effects of added phytate upon iron-mediated OH production and arachidonic acid peroxidation. Substantial amount of OH is produced by a superoxide-generating system in the presence of iron alone. Even greater amounts of OH are evolved if adenosine diphosphate (ADP) is added to chelate the iron. Generation of this oxyradical, however, is completely blocked by the addition of micromolar amounts of phytic acid. It is important to note that the inhibition of OH generation is found over a wide range of phytate:iron ratios from 1:4 to 20:1 (Graf et al., 1984). The effect is due to occupation of all iron coordination sites by phytate; all iron-phytate chelates prepared were completely soluble. Similarly, phytate prevents the peroxidation of arachidonic acid driven by AA and iron. Substantial amount of malondialdehyde arises from arachidonic acid in the presence of free iron or of an iron–ADP chelate. However, the addition of phytate prevents this iron-dependent generation of malondialdehyde. The magnitude of the effect of chelating agents on OH formation does not directly correspond to that on lipid peroxidation, suggesting that different reactions may be involved in the two processes and that, during lipid peroxidation, iron may catalyze several steps, for example, OH-dependent hydrogen abstraction, OH-independent formation of lipid peroxides, and catalysis of the formation of the final aldehydic cleavage products. Phytic acid is an antioxidant and chelating agent. It suppresses oxidative reactions catalyzed by iron. In plant seeds phytic acid helps to reduce the oxidation of its components but when ingested by humans it may reduce the risk of colon cancer and some other IBDs. The addition of phytic acids to foods improves its shelf life. It is also used as an antioxidant in many industrial applications. Toxicity studies of phytic acid revealed that single-dose test acute oral LD_{50} is 0.9 g/kg in the case of mouse and is 0.41 g/kg in the case of rat. Repeat-dose studies for 12 weeks, a non-toxic amount is 300 mg/kg/day in the case of rat. Reverse mutation test, chromosome aberration examination test, micronucleus test, all were found to be negative.

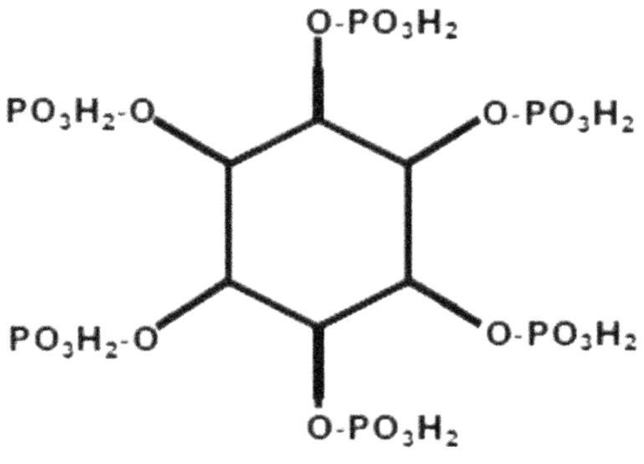

FIGURE 2.8   Structure of phytic acid.

2.13.2   FOOD APPLICATIONS OF PHYTIC ACID

Food applications of phytic acid includes, preservation of oils and fats in tofu and deep-fried tofu; chelate action in miso, soy sauce, pickle, meat industry products, canned foods, and soft drinks; browning prevention of fruit juice; sterilization/bacteriostatic action in boiling noodles; deodorant action in mutton meat; growth promotion action of lactic acid bacterium in fermented foods; acidulant action in soft drinks and pickled plums; struvite production prevention of canned foods; return prevention of bleaching; and brightness improvement of bean jam. Recommended dosage (%) for various food products are: soft drink 0.02~0.1; agriculture and fishery canned foods 0.02~0.2; pickle 0.02~0.1; bean jams 0.02~0.1; and boiling noodle 0.5~0.7.

2.14 SESAME LIGNANS

Crude or virgin sesame oil has these unique bioactive lignans namely sesamin, sesamolin, sesaminol, and sesamolinol which occur with their breakdown products like sesamol (from sesamolin) (Bhatnagar et al., 2015). Sesame oil lignans are reported to have unique bioactive, functional, physiological, and nutritional properties (Moazzami & Kamal-Eldin, 2006; Smeds et al., 2007; Namiki, 2007). Sesame seeds contain 0.26–1.16% of lignans mainly as sesamin, sesamolin, sesaminol, and sesamolinol (Moazzami & Kamal-Eldin, 2006) and sesamol is a minor component of the total lignans. Sesamin and sesamolin are usually present to an extent of 0.4 and 0.3% in sesame oil, respectively (Namiki, 2007). Sesame seeds and its oil have unique physiological and nutritional properties, which are attributed to the presence of oil soluble lignans such as sesamin and sesamolin and oil insoluble lignans present as lignan glucosides namely sesaminol di- and triglucosides, sesamolinol diglucoside, pinoresinol mono-, di-, and triglu- cosides, and other glucosides of lariciresinol, 7-hydroxy matairesinol, and medioresinol in minor amounts (Figs. 2.9 and 2.10) (Smeds et al., 2007); (Namiki, 2007); (Milder et al., 2005); (Katsuzaki et al., 1992). Sesame seeds contain 0.26–1.16% of lignans mainly as sesamin, sesamolin, sesaminol, and sesamolinol (Moazzami & Kamal-Eldin, 2006). Lignans are a group of natural compounds which are defined as an oxidative coupling product of β-hydroxyphenylpropane. Sesamin has a typical lignan structure of β-β′ (8-8′) linked product of two coniferyl alcohol radicals. Sesamolin has a unique structure involving one acetal oxygen bridge in a sesamin type struc- ture. Both sesamin and sesamolin are characteristic lignans of sesame seed (Namiki, 2007). Hydrolysis of sesamolin produces two breakdown prod- ucts namely sesamol and samin (Fukuda et al., 1986a). Samin and sesamol further combine to form sesaminol, another major sesame lignan (Nagata et al., 1987).

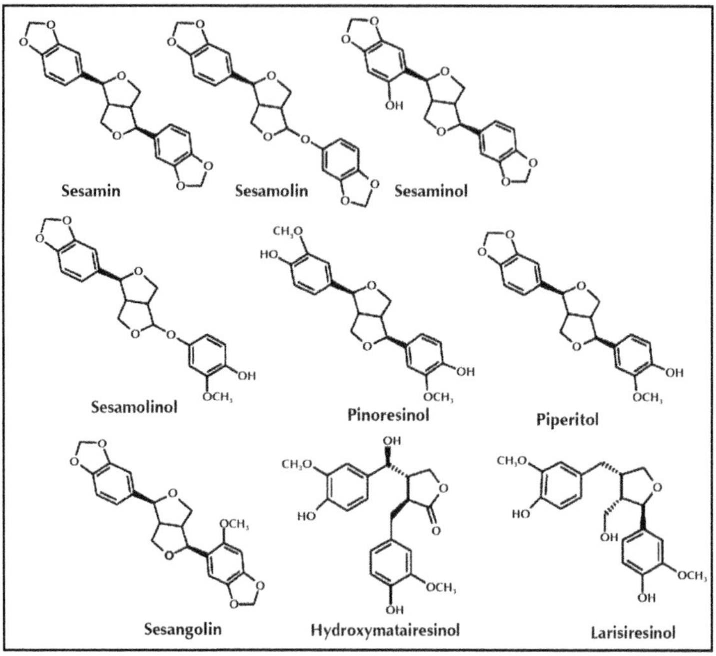

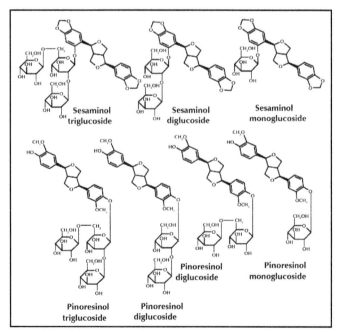

FIGURE 2.9 Oil soluble lignans of sesame seed.

FIGURE 2.10 Oil insoluble lignans of sesame seed.

2.14.1   ANTIOXIDATIVE EFFECT OF SESAME LIGNANS

Sesaminol has sesamol as a moiety and has far stronger antioxidative activity than sesamol because sesamol is easily dimerized and its products have lower activity (Fukuda et al., 1986a). The markedly strong and stable antioxidative property may be provided by the presence of a bulky samin group at the ortho position of the phenol group in sesamol similar to the BHT molecule (Namiki, 2007). Sesamin is the main characteristic lignan of sesame seed with a content of about 0.4% in seed oil, but it has no free phenol group and showed very weak or no antioxidative effect in conventional *in vitro* tests. However, sesamin exhibits significant physiological activities assumed to be due to antioxidative activity *in vivo*. Another important issue concerning the stereochemical structure of sesame lignans is the fact that the artifact episesamin, which is produced during food processing, has stronger physiological activities than native sesamin (Namiki, 2007). Sesame lignans have been found to exhibit various unique functions, such as a synergistic effect with tocopherols on vitamin E activity and the specific inhibition of fatty acid metabolism. These are quite different from the activities of other polyphenolic antioxidants, including sesamol, and they do not always appear to be based upon their antioxidative activity. These facts suggest the existence of some unique biochemical activity in sesame lignans due to their characteristic structures in addition to their antioxidative activities (Namiki, 2007).

2.14.2   SESAME LIGNANS IN FOOD PRESERVATION

Sesame oil is highly resistant to oxidative deterioration. In ancient Egypt it was used for making mummies, and in Japan it has been evaluated as the best oil for deep frying tempura because of its superior stability against deterioration by heating. There are two different kinds of sesame oil, roasted and unroasted. The antioxidative activities of these oils were demonstrated in experiments with other common vegetable oils that were stored at 60 °C. Soybean oil, rapeseed oil, and others showed rapid increase after about 10 days, whereas both roasted and unroasted sesame oils were very stable. The unroasted oil remained unchanged for 30 days, while no oxidation was observed even after 50 days in the roasted oil (Fukuda & Namiki, 1988). Roasted sesame seed oil has a characteristic flavor and red-brown color probably caused by the Maillard-type reaction during roasting. The antioxidative activity increases mainly in proportion to the roasting temperature along with the brown color, indicating that some products of the roast reaction contribute to the antioxidative activity. Thus, the very strong anti-oxidative activity of the roasted oil might result from the synergistic effect of the combination of such effective factors as sesamol produced from sesa-molin, γ-tocopherol, sesamin, and roasted products like melanoidin (Fukuda et al., 1986b; Koizumi et al., 1996; Fukuda et al., 1996). It has also been shown that when foodstuff covered with wet material is fried in roasted seed oil, as in the case of Japanese tempura, sesamol is produced by the splitting of sesamolin, resulting in the formation of a strong antioxidative coating on the fried food (Fukuda et al., 1986b).

2.14.3   FOOD APPLICATIONS OF SESAME LIGNANS

There are various forms and methods of using sesame seed and oil in Asian countries, particularly in China, Korea, and Japan, and people in these coun-tries enjoy many kinds of foods containing sesame seed and oil with superb taste and flavor. However, the use of sesame in the Western countries is limited in variety and sesame is utilized mostly as topping on bread and biscuits, with low consumption. In this respect, it will be necessary to develop various sesame foods which will suit many people's tastes throughout the world. For example, one recommended form of sesame may be used in salad dressing or seasoning containing ground sesame seed and oil, which can be used with various vegetables. This use of sesame is delicious in taste and has good digestibility with high nutritional value in combination with sesame lignans and various vegetable components (Namiki, 2007).

2.15   ORYZANOL

Rice has been widely cultivated as one of the major food resources and remains as staple food. With the advancement in rice milling technology, by-products of rice milling such as rice bran is being produced. Rice bran contains about 10–24% of oil which can be commercially utilized to produce rice bran oil for edible as well as cosmetic purposes. Rice bran oil is loaded with bioactive compounds such as γ-OZ, tocopherols, tocotrienols, PS, and so forth, which contribute to the excellent stability and functionality of rice bran oil. γ-OZ is a lipid soluble antioxidant/nutraceutical/bioactive compound uniquely present in rice bran and rice bran oil. It has been widely used in foods and cosmetics around the world. It is registered as a medicine in Japan and South Korea. γ-OZ is a naturally occurring component in rice bran and rice germ which consists of a mixture of ferulic acid esters of PS and triterpene alcohols (Fig. 2.11). There are numerous reports indicating the benefits, efficacy, and safety of γ-OZ (Rukumini & Raghuram, 1991).

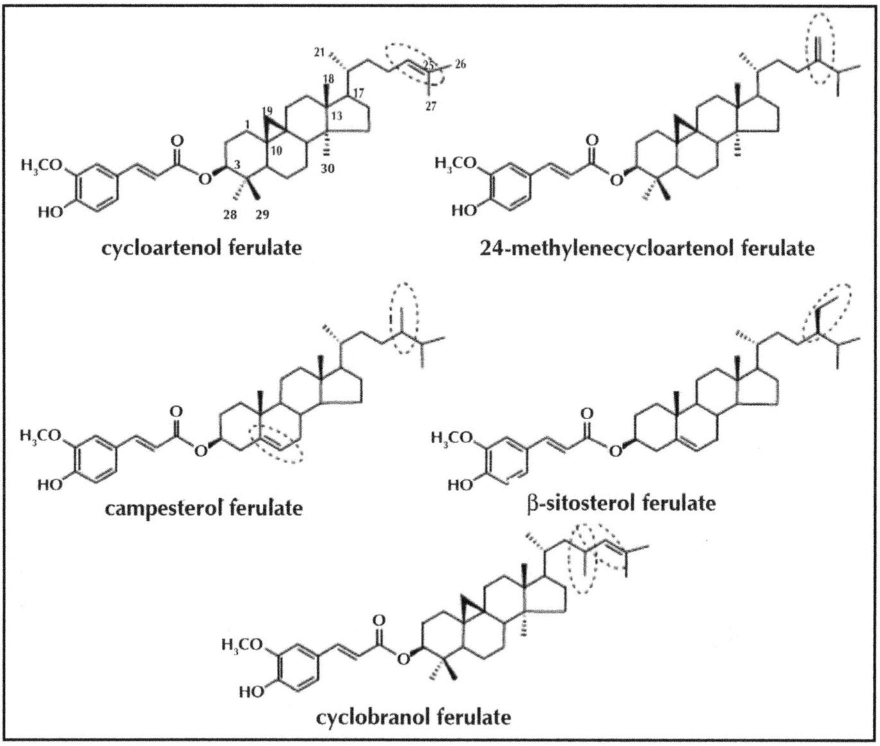

FIGURE 2.11   Different components of γ-oryzanol.

2.15.1   HEALTH BENEFITS OF ORYZANOL

Wilson et al. (2007) reported that γ-OZ reduced plasma cholesterol in hypercholesterolemic hamsters. Clinically, oral intake of rice bran oil (containing naturally occurring γ-OZ) has been shown to alleviate hypercholesterolemia and hyperlipidemia. γ-OZ has been advocated as treatment for relieving menopausal symptoms. Besides, Oka et al. (2010) reported that cycloartenyl ferulate, a component of rice bran oil-derived γ-OZ, inhibits mass cells degranulation. The anti-inflammatory effects of γ-OZ in ulcerative colitis induced in mice have also been reported (Rukumini & Raghuram, 1991).

2.15.2 SAFETY STUDIES OF ORYZANOL

Oral and intraperitoneal administration of γ-OZ (10,000 ppm) showed no abnormality generally and upon autopsy (Rukumini & Raghuram, 1991). Similarly, no abnormalities were observed on subcutaneous administration of γ-OZ (500 ppm). It has been reported that no abnormal finding observed in rats after six months of continuous oral administration of γ-OZ (30–1000 ppm) (Rukumini & Raghuram, 1991). No fetal teratogenicity observed in mouse with the administration of γ-OZ (6–600 ppm) during pregnancy (Rukumini & Raghuram, 1991). Oral administration of γ-OZ (2000 ppm) was given to mouse for 72 weeks and rat for two years, respectively. No carcinogenicity observed at the above dosage (Rukumini & Raghuram, 1991).

2.15.3 ANTIOXIDANT EFFECT OF ORYZANOL

The antioxidant effect of γ-OZ is well documented and has been reported to be excellent in inhibiting lipid peroxidation. Kanno et al. (1985) reported that γ-OZ (0.5~1%) inhibited thermal oxidative polymerization of soybean oil. The antioxidant effect of γ-OZ is contributed by ferulic acid entity, meanwhile, BHT and α-T have been revealed to be heat resistant. In addition, Oryza Oil & Fat Chemical Co. Ltd. showed that the antioxidant effect of γ-OZ was potentiated with rice bran/germ amino acid and showed a synergistic increase in antioxidant effect of γ-OZ (Rukumini & Raghuram, 1991). The excellent heat resistance property of γ-OZ is highly suitable for its incorporation in heat-processed foods. Currently in Japan, γ-OZ is approved and listed as "antioxidant" under the list of chemical composition of food additives. In addition, γ-OZ is being incorporated as antioxidant in cosmetic products (Rukumini & Raghuram, 1991).

2.15.4 ORYZANOL IN FOOD PRESERVATION

The antioxidant effect of natural OZ concentrate (15.5% OZ), purified OZ (80% OZ), and α-T (0.1%) on oxidative and thermal stability of sunflower oil was studied. It was concluded that sunflower oil containing a combination of 1% OZ (80% purity) and 0.1% α-T exhibited a synergistic effect in inhibiting primary and secondary oxidation products and also showed a very high thermal stability (Sunil et al., 2015).

To provide nutraceutical such as OZ through food, two instant mixes based on the Indian traditional food cuisine Bisibele bhath and Upma (Bhath-OZ and Upma-OZ) were developed and evaluated for shelf life. The formulations contained cereals, pulses, and spices along with OZ enriched oil and were packed in 200 gauge/50 gauge metallized polyester packaging material and stored under ambient (27 °C 65% relative humidity (RH)) and accelerated conditions (37 °C/92% RH). Samples were withdrawn periodically and PV, free fatty acid value (FFA), fatty acid composition, OZ, and total tocopherols content were estimated. Sensory evaluation of reconstituted products was also carried out. OZ content (610 and 550 mg%) did not change appreciably in Bhath-OZ and Upma-OZ, respectively. The PV under ambient condition increased from 1.1 to 9.3 meq.O_2/kg and 2.24 to 9.02 meq.O_2/kg during the six-month storage study at 27 °C and 65% RH, while under accelerated conditions at 37 °C and 92% RH, it increased from 1.12 to 8.54 meq.O_2/kg and 2.24 to 6.96 meq.O_2/kg during two-month storage period. Bhath-OZ and Upma-OZ packed in metallized polyester pouches stored at 27 °C and 65% RH had a shelf life of four months without affecting the OZ content and quality of instant mixes during the storage period (Baby Latha et al., 2014).

Biscuit is a well-known cereal based processed food and the fortification of OZ into the biscuits will go a long way to provide antioxidant rich, highly stable, and acceptable functional food to the consumers. Biscuits were prepared with commercially available fat (CF) and oryzanol fortified fat (OFF). The control biscuits (CB) and oryzanol fortified biscuits (OFB) were packed in 200 gauge polypropylene pouches, stored at 27 °C with different relative humidity (RH 11, 22, 32, 44, and 56%) and analyzed for its stability during storage of 120 days. Critical moisture content of OFB (4.8%) was slightly less than that of CB (5.3%). The fat content of the CB (12.2%) and OFB (12.5%) did not change during storage while free fatty acid content (0.36 and 0.60%) and PV (0.08 and 0.17 meq.O_2/100 g biscuit), respectively, for CB and OFB were showed small but significant changes during storage. OZ content (292 mg) and RSA (81.1%) of OFB did not change during storage. The biscuits had a shelf life of minimum three months at 27 °C. OZ in OFB showed good stability during baking and storage of biscuits (Prasanth Kumar et al., 2014).

2.16 APPLICATIONS OF NATURAL ANTIOXIDANTS/EXTRACTS IN FOOD PRODUCTS PRESERVATION

Foods of plant origin are stabilized by addition of antioxidants less frequently than foods of animal origin, perhaps with the exception of edible and essential oils. In contrast to animal foods, foods of vegetable origin usually contain natural antioxidants, such as tocopherols, carotenoids, or flavonoids in sufficient amounts. The pro-oxidative activity of iron and other heavy metals is less dangerous in plant materials than that of heme derivatives in animal products, as plant materials usually also contain metal-chelating agents. The only important oxidation catalyst in raw materials and foods of vegetable origin is a group of lipoxygenases and related enzymes. Synthetic antioxidants prevailed for the stabilization of foods of plant origin in earlier applications, but in the last decade or two, natural antioxidants have been intensively applied, following consumers' wishes.

Lipids in foods of vegetable origin are usually more unsaturated than that of animal origin; therefore, the initiation rate of oxidation reactions is higher and natural antioxidants, originally present in foods are more rapidly consumed than in lard or tallow and other animal fats. The stabilization of products of vegetable origin against autoxidation is thus less efficient than the stabilization of animal products. Protection factors of natural antioxidants are several times higher in lard than in edible oils.

The initial concentration of natural antioxidants in plant foods is already near the optimum so that a further addition of antioxidants has only a small effect, but it is useful for those cases when rapid decomposition of antioxidants is expected. For instance, additional natural antioxidants can be added to foods heated to high temperature or stored for a long time (Löliger, 1991; Pokorný &Trojáková, 2001).

Cereal products such as dehulled rice, white flour, or grits, are not usually stabilized. In whole grain flours, enzymes have to be inactivated to increase shelf life. After heating, natural antioxidants from brans are sufficient for lipid stabilization (Table 2.9). Natural antioxidants may be added to breakfast cereals, the shelf life of which should be long. Rice bran, stabilized by extrusion, has high natural antioxidant content, and thus it was found suitable as a component for breakfast cereals with high stability (Saunders, 1989; Pokorný &Trojáková, 2001).

TABLE 2.9 Natural Antioxidants Present in Cereal Brans (Rosa et al., 1999; Hídvégi and Lásztity, 2003).

Nutrients (values/100 g)	Rice bran	Corn bran	Oat bran	Wheat bran
Vitamin E				
Tocopherols (mg)	12.0	0.4	1.0	1.5
Tocotrienols (mg)	13.6	–	–	–
Total carotenoids (mg)	129.3	–	–	–
Gamma oryzanol (mg)	300.0	–	–	–
Phytosterols (mg)	341.1	–	–	–
Phytic acid (mg)	9500	620–1170	900–1420	520–1050

Aqueous extracts of natural antioxidants from other whole grains or brans, tea extracts and fruit extracts may be used with good results (Baublis et al., 2000a; Pokorný & Trojáková, 2001). The catalytic effect of iron is eliminated by phytic acid (Baublis et al., 2000b; Pokorný & Trojáková, 2001). Natural amino acids—methionine and cystine, PL and uric acid—are as active as synthetic antioxidants in breakfast cereals (Maestro-Duran & Borha-Padilla, 1993; Pokorný & Trojáková, 2001). The nutritional value of breakfast cereals is increased by the addition of flavonoids and related plant antioxidants, which extend shelf life (Shukla, 1993; Pokorný & Trojáková, 2001). Breakfast cereals, fortified with vitamin A (a very unstable compound) were efficiently stabilized with commercial phenolic natural antioxidants (Fritsch et al., 1975; Pokorný & Trojáková, 2001). Browning products, often present to improve the flavor of food products, may also help in their stabilization. Another kind of cereal products are extruded products, such as flat bread. Natural antioxidants may be added with flour and other additives to the extruder barrel. They are thus uniformly distributed in the extruded product. Spices are useful for stabilization as they impart interesting flavor notes to the final product. The agreeable color of extruded snack products, mainly due to carotenoids, rapidly disappears on storage. Therefore, it needs to be stabilized. An oil-soluble liquid rosemary extract (4942 Rosmanox) and its mixture with tocopherols (4993 Rosmanox E) preserved the natural coloration for more than seven months (Marcus, 1994; Pokorný & Trojáková, 2001). Various snacks are easily stabilized with salt-containing natural antioxidants (Sharma et al., 1997; Pokorný &Trojáková, 2001). Some cereal products contain added fat, mostly hydrogenated edible oil and/or fillings also rich in fat. Even when hydrogenated oils are rather stable against oxidation, off-flavors may arise on storage. Application of natural antioxidants may be useful in such products as their shelf life is expected to be long. Both synthetic and natural antioxidants or mixtures of both additives are available for these specialty products. Certain spices, Maillard products and essential oils could also be tested for this purpose. Sugar-snap cookies usually stabilized by BHA, are being stabilized with natural antioxidants as a replacement. Ferulic acid and sodium phytate were found to be suitable as natural antioxidants (Hix et al., 1997; Pokorný & Trojáková, 2001). Cookies containing phytate were sensorially fully acceptable. In sugar cookies, BHT may be replaced by casein, whey proteins, or Maillard reaction products without any loss of storage stability (Ferreira et al., 1996; Pokorný & Trojáková, 2001). Active natural antioxidants are formed during Maillard reactions in butter cookies (Bressa et al., 1996; Pokorný & Trojáková, 2001). Coffee bean components, such as chlorogenic acid, caffeic acid, and roasted coffee bean powder or extract, have been added to butter cookies to good effect. The last two natural additives are more active than tocopherol (Ochi et al., 1994; Pokorný & Trojáková, 2001). Ascorbic and erythorbic acids, citric acid and its isopropyl ester act as synergists of tocopherols (Ochi et al., 1993; Pokorný & Trojáková, 2001). The addition of spices, such as extracts from lemongrass, clove leaves, black pepper leaves, and turmeric increased the shelf life of cakes and also contributed to their characteristic flavor (Lean & Mohamed, 1999; Pokorný & Trojáková, 2001). The keeping quality of crackers and cookies is of great economic importance since these products are often stored for extended periods before they are consumed (sometimes after opening the packaging) and they are not protected from oxidation. A soda cracker biscuit was processed using a fine powder of marjoram, spearmint, peppermint, and basil, and their purified diethyl ether extracts as natural antioxidants. Addition of ether extract from each of the above four plant materials gave an excellent antioxidative effect compared with the effect of BHA at concentrations of 0.01, 0.02, and 0.03%. Addition of fine powder of all plant materials at 0.5% level gave an antioxidant effect compared to the control sample. Addition of a 1% mixture of equal amounts of the four plant powders caused a pro-oxidant effect (Bassiouny et al., 1990; Pokorný & Trojáková, 2001). Carotene in bread and crackers is stabilized against oxidative bleaching by α-T and ascorbyl palmitate (Ranhotra et al., 1995; Pokorný & Trojáková, 2001). Large losses of coloration, otherwise observed during baking, were thus reduced. The shelf life of fruits and vegetables is limited by factors other than lipid oxidation, for example, antioxidants are added to fruit and mushrooms to prevent oxidation of polyphenols, resulting in the enzymic browning (Nisperos-Carriedo et al., 1991; Pokorný & Trojáková, 2001). The lipid content in fruit and vegetables is about 1%

or less, so that the effect of their rancidification may be masked by other, sensorily more active substances. Fruits contain natural essential oils, which possess antioxidant activities but are also easily oxidized. They are protected by similar antioxidants as glyceridic oils. If lipoxygenases are deactivated by blanching, the content of natural antioxidants (mostly flavonoids) would be sufficient to protect the lipid fraction against oxidation. Natural antioxidants are applied only exceptionally, for example to protect carotenoids or anthocyanins against oxidation. Pigmented orange juice was stabilized with AA and phenolic acids and pasteurized (Maccarone et al., 1988; Pokorný & Trojáková, 2001). The use of natural antioxidants is more justified for the stabilization of dried products. The stability of dehydrated mashed potatoes was achieved with α-T or ascorbyl palmitate or with Prolong P (a mixture of rosemary, thyme, and marjoram) with more success than with TBHQ (Baardseth, 1989; Pokorný & Trojáková, 2001).

The application of synthetic antioxidants will probably be reduced further. They could be replaced by natural or nature-identical antioxidants. We believe that only few more new natural antioxidants will be introduced to the market in the near future in addition to the rosemary extract currently being used. Green tea extracts (prepared from dust, old leaves, and other tea wastes) also have a fair prospect of market success. Tocopherols and β-carotene will probably be increasingly used. Prolongation of the shelf life of complex foods will be achieved mainly by modifying recipes, introducing herbs and spices with a high concentration of natural antioxidants, using high-oleic edible oils requiring lower antioxidant levels and using protein hydrolysates, which act as good synergists.

2.17 REGULATORY STATUS OF NATURAL ANTIOXIDANT EXTRACTS, CONCENTRATES, AND RESINS

Synthetic antioxidants (BHA, BHT, PG, TBHQ, and EDTA) are regulated by the FDA as direct food additives. They may be used alone or in combination not to exceed 0.02% (200 ppm) of the final product in specified food products (21CFR172.110). These antioxidants are considered to be safe and suitable ingredients for use in meat, poultry, and egg products, alone or in combination, not to exceed 0.02% of the fat content (FSIS Directive 7120.1. revision 5). Some herbs, spice extracts, and oleoresins are GRAS. Some are considered to be indirect additives (21 CFR Vol. 3. Part 101); as such, solvents permitted for the extraction process and solvent residues allowed are specified. Some extracts, concentrates, and resins are regulated by the FDA

"Dietary Supplement Health and Education Act of 1994" and are considered to be one (or more) of several defined dietary ingredients a vitamin, a mineral, an herb or other botanical, amino acid, a dietary substance for use by man to supplement the diet by increasing the total dietary intake, or a concentrate, metabolite, constituent, extract, or combination of any ingredient described in clause (A), (B), (C), (D), or (E) and is excluded from regulation as a food additive. Extracts, concentrates, and resins are also regulated under the Food Labeling Regulation, Amendments; Food Regulation Uniform Compliance Date; and New Dietary Ingredient Premarket Notification Final Rule (1997). If they are added to cause flavor or color changes, they are regulated as such and specific quantities allowable for use in various foods are set forth. Based on the number of various classifications under which an extract, concentrate, or resin could be covered, allowable use levels vary widely (Brewer, 2011).

2.18 CONCLUSION

Plant and animal tissues contain unsaturated fatty acids, primarily in the PL fraction of cell membranes. These lipids are especially susceptible to oxidation because of their electron deficient double bonds. The breakdown products of oxidation can produce off-odors, new flavors, loss of nutrient content, and color deterioration. To manufacture high-quality, stable food products, the most effective solution is often the addition of antioxidants, either synthetic or natural, which can serve as "chain breakers," by intercepting the free radicals generated during various stages of oxidation or to chelate metals. Chain-breaking antioxidants are generally the most effective. A common feature of these compounds is that they have one or more aromatic rings (often phenolic) with one or more $-OH$ groups capable of donating H to the oxidizing lipid. Synthetic antioxidants, such as BHA, BHT, and PG, have one aromatic ring. The natural antioxidants AA and α-T each have one aromatic ring as well. However, many of the natural antioxidants (flavonoids and anthocyanins) have more than one aromatic ring. The effectiveness of these aromatic antioxidants is generally proportional to the number of $-OH$ groups present on the aromatic ring(s). Depending on the arrangement of the $-OH$ groups, these compounds may also chelate pro-oxidative metals. The facts that they are natural, and have antioxidative activity that is as good or better than the synthetic antioxidants, make them particularly attractive for commercial food processors because of consumer demand for natural ingredients.

KEYWORDS

- **antioxidants**
- **free radicals**
- **preservation**
- **applications**

REFERENCES

Abdulkarim, S. M.; Long, K; Lai, O. M.; Muhammad, S. K. S.; Ghazali, H. M. Frying Quality and Stability of High-Oleic Moringa Oleifera Seed Oil in Comparison with Other Vegetable Oils. *Food Chem.* **2007,** *105,* 1382–1389.

Agati, G.; Matteini, P.; Goti, A; Tattini, M. Chloroplast-Located Flavonoids can Scavenge Singlet Oxygen. *New Phytol.* **2007,** *174* (1), 77–81.

Akoh, C. C.; Min, D. B. *Food Lipids: Chemistry, Nutrition and Biotechnology;* Marcel Dekker Inc.: New York, 2002; p 54. ISBN: 0-8247-0749-4.

Baardseth, P. Effect of Selected Antioxidants on the Stability of Dehydrated Mashed Potatoes. *Food Addit. Contam.* **1989,** *6,* 201–207.

Baby Latha, R.; Debnath, S.; Sarmandal, C. V.; Hemavathy, J.; Khatoon, S.; Gopala Krishna, A. G.; Lokesh, B. R. Shelf-Life Study of Indian Traditional Food Based Nutraceutical (*oryzanol*) Enriched Instant Mixes Bhath-OZ and Upma-OZ. *J. Food Sci. Technol.* **2014,** *51* (1), 124–129.

Bassiouny, S. S.; Hassanien, F. R.; Abd-El-Razik, A. F.; El-Kayati, S. M. Efficiency of Antioxidants from Natural Sources in Bakery Products. *Food Chem.* **1990,** *37* (4), 297–305.

Baublis, A. J.; Clydesdale, F. M.; Decker, E. A. Antioxidants in Wheat-Based Breakfast Cereals. *Cereal Food World.* **2000b,** *45,* 71–74.

Baublis, A. J.; Decker, E. A.; Clydesdale, F. M. Antioxidant Effect of Aqueous Extracts from Wheat Based Ready-to-Eat Breakfast Cereals. *Food Chem.* **2000a,** *68,* 1–6.

Bauernfeind, J. C. The Use of Ascorbic Acid in Processing Foods. In *Advances in Food Research;* Mrak, E. M.; Stewart G. F., Ed.; Academic Press, Inc: New York, 1953; Vol. 4, 359–431.

Bendini, A.; Cerretani, L.; Vecchi, S.; Carrasco-Pancorbo, A.; Lercker, G. Protective Effects of Extra Virgin Olive Oil Phenolics on Oxidative Stability in the Presence or Absence of Copper Ions. *J. Agric. Food. Chem.* **2006,** *54* (13), 4880–4887.

Berger, R. G. Biotechnology of Flavours-the Next Generation. *Biotech. Lett.* **2009,** *31* (11), 1651–1659.

Bhatnagar, A. S.; Gopala Krishna, A. G. Bioactives Concentrate from Commercial Indian Niger (*Guizotia Abyssinica* (L.f.) Cass.) Seed and Its Antioxidant and Antiradical Activity. *Am. J. Nutr. Food Sci.* **2015,** *1* (1), 10–20.

Bhatnagar, A. S.; Hemavathy, J.; Gopala Krishna, A. G. Development of a Rapid Method for Determination of Lignans Content In Sesame Oil. *J. Food Sci. Technol.* **2015,** *52* (1), 521–527.

Bhatnagar, A. S.; Prasanth Kumar, P. K.; Hemavathy, J.; Gopala Krishna, A. G. Fatty Acid Composition, Oxidative Stability, and Radical Scavenging Activity of Vegetable Oil Blends with Coconut Oil. *J. Am. Oil Chem. Soc.* **2009,** *86* (10), 991–999.

Booth, S. L.; Suttie, J. W. Dietary Intake and Adequacy of Vitamin K. *J. Nutr.* **1998,** *128,* 785–788.

Boskou, D. *Olive Oil, Chemistry and Technology;* 2nd ed.; AOCS Press: Champaign, IL, 2006; p 58.

Botelho, P. B.; Galasso, M.; Dias, V.; Mandrioli, M.; Lobato, L. P.; Rodriguez-Estrada, M. T.; Castro, I. A. Oxidative Stability of Functional Phytosterol-Enriched Dark Chocolate. *LWT-Food Sci. Technol.* 2014, *55* (2), 444–451.

Bressa, F.; Tesson, N.; Rosa, M.; Sensidoni, A.; Tubaro, I. Antioxidant Effect of Maillard Reaction Product: Application to a Butter Cookie of a Competition Kinetics Analysis. *J. Agric. Food Chem.* **1996,** *44,* 692–695.

Brewer, M. S. Natural Antioxidants: Sources, Compounds, Mechanisms of Action, and Potential Applications. *Compr. Rev. Food Sci. Food Saf.* **2011,** *10* (4), 221–247.

Brewer, M. S.; Prestat, C. Consumer Attitudes toward Food Safety Issues. *J. Food Saf.* **2002,** *22* (2), 67–83.

Brewer, M. S.; Sprouls, G. K.; Russon, C. Consumer Attitudes toward Food Safety Issues. *J. Food Saf.* **1994,** *14,* 63–76.

Britton, G. Structure and Properties of Carotenoids in Relation to Function. *FASEB J.* **1995,** *9,* 1551–1558.

Brown, J. E.; Kelly, M. F. Inhibition of Lipid Peroxidation by Anthocyanins, Anthocyanidins and Their Phenolic Degradation Products. *Eur. J. Lipid Sci. Technol.* **2007,** *109* (1), 66–71.

Cabelli, D. E.; Bielski, B. H. J. Kinetics and Mechanism for the Oxidation of Ascorbic Acid/Ascorbate by HO_2/O_2^- Radicals. A Pulse Radiolysis and Stopped-Flow Photolysis Study. *J. Phys. Chem.* **1983,** *87,* 1809–1812.

Chen, Z. Research of Antioxidative Capacity in Essential Oils of Plants. *Chin. Cond.* **2008,** *11,* 40–43.

Choe, E.; Min, D. B. Chemistry of Deep-Fat Frying Oils. *J. Food Sci.* **2007,** *72* (5), R77–R86.

Christie, W.W. The Lipid Library. A Lipid Primer Structures, Occurrence, Basic Biochemistry and Function. 2013. http://lipidlibrary.aocs.org/ (accessed Mar 20, 2013).

Christine, M. S.; Song, Q.; Csallany, A. S. The Antioxidant Functions of Tocopherol and Tocotrienol Homologues in Oils, Fats, and Food Systems. *J. Am. Oil Chem. Soc.* **2010,** *87,* 469–481.

Clifton, P. Plant Sterols and Stanols-Comparison and Contrasts. Sterols Versus Stanols in Cholesterol Lowering: Is there a Difference? *Atheroscler. Suppl.* **2002,** *3* (3), 5–9.

CODEX STAN 210-1999; *Codex Standard for Named Vegetable Oil,* Adopted in 1999. Revisions: 2001, 2003, 2009, Amendment 2005, 2011, 2013 and 2015, FAO, WHO: Geneva, 2011.

Cort, W. M. Antioxidant Activity of Tocopherols, Ascorbyl Palmiatte, and Ascorbic Acid and Their Mode of Action. *J. Am. Oil Chem. Soc.* **1974,** *51,* 321–325.

Dai, J.; Mumper, R. J. Plant Phenolics: Extraction, Analysis and Their Antioxidant and Anticancer Properties. *Molecules.* **2010,** *15* (10), 7313–7352.

Damon, M.; Zhang, N. Z.; Haytowitz, D. B.; Booth, S. L. Phylloquinone (vitamin K1) Content of Vegetables. *J. Food Compost. Anal.* **2005,** *18* (8), 751–758.

Dost, K.; Tokul, O. Determination of Phytic Acid in Wheat and Wheat Products by Reverse Phase High Performance Liquid Chromatography. *Anal. Chim. Acta.* **2006,** *558,* 22–27.

Dougherty, M. E. Tocopherols as Food Antioxidants. *Cereal Food World.* **1988,** *33* (2), 222–223.

Elez-Martínez, P.; Soliva-Fortuny, R.; Martín-Belloso, O. Oxidative Rancidity in Avocado Puree as Affected by A-Tocopherol, Sorbic Acid and Storage Atmosphere. *Eur. Food Res. Technol.* **2007,** *226,* 295–300.

Erkkilä, A. T.; Booth, S. L.; Hu, F. B.; Jacques, P. F.; Lichtenstein, A. H. Phylloquinone Intake and Risk of Cardiovascular Diseases in Men. *Nutr. Metab. Cardiovasc. Dis.* **2007,** *17,* 58–62.

Faustman, C.; Cassens, R. G.; Schaefer, D. M.; Buege, D. R.; Williams, S. N.; Scheller, K. K. Improvement of Pigment and Lipid Stability in Holstein Steer Beef by Dietary Supplementation with Vitamin E. *J. Food Sci.* **1989,** *54* (4), 858–862.

Ferreira, M. T.; Harris, N. D. In *Effect of Casein, Whey Proteins, Their Maillard Reaction Products, and Cysteine on Storage Stability of Cookies,* IFT Annual Meeting, New Orleans, Louisiana, 1996; 49–50.

Fessenden, R. W.; Verma, N. C. A Time-Resolved Electron Spin Resonance Study of the Oxidation of Ascorbic Acid by Hydroxyl Radical. *Biophys. J.* **1978,** *24,* 93–100.

Fleming, J. E.; Bensch, K. G. Effect of Amino Acids, Peptides, and Related Compounds on the Autoxidation of Ascorbic Acid. *Int. J. Pept. Protein Res.* **1983,** *22,* 355–361.

Frankel, E. N. Antioxidants in Lipid Foods and Their Impact on Food Quality. *Food Chem.* **1996,** *57* (1), 51–55.

Fritsch, C. W.; Maxwell, D. S.; Andersen, S. H. Effect of Antioxidants upon the Stability of a Vitamin Fortified Breakfast Cereals. *J. Am. Oil Chem. Soc.* **1975,** *52,* 122A.

Fukuda, Y.; Isobe, M.; Nagata, M.; Osawa, T.; Namiki, M. Acidic Transformation of Sesamolin of Sesame Oil Constituent into an Antioxidant Bisepoxylignan, Sesaminol. *Heterocycles.* **1986a,** *24,* 923–926.

Fukuda, Y.; Koizumi, T.; Ito, R.; Namiki, M. Synergistic Action of the Antioxidative Components in Roasted Sesame Seed Oil. *J. Jpn. Soc. Food Sci. Technol.* **1996,** *43,* 1272–1277.

Fukuda, Y.; Nagata, M.; Osawa, T.; Namiki, M. Chemical Aspects of the Antioxidative Activity of Roasted Sesame Seed Oil and the Effect of Using the Oil for Frying. *Agric. Biol. Chem.* **1986b,** *50,* 857–862.

Fukuda, Y.; Namiki, M. Recent Studies on Sesame Seed and Oil. *J. Jpn. Soc. Food Sci. Technol.* **1988,** *35,* 552–562.

Gao, Z. H.; Ackman, R. G. Determination of Vitamin K_1 in Canola Oils by High Performance Liquid Chromatography with Menaquinone-4 as Internal Standard. *Food Res. Int.* **1995,** *28,* 61–69.

Garcia, A.; Brenes, M.; Martinez, F.; Alba, J.; Garcia, P.; Garrido, A. High Performance Liquid Chromatography Evaluation of Phenols in Virgin Olive Oil During Extraction of Laboratory and Industrial Scale. *J. Am. Oil Chem. Soc.* **2001,** *78* (6), 625–629.

Golan-Goidhirsh, A.; Whitaker, J. R.; Kahn, V. Relation between Structure of Polyphenol Oxidase and Prevention of Browning. *Adv. Exp. Med. Biol.* **1984,** *177,* 437–456.

Golan-Goldhirsh, A.; Osuga, D. T.; Chan, A. O.; Whitaker, J. T. In *Effects of Ascorbic Acid and Copper on Proteins,* Abstract HIST No. 26, 193rd National Meeting of the American Chemical Society, Denver, CO, Apr 5–10, 1987.

Gorman, J. E.; Clydesdale, F. M. The Behaviour and Stability of Iron-Ascorbate Complexes in Solution. *J. Food Sci.* **1983,** *48,* 1217–1225.

Graf, E.; Eaton, J. W. Antioxidant Functions of Phytic Acid. *Free Radic. Biol. Med.* **1990,** *8* (1), 61–69.

Graf, E.; Empson, K. L.; Eaton, J. W. Phytic Acid. A Natural Antioxidant. *J. Biol. Chem.* **1987,** *262,* 11647–11650.

Graf, E.; Mahoney, J. R.; Bryant, R. G.; Eaton, J. W. Iron-Catalyzed Hydroxyl Radical Formation: Stringent Requirement for Free Iron Coordination Site. *J. Biol. Chem.* **1984,** *259,* 3620–3624.

Grases, F.; Simonet, B. M.; Perelló, J.; Costa-Bauz, A.; Prieto, R. M. Effect of Phytate on Element Bioavailability in the Second Generation of Rats. *J. Trace Elem. Med. Biol.* **2004,** *17* (4), 229–234.

Gülcin, İ. Antioxidants Activity of Food Constituents: An Overview. *Arch Toxicol.* **2012,** *86* (3), 345–391.

Gülcin, İ.; Elias, R.; Gepdiremen, A.; Boyer, L. Antioxidant Activity of Lignans from Fringe Tree (Chionanthus virginicus L.). *Eur. Food Res. Technol.* **2006,** *223* (6), 759–767.

Gunstone, F. D. *Vegetable Oils in Food Technology: Composition, Properties and Uses;* Blackwell publishing, CRC Press: Boca Raton, FL, 2002; p 62, 103, 162, 216, 244–274, 284. ISBN: 1-84127-331-7.

Gunstone, F. D. Composition and Properties of Edible Oils. In *Edible Oil Processing;* Hamm W., Hamilton R. J., Eds.; The Oily Press: Bridgwater, UK, 2000; p 1–33.

Halliwell, B. Antioxidants in Human Health and Disease. *Annu. Rev. Nutr.* **1996,** *16,* 33–50.

Halliwell, B.; Gutteridge, J. M. C.; Aruoma, O. The Deoxyribose Method: A Simple "Test Tube" Assay for Determination of Rate Constants for Reactions of Hydroxyl Radicals. *Anal. Biochem.* **1987,** *165,* 215–219.

Halliwell, B.; Murcia, M. A.; Chirico, S.; Aruoma, O. I. Free Radicals and Antioxidants in Food and *In Vivo:* What They Do and How They Work. *Crit. Rev. Food Sci. Nutr.* 1995, *35,* 7–20.

Hawkins, P. T.; Poyner, D. R.; Jackson, T. R.; Letcher, A. J.; Lander, D. A.; Irvine, R. F. Inhibition of Iron-Catalysed Hydroxyl Radical Formation by Inositol Polyphosphates: A Possible Physiological Function for Myo-Inositol Hexakisphosphate. *Biochem. J.* **1993,** *294,* 929–934.

Hay, G. W.; Lewis, B. A.; Smith, F. Ascorbic Acid. II. Chemistry. In *The Vitamins. Chemistry, Physiology, Pathology, Methods;* Sebrell, Jr. W. H., Harris, R. S., Eds.; Academic Press Inc.: New York, 1967; Vol. 1, pp 307–336.

Hídvégi, M.; Lásztity, R. Phytic Acid Content of Cereals and Legumes and Interaction with Proteins. *Period. Polytech. Chem. Eng.* **2003,** *46* (1–2), 59–64.

Hillmann, J. Reformulation Key for Consumer Appeal into the Next Decade. *Food Rev.* **2010,** *37* (1), 14, 16, 18–19.

Hix, D. K.; Klopfenstein, C. F.; Walker, C. E. Physical and Chemical Attributes and Consumer Acceptance of Sugar-Snap Cookies Containing Naturally Occurring Antioxidants. *Cereal Chem.* **1997,** *74,* 281–283.

Iacopini, P.; Baldi, M.; Storchi, P.; Sebastiani, L. Catechin, Epicatechin, Quercetin, Rutin, and Resveratrol in Red Grapes: Content, In Vitro Antioxidant Activity and Interactions. *J. Food Compost. Anal.* **2008,** *21* (8), 589–598.

Inagaki, C.; Igarashi, O.; Arakawa, N.; Ohta, T. Antioxidative Components in the Flavedo Oil from Unshu-Orange. *J. Agric. Chem. Soc. Jpn.* **1968,** *42,* 731–734.

Jakob, E.; Elmadfa, I. Rapid and Simple HPLC Analysis of Vitamin K in Food, Tissue and Blood. *Food Chem.* **2000,** *68* (2), 219–221.

Jalani, B. S.; Cheah, S. C.; Rajanaidu, N.; Darus, A. Improvement of Oil Palm Through Breeding and Biotechnology. *J. Am. Oil Chem. Soc.* **1997,** *47* (11), 1451–1455.

Jonker, D.; van der Hoek, G. D.; Glatz, J. F. C.; Homan, C.; Posthumus, M. A.; Katan, M. B. Combined Determination of Free, Esterified and Glycosylated Plant Sterols in Foods. *Nutr. Rep. Int.* **1985,** *32* (4), 943–951.

Joppen, L. Taking Out the Chemistry. *Food Eng. Ingred.* **2006,** *31* (2), 38–39, 41.

Kamal-Eldin, A.; Andersson, R. A Multivariate Study of the Correlation between Tocopherol Content and Fatty Acid Composition in Vegetable Oils. *J. Am. Oil Chem. Soc.* **1997,** *74,* 375–380.

Kamal-Eldin, A.; Appelqvist, L. A. Variations in the Composition of Sterols, Tocopherols and Lignans in Seed Oils from Four *Sesamum* species. *J. Am. Oil. Chem. Soc.* **1994,** *71,* 149–156.

Kanematsu, H.; Morise, E.; Niiya, I.; Imamura, M.; Matsumoto, A.; Katsui, G. Influence of Tocopherols on Oxidative Stability of Margarines. *J. Jpn. Soc. Food Nutr.* **1972,** *25,* 343–348.

Kanno, H.; Usuki, R.; Kaneds, T. Antioxidative Effects of Oryzanol on Thermal Oxidation of Oils. *J. Jpn. Soc. Food Sci. Technol.* **1985,** *32* (3), 170.

Katsuzaki, H.; Kawasumi, M.; Kawakishi, S.; Osawa, T. Structure of Novel Antioxidative Lignin Glucosides Isolated from Sesame Seed. *Biosci. Biotechnol. Biochem.* **1992,** *56,* 2087–2088.

Kefford, J. F.; McKenzie, H. A.; Thompson, P. C. O. Effects of Oxygen on Quality and Ascorbic Acid Retention in Canned and Frozen Orange Juices. *J. Sci. Food Agric.* **1959,** *10* (1), 51–63.

Keypour, H.; Silver, J.; Wilson, M. T.; Hamed, M. Y. Studies on the Reactions of Ferric Ion with Ascorbic Acid. A Study of Solution Chemistry with Mössbauer Spectroscopy and Stopped-Flow Techniques. *Inorg. Chim. Acta,* **1986,** *125,* 97–106.

Khanduja, K. L. Stable Free Radical Scavenging and Antiperoxidative Properties of Resveratrol *In Vitro* Compared with Some Other Bioflavonoids. *Indian J. Biochem. Biophys.* **2003,** *40,* 416–422.

King, C. C. The Effect of Antioxidants and Modified Atmosphere on the Storage Stability of the Pecan Kernel. *Diss. Abstr. Int., B.* **1986,** *47,* 1347.

Koivu, T.; Piironen, V.; Lampi, A. M.; Mattila, P. Dihydrovitamin K1 in Oils and Margarines. *Food Chem.* **1999,** *64* (3), 411–414.

Koizumi, Y.; Fukuda, Y.; Namiki, M. Effect of Roasting Conditions on Antioxidative Activity of Seed Oils Developed by Roasting of Sesame Seeds. I. Marked Antioxidative Activity of Sesame Oils Developed by Roasting of Sesame Seeds. *J. Jpn. Soc. Food Sci. Technol.* **1996,** *43,* 689–694.

Kuwahara, M.; Uno, H.; Fujiwara, A.; Yoshikawa, T.; Uda, I. Antioxidative Effect of Natural Vitamin E for Lard Used for Frying Instant Ramen Part I. On the Effect of Natural Vitamin E of Various Concentration in the Comparison with Synthetic Antioxidants. *J. Food Sci. Technol. Tokyo, Jpn.* **1971,** *18,* 64–69.

Laguerre, M.; Lecomte, J.; Villeneuve, P. The Physico-Chemical Basis of Phenolic Antioxidant Activity. *Lipid Technol.* **2014,** *26,* 59–62.

Larson, S. R.; Rutger, N. J.; Young, K. A.; Raboy, V. Isolation and Genetic Mapping of a Non-Lethal Rice (*Oryza sativa L.*) Low Phytic Acid 1 Mutation. *Crop Sci.* **2000,** *40,* 1397–1405.

Lean, L. P.; Mohamed S. Antioxidative and Antimycotic Effects of Turmeric, Lemon-Grass, Betel Leaves, Clove, Black Pepper Leaves and Garcinia Atriviridis on Butter Cakes. *J. Sci. Food Agric.* **1999,** *79,* 1817–1822.

Lee, K. H.; Jung, M. Y.; Kim, S. Y. Effects of Ascorbic Acid on the Light-Induced Riboflavin Degradation and Color Changes in Milks. *J. Agric. Food Chem.* **1998,** *46* (2), 407–410.

Lee, S. H.; Park, H. J.; Cho, S. Y.; Jung, H. J.; Cho, S. M, Cho, Y. S.; Lillehoj, H. S. Effects of Dietary Phytic Acid on Serum and Hepatic Lipid Levels in Diabetic KK Mice. *Nutr. Res.* **2005,** *25,* 869–876.

Li, S. X.; Cherian, G.; Ahn, D. U.; Hardin, R. T.; Sim, J. S. Storage, Heating, and Tocopherols affect Cholesterol Oxide Formation in Food Oils. *J. Agric. Food Chem.* **1996,** *44* (12), 3830–3834.

Loewus, F. A.; Loewus, M. W. Biosynthesis and Metabolism of Ascorbic Acid in Plants. *Crit. Rev. Plant Sci.* **1987,** *5* (1), 101–119.

Löliger, J. The Use of Antioxidants in Foods. In *Free Radicals and Food Additives;* Aruoma, O. I., Halliwell, B., Eds.; Taylor & Francis: London, 1991; pp 121–150.

Lugasi, A.; Dworschak, E.; Hovari, J. In *Characterization of Scavenging Activity of Natural Polyphenols by Chemiluminescence Technique,* Proceedings of the European Food Chemists. VIII, Vienna, Austria, Sept 18–20, 1995; Federation of the European Chemists' Society: Vienna, Vol. 3, pp 639–643.

Maccarone, E.; Longo, M. L.; Leuzzi, U.; Maccarone, A.; Passerini, A. Stabilizzazione del Succo D'arancia Pigmentata con Trattamenti Fisici e Additivi Fenolici. *Chim. Ind.* **1988,** *70,* 95–98.

Maestro-Durán, R.; Borja-Padilla, R. Antioxidant Activity of the Nitrogeneous Natural Compounds. *Grasas Aceites.* **1993,** *44* (3), 204–207.

Marcus, F. K. Improved Light Stability with Natural Color Formulations. *Food Market. Technol.* **1994,** *8* (3), 8–10.

Mateos, R.; Espartero, J. L.; Trujillo, M.; Rios, J. J.; Leòn-Camacho, M.; Alcudia, F.; Cert, A. Determination of Phenols, Flavones and Lignans in Virgin Olive Oils by SPE and HPLC with Diode Array Ultraviolet Detection. *J. Agric. Food Chem.* **2001,** *49,* 2185–2192.

Miková, K. The Regulation of Antioxidants in Food. In *Food Chemical Safety;* Watson D. H., Ed.; Woodhead Publishing Ltd.: Cambridge, UK, 2003; Vol. 2, p 275.

Milder, I. E.; Arts, I. C.; van de Putte, B.; Venema, D. P.; Hollman, P. C. Lignan Contents of Dutch Plant Foods: A Database Including Lariciresinol, Pinoresinol, Secoisolariciresinol and Matairesinol. *Br. J. Nutr.* **2005,** *93,* 393–402.

Min, D. B.; Boff, J. M. Lipid Oxidation of Edible Oil. In *Food Lipids;* Akoh, C. C.; Min, D. B., Eds.; 2nd ed.; Marcel Dekker Inc: New York, 2002; p 344.

Ming-Long, L; Paul, A. S. Chemistry of L-Ascorbic Acid Related to Foods. *Food Chem.* **1988,** *30,* 289–312.

Moazzami, A. A.; Kamal-Eldin, A. Sesame Seed is a Rich Source of Dietary Lignans. *J. Am. Oil Chem. Soc.* **2006,** *83* (8), 719–723.

Morales, M. T.; Tsimidou, M. The Role of Volatile Compounds and Polyphenols in Olive Oil Sensory Quality. In *Handbook of Olive Oil;* Harwood, J.; Aparicio, R., Eds.; Aspen Publishers: Gaithersburg, MD, 2000; pp 393–458.

Moreau, R. A.; Whitaker, B. D.; Hicks, K. B. Phytosterols, Phytostanols and Their Conjugates in Foods: Structural Diversity, Quantitaitve Analysis, and Health-Promoting Uses. *Prog. Lipid Res.* **2002,** *41* (6), 457–500.

Muchuweti, M.; Kativu, E.; Mupure, C. H.; Chidewe, C.; Ndhlala, A. R.; Benhura M. A. N. Phenolic Composition and Antioxidant Properties of Some Spices. *Am. J. Food Technol.* **2007,** *2* (5), 414–420.

Nagata, M.; Osawa, T.; Namiki, M.; Fukuda, Y.; Ozaki, T. Stereochemical Structures of Antioxidative Bisepoxylignans, Sesaminol and Its Isomers, Transformed from Sesamolin. *Agric. Biol. Chem.* **1987,** *51* (5), 1285–1289.

Nakatani, N. Biologically Functional Constituents of Spices and Herbs. *J. Jpn. Soc. Nutr. Food Sci.* **2003,** *56* (6), 389–395.

Namiki, M. Nutraceutical Functions of Sesame: A Review. *Crit. Rev. Food Sci. Nutr.* **2007,** *47* (7), 651–673.

Nanni, E. J. Jr.; Stallings, M. D.; Sawyer, D. T. Does Superoxide Ion Oxidize Catechol, α-Tocopherol, and Ascorbic Acid by Direct Electron Transfer? *J. Am. Chem. Soc.* **1980,** *102,* 4481–4485.

Nawar, W. F. Lipids. In *Food Chemistry;* Fennema, O. R., Ed.; Marcel Dekker Inc: New York, 1996; pp 225–320.

Nes, W. R. Multiple Roles for Plant Sterols. In *The Metabolism, Structure and Function of Plant Lipids;* Stumpf, P. K.; Mudd, B. J.; Nes, W. R., Eds.; Springer: New York, 1987; pp 3–9.

Nisperos-Carriedo, M. O.; Baldwin, E. A.; Shaw, P. E. Development of an Edible Coating for Extending Postharvest Life of Selected Fruits and Vegetables. *Proc. Fla. State Hort. Soc.* **1991,** *104,* 122–125.

Ochi, T.; Otsuka, Y.; Aoyama, M.; Maruyama, T.; Niiya, I. Studies on the Improvement of Antioxidant Effect of Tocopherols. XXV. Synergistic Effects of Several Components of Coffee Beans in Cookies. *J. Jpn. Oil Chem. Soc.* **1994,** *43,* 719–723.

Ochi, T.; Tsuchiya, K.; Aoyama, M.; Maruyama, T.; Niiya, I. Effects of Tocopherols on Qualitative Stability of Cookies and Influence of Powdered Milk and Egg. *J. Jpn. Soc. Food Sci. Technol.* **1988,** *35* (4), 259–264.

Ochi, T.; Tsuchiya, K.; Ohtsuka, Y.; Aoyama, M.; Maruyama, T.; Niiya, I. Synergistic Antioxidant Effects of Organic Acids and Their Derivatives with Tocopherols on Cookies. *J. Jpn. Soc. Food Sci. Technol.* **1993,** *40,* 393–399.

Oka, T.; Fujimoto, M.; Nagasaka, R.; Ushio, H.; Hori, M.; Ozaki, H. Cycloartenyl Ferulate, a Component of Rice Bran Oil-Derived Γ-Oryzanol, Attenuates Mast Cell Degranulation. *Phytomedicine.* **2010,** *17* (2), 152–156.

Ostlund, R. E. Jr. Phytosterols in Human Nutrition. *Annu. Rev. Nutr.* **2002,** *22,* 533–549.

Otles, S.; Cagindi, O. Determination of Vitamin K1 Content in Olive Oil, Chard and Human Plasma by RP-HPLC Method with UV-Vis Detection. *Food Chem.* **2007,** *100,* 1220–1222.

Ozsoy, N.; Candoken, E.; Akev, N. Implications for Degenerative Disorders: Antioxidative Activity, Total Phenols, Flavonoids, Ascorbic Acid, Beta-Carotene and Beta-Tocopherol in Aloe Vera. *Oxid. Med. Cell. Longev.* **2009,** *2* (2), 99–106.

Piironen, V.; Koivu, T.; Tammisalo, O.; Mattila, P. Determination of Phylloquinone in Oils, Margarines and Butter by High-Performance Liquid Chromatography with Electrochemical Detection. *Food Chem.* **1997,** *59,* 473–480.

Pokorný, J.; Trojáková, L.; Takácsová, M. The Use of Natural Antioxidants in Food Products of Plant Origin. In *Antioxidants in Food: Practical Applications;* Pokorný, J., Yanishlieva, N., Gordon, M., Eds.; Wood Head Publishing Ltd.: Cambridge, UK. 2001; pp 355–372.

Prasanth Kumar, P. K.; Sai Manohar, R.; Indiramma, A. R.; Gopala Krishna, A. G. Stability of Oryzanol Fortified Biscuits on Storage. *J. Food Sci. Technol.* **2014,** *51* (10), 2552–2559.

Psomiadou, E.; Tsimidou, M. Simultaneous HPLC Determination of Tocopherols, Carotenoids and Chlorophylls for Monitoring Their Effect on Virgin Olive Oil Oxidation. *J. Agric. Food Chem.* **1998,** *46* (12), 5132–5138.

Ramadan, M. F. Functional Properties, Nutritional Value and Industrial Applications of Niger Oilseeds (*Guizotia abyssinica* Cass.). *Criti. Rev. Food Sci. Nutr.* **2012**, *52* (1), 1–8.

Ramadan, M. F.; Kroh, L. W.; Moersel, J, T. Radical Scavenging Activity of Black Cumin (*Nigella sativa L.*), Coriander (*Coriandrum sativum L.*), and Niger (*Guizotia abyssinica* Cass.) Crude Seed Oils and Oil Fractions. *J. Agric. Food Chem.* **2003**, *51* (24), 6961–6969.

Ramadan, M. F.; Moersel, J. T. Direct Isocratic Normal-Phase HPLC Assay of Fat-Soluble Vitamins and Beta-Carotene in Oilseeds. *Eur. Food Res. Technol.* **2002**, *214,* 521–527.

Ramadan, M. F.; Moersel, J. T. Oxidative Stability of Black Cumin (*Nigella sativa L.*), Coriander (*Coriandrum sativum L.*), and Niger (*Guizotia abyssinica* Cass.) Crude Seed Oils upon Stripping. *Eur. J. Lipid Sci. Technol.* **2004**, *106,* 35–43.

Ranhotra, G. S.; Gelroth, J. A.; Langemeier, J.; Rogers, D. E. Stability and Contribution of B-Carotene Added to Whole Wheat Bread And Crackers. *Cereal Chem.* **1995**, *72,* 139–141.

Rojas, M. C.; Brewer, M. S. Consumer Attitudes towards Issues in Food Safety. *J. Food Saf.* **2008**, *28* (1), 1–22.

Rosa, M.; Garcia-Estepa, E.; Guerra-Hernandez, E. Belen Garcia-Villanova, Phytic Acid Content in Milled Cereal Products and Breads. *Food Res. Int.* **1999**, *32,* 217–221.

Rukumini, C.; Raghuram, T. C. Nutritional and Biochemical Aspects of the Hypolipidemic Action of the Rice Bran Oil. *J. Am. Coll. Nutr.* **1991**, *10* (6), 593–601.

Saad, N.; Esa, N. M.; Ithnin, H.; Shafie, N. H. Optimization of Optimum Condition for Phytic Acid Extraction from Rice Bran. *Afr. J. Plant Sci.* **2011**, *5* (3), 168–176.

Sapper, H., Kang, S. O., Paul, H. H.; Lohmann, W. The Reversibility of the Vitamin C Redox System: Electrochemical Reasons and Biological Aspects. *Z. Naturforsch. C.* **1982**, *37C,* 942–946.

Saunders, R. M. Development of New Rice Products as a Consequence of Bran Stabilization. *Bull. Assoc. Oper. Millers.* **1989**, 5559–5561.

Schwarz, K.; Bertelsen, G.; Nissen, L. R.; Gardner, P. T.; Heinonen, M. I.; Huynh-Ba, A. H. T.; Lambelet, P.; McPhail, D.; Skibsted, L. H.; Tijburg, L. Investigation of Plant Extracts for the Protection of Processed Foods against Lipid Oxidation. Comparison of Antioxidant Assays Based on Radical Scavenging, Lipid Oxidation and Analysis of the Principal Antioxidant Compounds. *Eur. Food Res. Technol.* **2001**, *212,* 319–328.

Senesi, E.; Galvis, A.; Fumagalli, G. Quality Indexes and Internal Atmosphere of Packaged Fresh-Cut Pears (Abate Fetel and Kaiser Varieties). *Ital. J. Food Sci.* **1999**, *2* (11), 111–120.

Shahidi, F.; Wanasundara, J. P. K. P. D. Phenolic Antioxidants. *Crit. Rev. Food Sci. Nutr.* **1992**, *32* (1), 67–103.

Shamsuddin, A. K. M.; Vucenik, I. IP6 and Inositol in Cancer Prevention and Therapy. *Curr. Cancer Ther. Rev.* **2005**, *1* (11), 259–269.

Shamsuddin, A. M. Anti-Cancer Function of Phytic Acid. *Int. J. Food Sci. Technol.* **2002**, *37* (7), 769–782.

Sharma, G. K.; Semwal, A. D.; Narashima-Murthy, M. C.; Arya, S. S. Suitability of Antioxygenic Salts for Stabilization of Fried Snacks. *Food Chem.* **1997**, *60,* 19–24.

Shearer, M. J. Vitamin K Metabolism and Nutriture. *Blood Rev.* **1992**, *6,* 92–104.

Shearer, M. J.; Bach, A.; Kohlmeier, M. Chemistry, Nutritional Sources, Tissue Distribution and Metabolism of Vitamin K with Special Reference to Bone Health. *J. Nutr.* **1996**, *126,* 1181S–1186S.

Shukla, T. P. High Nutrition Breakfast Cereals. *Cereal Food World.* **1993**, *38,* 437–438.

Siger, A.; Nogala-Kalucka, M.; Lampart-Szczapa, E. The Content and Antioxidant Activity of Phenolic Compounds in Cold-Pressed Plant Oils. *J. Food Lipids.* **2008**, *15,* 137–149.

Sloan, A. E. Top Ten Trends to Watch and Work on for the Millennium. *Food Technol.* **1999,** *53* (8), 40–48, 51–58.

Smeds, A. I.; Eklund, P. C.; Sjoholm, R. E.; Willfor, S. M.; Nishibe, S.; Deyama, T.; Holmbom, B. R. Quantification of a Broad Spectrum of Lignans in Cereals, Oilseeds and Nuts. *J. Agric. Food Chem.* **2007,** *55* (4), 1337–1346.

Smoot, J. M.; Nagy, S. Effects of Storage Temperature and Duration on Total Vitamin C Content of Canned Single-Strength Grapefruit Juice. *J. Agric. Food Chem.* **1980,** *28* (2), 417–421.

Soliva-Fortuny, R. C.; Grigelmo-Miguel, N.; Odriozola-Serrano, I.; Gorinstein, S.; Martín-Belloso, O. Browning Evaluation of Ready-to-Eat Apples as Affected by Modified Atmosphere Packaging. *J. Agri. Food Chem.* **2001,** *49* (8), 3685–3690.

Solomons, N. W.; Viteri, F. E. Biological Interaction of Ascorbic Acid and Mineral Nutrients. In *Ascorbic Acid: Chemistry, Metabolism and Uses;* Seib, P. A.; Tolbert, B. M., Eds.; American Chemical Society: Washington, DC, 1982; Vol. 200, p 551–569.

Soobrattee, M. A.; Neergheen, V. S.; Luximon-Ramma, A.; Aruoma, O. I.; Bahorun, T. Phenolics as Potential Antioxidant Therapeutic Agents: Mechanism and Actions. *Mutat. Res.* **2005,** *579* (2), 200–213.

Steinberg, F. M.; Rucker, R. B. Vitamin C. *Reference Module in Biomedical Sciences Encyclopedia of Biological Chemistry;* 2nd ed.; Elsevier: Amsterdam, 2013; pp 530–534.

Sunil, L.; Srinivas, P.; Prasanth Kumar, P. K.; Gopala Krishna, A. G. Oryzanol as Natural Antioxidant for Improving Sunflower Oil Stability. *J. Food Sci. Technol.* **2015,** *52* (6), 3291–3299.

Trichopoulou, A.; Vasilopoulou, E. Mediterranean Diet and Longevity. *Br. J. Nutr.* **2000,** *84* (2), S205–S209.

Trojáková, L.; Reblova, Z.; Nguyen, H. T. T.; Pokornya, J. Antioxidant Activity of Rosemary and Sage Extracts in Rapeseed Oil. *J. Food Lipids.* **2001,** *8* (1), 1–13.

Verghese, M.; Rao, D. R.; Chawan, C. B.; Walker, L. T.; Shackeleford, L. Anticarcinogenic Effects of Phytic Acid (IP$_6$): Apoptosis as a Possible Mechanism of Action. *LWT-Food Sci. Technol.* **2006,** *39* (10), 1093–1098.

Visioli, F. Antioxidants in Mediterranean Diets. In *Mediterranean Diets;* Simopoulos, A. P., Visioli, F., Eds.; World Review of Nutrition and Dietetics, Karger: Basel, Switzerland, 2000; Vol. 87, pp 43–53.

Vucenik, I.; Shamsuddin, A. M. Cancer Inhibition by inositol Hexaphosphate (IP6) and Inositol: From Laboratory to Clinic. *J. Nutr.* **2003,** *133,* 3778S–3784S.

Wales, N. S. The Antioxidant Properties of Ascorbic Acid and Its Use for Improving the Shelf-Life of Beer. *Wallerstein Lab. Commun.* **1956,** *19,* 193–207.

Wanasundara, P. K. J. P. D.; Shahidi, F. Antioxidants: Science, Technology, and Applications. In *Bailey's Industrial Oil and Fat Products;* Shahidi, F., Ed.; Wiley: New York, 2005; Vol. 1, pp 431–489.

Wanatabe, Y.; Nakanashi, H.; Goto, N.; Otsuka, K.; Kimura, T.; Adachi, S. Antioxidative Properties of Ascorbic Acid and Acyl Ascorbates in ML/W Emulsion. *J. Am. Oil Chem. Soc.* **2010,** *85,* 1475–1480.

Weihrauch, J. L.; Gardner, J. M. Strerol Content of Foods of Plant Origin. *J. Am. Diet. Assoc.* **1978,** *73* (1), 39–47.

Wiley, R. C. Preservation Methods for Minimally Processed Refrigerated Fruits and Vegetables. In *Minimally Processed Refrigerated Fruits and Vegetables*; Wiley, R. C., Ed.; Springer: New York, 1994; pp 66–134.

Wilson, T. A.; Nicolosi, R. J.; Woolfrey, B.; Kritchevsky, D. Rice Bran Oil and Oryzanol Reduce Plasma Lipid and Lipoprotein Cholesterol Concentrations and Aortic Cholesterol Ester Accumulation to a Greater Extent than Ferulic Acid in Hypocholesterolemic Hamsters. *J. Nutr. Biochem.* **2007,** *18,* 105–112.

Yap, S. C.; Choo, Y. M.; Ooi, C. K.; Ong, A. S. H.; Goh, S. H. Quantitative Analysis of Carotenes in the Oil from Different Palm Species. *J. Oil Palm Res.* **1991,** *3,* 369–378.

Zeb, A.; Mehmood, S. Carotenoids Contents from Various Sources and Their Potential Health Applications. *Pak. J. Nutr.* **2004,** *3* (3), 199–204.

CHAPTER 3

POTENTIAL APPLICATIONS OF NATURAL ANTIOXIDANTS IN MEAT AND MEAT PRODUCTS

RITUPARNA BANERJEE[1,*], ARUN K VERMA[2],
MOHAMMED WASIM SIDDIQUI[3], B. M. NAVEENA[1], and
V. V. KULKARNI[1]

[1]ICAR-National Research Centre on Meat, Chengicherla, Hyderabad 500092, Telangana, India

[2]ICAR-Central Institute of Research on Goats, Makhdoom, Farah, Mathura 281122, Uttar Pradesh, India

[3]Department of Food Science and Postharvest Technology, Bihar Agricultural University, Sabour, Bhagalpur 813210, Bihar, India

*Corresponding author. E-mail: rituparnabnrj@gmail.com

CONTENTS

ABSTRACT

In recent years, there is a lot of buzz about natural antioxidants. Scientific advances, awareness of personal health, increasing healthcare costs, busy lifestyles, and technical advances in the meat industry have stimulated the "green consumerism." Demands for the natural ingredients have forced the researchers as well as meat industry to go for natural alternatives for synthetic antioxidants. In this journey numerous plant materials have been screened for their potential to prevent protein and lipid peroxidation. The extracts of these plant materials have also been screened for their active principles and have been attempted in different meat and meat products at various concentrations or levels and the quality, acceptability of the products have been assessed. In future, we can see many more natural alternatives of synthetic antioxidants with better potency and functionality for meat products.

3.1   INTRODUCTION

Meat as a food has a complex physical structure and chemical composition that is very prone to oxidation (Wood et al., 2008). The oxidative stability of meat depends upon the interaction between endogenous anti- and pro-oxidant substances and the substrates prone to oxidation including polyunsaturated fatty acids (PUFA), cholesterol, proteins, and pigments (Bertelsen et al., 2000). Additionally, a variety of intrinsic properties and processing steps can pre-dispose meat to lipid oxidation. For example, meat from non-ruminants is more prone to lipid oxidation than that of ruminants (Tichivangana & Morrissey, 1985) due to greater concentrations of unsaturated fatty acids (Enser et al., 1996); muscle with red fibers are more susceptible than white fibers because they contain more iron and phospholipid (Wood et al., 2004); processing of meat will accelerate lipid oxidation as the comminution or grinding process will incorporates oxygen and increase surface area as a result of particle size reduction (Gray et al., 1996).

Oxidation of lipids is a complex chemical process which involves the development of off-flavor, decreases the acceptability of meat and meat products by deteriorating their color, texture, and nutritive value, and can ultimately precipitate health hazards and economic losses in terms of inferior product quality (Naveena et al., 2008b). The oxidation process can be reduced or inhibited through application of antioxidants. These antioxidants could be either from synthetic or from natural sources; the latter includes both endogenous natural antioxidants (present in meat itself) and exogenous natural antioxidants (present in plant materials and other natural sources). Several endogenous antioxidants (including ubiquinone, glutathione, lipoic acid, spermine, carnosine, and anserine) have been studied in skeletal muscle (Decker et al., 2000). Both carnosine and anserine are histidyl dipeptides and the most abundant antioxidants in meat. Carnosine is present at around 365 mg/100 g in beef (Purchas & Busboom, 2005) and 400 mg/100 g in lamb (Purchas et al., 2004). Anserine is especially abundant in chicken muscle. The antioxidant activity of these dipeptides may result from their ability to chelate transition metals (Brown, 1981) and form complexes with copper, zinc, and cobalt. Levels of Coenzyme Q10 (ubiquinone) in meat has been estimated to be around 2 mg/100 g in both beef and mutton (Purchas & Busboom, 2005). Glutathione, a component of glutathione peroxidase enzymes, has an important antioxidant function. Glutathione levels in red meat are estimated to be 12–26 mg/100 g in beef (Jones et al., 1992). In addition, numerous Maillard reaction products formed during cooking have also been shown to have antioxidant activities (Bailey, 1988). However, none of these antioxidant systems, individually or combined, have been shown to sufficiently delay oxidation in meat or meat products under common processing conditions (Decker & Mei, 1996). Therefore, the best strategy to combat this problem is either supplementation of exogenous antioxidants through dietary manipulation or incorporation during processing of products. This chapter will deal with sources of natural antioxidants such as fruits, vegetables, herbs, and spices as well as marine macroalgae, their application in meat and meat products and effects on various quality and acceptability.

3.2   ANTIOXIDANTS: MECHANISM OF ACTION

Antioxidants are substances that at low concentrations retard the oxidation of easily oxidizable biomolecules, such as lipids and proteins in meat products, thus improving shelf life of products by protecting them against deterioration caused by oxidation (Karre et al., 2013).

According to their mechanism of action, antioxidants can be classified into different groups:

- Antioxidants that act as radical scavengers (react with free radicals): Two basic mechanisms are involved in this free radical scavenging activity of antioxidants: (a) A chain breaking mechanism in which the antioxidants break the chain reaction of radicals and avoid the propagation step of lipid peroxidation process by donating electrons to the free radicals present in the system (Masuda et al., 2002; Amakura et al., 2000). (b) The second mechanism involves the removal of reactive oxygen species (ROS) and reactive nitrogen species (RNS) initiators by quenching the chain initiator catalyst (Hamid et al., 2010). Examples of antioxidants which scavenge free radicals are phenolic compounds (tocopherols, butylated hydroxytoluene (BHT), butylated hydroxyanisole (BHA), tert-butylhydroquinone (TBHQ), propyl gallate (PG), lignans, flavonoids, and phenolic acids, carotenoids, and so forth.

- Antioxidants that react with transition metals to form complexes, and thus avoid the catalytic effect of the metals in the oxidation process. Metal chelators decrease oxidation by preventing metal redox cycling, forming insoluble metal complexes, or providing steric hindrance between metals and food components or their oxidation intermediates (Graf & Eaton, 1990). The most common metal chelators used in foods contain multiple carboxylic acid (e.g., ethylene diamine tetra acetic acid (EDTA) and citric acid) or phosphate groups (e.g., polyphosphates and phytate). Chelators are typically water soluble but many also exhibit lipid solubility (e.g., citric acid), thus allowing it to inactivate metals in the lipid phase. Lignans, polyphenols, ascorbic acid, and amino acids such as carnosine and histidine can also chelate metals (Decker et al., 2001). Phenolics, which possess hydroxyl and carboxyl groups are able to bind particularly well with metals like Fe or Cu (Jung et al., 2003).

- Antioxidants that decompose peroxides and produce stable substances which are unable to produce radicals, such as selenium (Se) containing glutathione peroxidase, an antioxidative enzyme, which inactivate free radicals and other oxidants, particularly hydrogen peroxide.

- Antioxidants which inactivate the singlet form of oxygen: In the presence of a photosensitizer, such as chlorophylls and pheophytins, singlet oxygen may be formed from ordinary triplet oxygen by the action of light. This singlet form of oxygen is very reactive; it is extremely important to deactivate it back to the triplet form very rapidly to prevent the photo-oxidation process. Tocopherols, carotenoids, curcumin, phenolics, urate, and ascorbate can quench singlet oxygen (Das & Das, 2002; Choe & Min, 2005).

- Antioxidants which prevent the enzymatic activity required for autooxidation. Examples are flavonoids, phenolic acids, and gallates, which deactivate the lipoxygenase.

3.3   SYNTHETIC OR NATURAL?

Though there are several different types of available antioxidants, they can be broadly grouped into two categories: natural and synthetic. While natural antioxidants are those that can be harvested directly from any organic source such as herbs, fruits, vegetables, and so forth, synthetic antioxidants are compounds produced artificially and added to processed or pre-packaged food to prevent rancidity, browning or to preserve the flavor and texture. Synthetic antioxidants such as BHA, BHT, TBHQ, and PG have been widely used in meat and meat products (Biswas et al., 2004; Formanek et al., 2001; Jayathilakan et al., 2007) by the food processors as they are cheaper than the natural ones. But the demand for natural antioxidants, especially of plant origin has increased in the recent years due to the growing concern among consumers about these synthetic antioxidants because of their potential toxicological effects (Naveena et al., 2008b). However, both of these antioxidants differ in performance level; the effectiveness can be measured by the number of peroxides formed in lipids over time and by their ability to provide stability under different processing conditions. Both natural and synthetic antioxidants act by donating electron density to fat and preventing their oxidation but synthetic antioxidants have shown to possess a higher performance than the natural ones. They differ in their fortification values also. The natural antioxidants are known to have higher additional health benefits in preventing cancer and heart diseases.

3.4   NATURAL ANTIOXIDANTS

Plants are persistently the generous source to supply man with valuable bioactive substances (Tayel & El-Tras, 2012) and thus different plant products are being evaluated as natural antioxidants to improve the overall quality of meat and meat products. The focus for using natural antioxidants for the effective preservation of meat or meat products has almost exclusively been on the use of plant phenolics or phenolic-containing extracts. Phenolic compounds are plant secondary metabolites commonly found in herbs and fruits, vegetables, grains and cereals, tea, coffee, and red and white wines. Phenolic acids are phenols that possess carboxylic acid functionality. Phenolic compounds can be broadly divided into two categories, flavonoids and non-flavonoid polyphenols. Among phenolic compounds found in plants, flavonoids are the most widely studied class of polyphenols with respect to their antioxidant and biological activities. Flavonoids may be divided into different subclasses according to the degree of oxidation of the heterocyclic ring: anthocyanins, flavonols, flavanones, flavanol, flavones, and isoflavones (Scalbert & Williamson, 2000). Flavonols are the most ubiquitous flavonoids in foods, and the main representatives are quercetin and kaempferol. Onions, kale, broccoli, and blueberries are the richest sources. Flavones are much less common than flavonols in fruit and vegetables. Celery, parsley, wheat, millet, and skin of citrus fruits are important sources of flavones. Anthocyanins are mainly found in red wine, certain leafy and root vegetables, and are most abundant in fruits. Flavanols exist as catechins or proanthocyanidins. Catechin and epicatechin are the main flavanols in fruit, whereas gallocatechin, epigallocatechin, and epigallocatechin gallate are found in certain seeds of leguminous plants, grapes, and more importantly in tea (Arts et al., 2000). Proanthocyanidins (condensed tannins) are responsible for the astringent character of grapes, peaches, apples, pears, berries, tea, wine, and bear. Flavanones are common in tomatoes, mint and to a considerable extent in citrus fruit. Soya and its processed products are the main source of isoflavones in the human diet. The importance of all these antioxidant constituents of plant materials and other natural sources in the maintenance of health and protection from coronary heart disease and cancer is raising interest among scientists, food processors, and consumers as the future trend is moving toward functional food with specific health effects.

The extraction of bioactive compounds is the first step in utilization of natural antioxidants as additives in meat products. Generally the plant material is cleaned, dried, and ground into fine powder followed by extraction. The drying process has some undesirable effects on the constituent profile of plant material; however, freeze-drying process retains higher levels of phenolic content in plant samples than air-drying (Abascal et al., 2005). There are many techniques to recover antioxidants from plants, such as Soxhlet extraction, maceration, supercritical fluid extraction, subcritical water extraction, and ultrasound assisted extraction but the solvent extraction process is the most commonly used procedure to prepare extracts from plant materials due to their ease of use, efficiency, and wide applicability. Table 3.1 represents the common solvents used for extraction of antioxidant compound from different plant parts. Extraction efficiency is affected by the chemical nature of phytochemicals, the extraction method used, sample particle size, the solvent used, as well as the presence of interfering substances (Stalikas, 2007) whereas the yield of extraction depends on the solvent with varying polarity, pH, temperature, extraction time, and composition of the sample (Turkmen et al., 2006). Solvents, such as methanol, ethanol, acetone, ethyl acetate, and their combinations have been used for the extraction of phenolics from plant materials, with different proportions of water. Methanol has been generally found to be more efficient in extraction of lower molecular weight polyphenols while the higher molecular weight flavanols are better extracted with aqueous acetone (Prior et al., 2001, Guyot et al., 2001). Ethanol is another good solvent for polyphenol extraction (Shi et al., 2005). Water is the safest solvent but it is less efficient than the organic solvents in extracting all the antioxidants. On the contrary, extra precaution should be taken to remove all the traces of organic solvent as it may not be acceptable for consumers if residues are left in the final product. In addition, the cost-effectiveness of extraction process should also be taken in account in order to reduce the cost of natural antioxidants and its wider exploitation in meat industry.

TABLE 3.1 Extraction of Antioxidant Components from Different Sources and Its Application in Meat Products.

Source	Part used	Extraction solvent	Reference
Fruits			
Bearberry	Leaf	95% ethanol and 50% acetone	Pegg et al. (2005)
Citrus paradisi (grape fruit)	Bark	Ethyl acetate, methanol, and water	Sayari et al. (2015)
Grape	Seed	80% ethanol	Shan et al. (2009)
	Pomace	Methanol	Garrido et al. (2011)
Kinnow	Peel	Water	Devatkal et al. (2010)
Pomegranate	Peel	Water	Devatkal et al. (2010)
		70% ethanol	Tayel and El-Tras (2012)
		80% ethanol	Shan et al. (2009)
Prunus mume	Fruit	Methanol	Jo et al. (2006)
Herbs and spices			
Clove	Bud	80% ethanol	Shan et al. (2009)
Cinnamon	Bark	70% ethanol	Tayel and El-Tras (2012)
Cinnamon stick	Cortex	80% ethanol	Shan et al. (2009)
Fenugreek	Seed	90% ethanol	Mansour and Khalil (2000)
Green tea	Leaf	Water	Rababah et al. (2011)
Rosemary & hissop	Leaf and secondary branches	Dimethyl sulfoxide	Fernandez-Lopez et al. (2003)
Rosemary	Leaf	Deionized water	Akarpat et al. (2008)
	Leaf	Acetone, hexane	Naveena et al. (2013)

TABLE 3.1   *(Continued)*

Source	Part used	Extraction solvent	Reference
Mint	Leaf	Water	Kanatt et al. (2007, 2008)
	Leaf	Water, ethanol, and 50% ethanol:50% water	Biswas et al. (2012)
Vegetables			
Broccoli	Flowering head	Water	Banerjee et al. (2012)
Cauliflower	Flowering head	Water	Banerjee et al. (2015)
		Acetone:water (1:1)	
Drumstick	Leaf	Water	Das et al. (2012); Muthu-kumar et al. (2014)
Green leafy vegetables	Leaf	70% ethanol	Kim et al. (2013a)
Butterbur, chamnamul, bok choy, Chinese chives, crown daisy, fatsia pump-kin, sesame stonecrop			
potato	Peel	90% ethanol	Mansour and Khalil (2000)

3.5   NATURAL ANTIOXIDANTS IN MEAT SYSTEM

The natural antioxidants from plants, in several forms, have been obtained from different sources such as fruits (grapes, pomegranate, date, and kinnow), vegetables (broccoli, potato, drumstick, and curry leaves), herbs, and spices (tea, rosemary, oregano, cinnamon, sage, thyme, mint, ginger, and clove) and explored to decrease the lipid oxidation (Akarpat et al., 2008; Banerjee et al., 2012; Das et al., 2012; Devatkal et al., 2010; Huang et al., 2011; Kanatt et al., 2007; Mansour & Khalil, 2000; McCarthy et al., 2001a, b; Rojas & Brewer, 2007, 2008; Shan et al., 2009).

3.5.1   FRUIT-BASED ANTIOXIDANTS

Fruits have gathered interest from the public and scientific groups because of their health-promoting properties. The benefits of fruits have been credited to their high phenolic content, which acts as antioxidants (Zuo et al., 2002). Numerous studies conducted on the antioxidant potential of fruits in meat and meat products are presented below.

3.5.1.1   BAEL

The bael fruit (*Aegle marmelos* L. Correa) is known in India since pre-historic times. This fruit is native to Northern India but widely found throughout the Indian Peninsula (Rahman & Pravin, 2014). Bael fruit pulp contains many functional and bioactive compounds such as dietary fiber, carotenoids, phenolics, alkaloids, coumarins, flavonoids, terpenoids, and other antioxidants (Suvimol & Anprung, 2008). Major antioxidants in bael fruit are phenolics, flavonoids, carotenoids, and vitamin C (Roy & Khurdiya, 1995). Bael fruit is rich in carbohydrates, fibers and is also a good source of protein, vitamins, and minerals (Ramulu & Rao, 2003). Kamalakkannan and Prince (2003a, 2003b) reported that the aqueous extract of the bael fruit pulp possesses potent antioxidant effects. Abdullakasim et al. (2007) also reported that the bael fruit drink was found to possess high quantities of total phenolic compounds (83.89/37.6 mg gallic acid equivalents/100 mL) and was also a good antioxidant in both 2,2-diphenyl-1-picrylhydrazyl (DPPH) and photochemiluminescence assays. Das et al. (2014) had reported the antioxidant potential of bael pulp residue (BPR), a by-product of bael fruit pulp in goat meat nuggets. BPR was found to be a rich source of phenolic compounds and contained 15.16 mg GAE/g dry weight (DW) total phenolics. Incorporation of BPR (0.25, 0.5%) in goat meat nuggets improved the lightness and redness values, whereas yellowness value remained unaffected. The lighter and redder goat meat nuggets looked very much appealing and could be helpful in attracting the consumers. Lower thiobarbituric acid reactive substances (TBARS) value was recorded in BPR treated nuggets; lowest value was observed in 0.5% treatment. Incorporation of BPR may enrich meat products with dietary fiber and antioxidants, and can be helpful in enhancing their physiological and functional values as well as oxidative stability.

3.5.1.2   BEARBERRY

Bearberries also known as *Uva Ursi*, is a member of genus *Arctostaphylos* and is one of the lesser-studied source of natural antioxidants. Traditionally, the astringent leaves of bearberry plant have been used in the treatment of bladder infections and other afflictions of the urinary tract. The plant contains arbutin, ursolic acid, tannic acid, gallic acid, some essential oils (EOs), hydroquinones, phenolic glycosides, and flavonoids (Hansel et al., 1992). O'Brien et al. (2006) investigated the antioxidant activity of several plant extracts under oxidative stress in cells and found bearberry to be a strong antioxidant. The total polyphenol content of bearberry extract was reported by Carpenter et al. (2007) as $57.4 \text{ g} \pm 1.73 \text{ GAE}/100 \text{ g}$. Bearberry extract reduced the lipid oxidation in raw and cooked pork patties during storage up to 12 days at 4 °C and sensory properties were affected by its addition. Pegg et al. (2005) had reported that bearberry leaf extract possesses marked antioxidant activity in model and meat systems. Crude leaf extract, and its fractions (acetone, ethanol) inhibited TBARS formation in cooked meat systems after seven days of refrigerated storage.

3.5.1.3   CAROB FRUIT

The Carob is the fruit of an evergreen *Ceratonia silique* L. cultivated in the Mediterranean area. Use of the whole carob fruit for consumption is rather limited, due to a high level of tannins causing astringency (Avallone et al., 1997). The two main carob pod constituents are pulp (90%) and seed (10%). The seed coat contains antioxidants (Batista et al., 1996). Phenolic contents of pulps and leaves from carob tree have been reported (Avallone et al., 1997; Corsi et al., 2002; Kumazawa et al., 2002; Owen et al., 2003; Makris & Kefalas, 2004). Flavonol glycoside, 4`-p-hydroxybenzoylisorhamnetin-3-O-α-L-rhamnopyranoside named ceratoside, together with the known kaempferol-3-O- α-L-rhamnopyranoside (afzelin), quercetin-3-O- α -L-arabi-nofuranoside (auriculain), quercetin-3-O- α-L-rhamnopyranoside, β-sitosterol, and β-sitosterol-3-O- β -D-glucoside were isolated from carob seeds (Gohar et al., 2009). Vaya and Mahmood (2006) observed that the carob leaves are rich in flavonoids; and more than nine compounds were identified. Researchers had isolated and identified the major polyphenols in carob fibers (Owen et al., 2003; Papagiannopoulos et al., 2004).

Bastida et al. (2009) evaluated the effect of adding condensed tannins in the form of non-purified (Liposterine®) or purified (Exxenterol®) extracts obtained from carob fruit to prevent oxidation in lipid-cooked pork meat systems during chilling and frozen storage. The antioxidant activity of these extracts was compared with that of α-tocopherol (TM). TBARS levels were significantly lower in samples containing Liposterine (LM), Exxenterol (EM), and TM than in control sample under chilled storage. TBARS formation was similar ($P > 0.05$) for LM and EM but significantly lower than that for TM. Thermal oxidation compounds were lower ($P < 0.05$) in EM than in LM or TM, which is also having nutritional importance as thermal oxidation products are potentially toxic. Therefore, Carob extract has the potential to improve the fat stability and toxicological safety of meat systems.

3.5.1.4   CITRUS FRUITS

Citrus fruits are an important source of bioactive compounds including antioxidants such as ascorbic acid, flavonoids, phenolic compounds, and pectins that are important to human nutrition (Fernandez-Lopez et al., 2005; Jayaprakasha & Patil, 2007). Flavanones, flavones, and flavonols are three types of flavonoids which are present in citrus fruit (Calabro et al., 2004). The main flavonoids found in citrus species are hesperidine, narirutin, naringin, and eriocitrin (Schieber et al., 2001). Epidemiological studies on dietary citrus flavonoids improved a reduction in risk of coronary heart disease (Di Majo et al., 2005) and are attracting more and more attention not only due to their antioxidant potential, but also as anti-carcinogenic and anti-inflammatory agents because of their lipid anti-peroxidation effects (Stavric, 1993; Elangovan et al., 1994; Martın et al., 2002). In addition, citrus by-products represent a rich source of naturally occurring flavonoids. The peel which constitutes almost one-half of the fruit mass, contains the highest concentrations of flavonoids in the citrus fruit (Anagnostopoulou et al., 2006; Manthley & Grohmann, 2001).

The antioxidant effects of orange and lemon extracts were investigated in cooked Swedish-style meatballs by Fernandez-Lopez et al. (2005). Antioxidant activities of each natural extract were expressed as stability index (SI). The SI of orange extract (1.30) was higher ($P < 0.05$) than that of lemon extract (1.19). TBARS data indicated that orange extract was superior in reducing the lipid oxidation of cooked products in comparison to lemon extract throughout the storage period (8 ± 1 °C, 12 days). Viuda-Martos et al. (2009) stated that the addition of citrus waste water (5, 10%) obtained as co-product during the extraction of dietary fiber to the bologna samples reduced the residual nitrite levels and the degree of lipid oxidation. The flavonoids hesperidin and narirutin were detected in all the samples. Viuda-Martos et al. (2010) studied the effect of adding orange dietary fiber (1%), rosemary EO (0.02%), or thyme EO (0.02%) and the storage conditions on the quality characteristics and the shelf life of *mortadella*, a bologna-type sausage. Color coordinates lightness (L^*) and yellowness (b^*) were affected by the fiber content. The treatments had lower level of residual nitrite; the extent of lipid oxidation was also reduced, and analysis of the samples revealed the presence of the flavonoids, hesperidin, and narirutin. Antioxidant and antibacterial properties of *Citrus paradisi* fruit barks (CPFB) extract was evaluated in turkey sausage formulation (Sayari et al., 2015). The CPFB water extract contained a high amount of total phenolics and flavonoids (118 ± 4 mg GAE/g dried extract and 794 ± 8.7 mg QE/g dried extract, respectively) and showed important antioxidant and antibacterial activities. Water extracts from CPFB showed a higher antioxidant activity than sodium lactate.

Kinnow or Tangerine (*Citrus reticulata*), a citrus fruit variety is grown in North Indian states, mainly Punjab and Rajasthan. In the process of juice extraction, 30–34% of kinnow peel is obtained as a major by-product, which is a very rich source of vitamin C, carotenoids, limonene, and polyphenolic antioxidants (Anwar et al., 2008). Kinnow rind powder (KRP), pomegranate rind powder (PRP), and pomegranate seed powder (PSP) extracts were investigated (Devatkal et al., 2010) in goat meat patties stored at 4 ± 1 °C. It was found that these extracts are rich in phenolic compounds and have free radical scavenging activity. Hunter Lab $L*$ value was significantly ($P < 0.05$) lower in PRP followed by PSP and KRP treated patties. Sensory evaluation indicated no significant differences among patties. Further, a significant ($P < 0.05$) reduction in TBARS values during storage of goat meat patties was observed in PRP, PSP, and KRP as compared to control patties. The overall antioxidant effect was in the order of PRP > PSP > KRP.

Effects of salt, kinnow, and pomegranate fruit by-product powders on color and oxidative stability of raw ground goat meat stored at 4 ± 1 °C was evaluated by Devatkal and Naveena (2010). Five treatments were evaluated including control (no treatment), meat with salt (MS) (meat + 2% salt), KRP (meat + 2% salt + 2% KRP), PRP (meat + 2% salt + 2% PRP), and PSP (meat + 2% salt + 2% PSP). TBARS values were higher ($P < 0.05$) in MS followed by control and KRP samples compared to PRP and PSP samples throughout storage. The PSP treated samples showed lowest TBARS values than others. Percent reduction of TBARS values was highest in PSP (443%) followed by PRP (227%) and KRP (123%). Salt accelerated the TBARS formation and by-products of kinnow and pomegranate fruits counteracted this effect. The overall antioxidant effect was in the order of PSP > PRP > KRP > control > MS.

3.5.1.5   CRANBERRY

Cranberries are a group of evergreen dwarf shrubs or trailing vines in the subgenus *Oxycoccus* of the genus *Vaccinium*. It is generally processed into three basic categories: fresh (5%); sauce products, concentrate, and various value-added applications (35%); and juice drinks (60%) (NASS, 2001). Extensive processing of cranberry for different products such as juice yields cranberry pomace as a by-product, a cheap source of natural antioxidants with potential health benefits as food ingredients (Yamaguchi et al., 1999; Koga et al., 1999). Cranberry contains various classes of polyphenolic compounds, including phenolic acids, flavonol glycosides, anthocyanins, and proanthocyanidins (Foo et al., 2000; Chen et al., 2001; Sun et al., 2002; Zuo et al., 2002). Studies have shown that cranberry phenolic compounds possess antioxidant activity against peroxyl (Wang & Stretch, 2001; Gunes et al., 2002; Zheng & Wang, 2003), superoxide (Wang & Jiao, 2000), hydroxyl (Wang & Jiao, 2000) and DPPH (Yan et al., 2002) radicals, hydrogen peroxide, and singlet oxygen (Wang & Jiao, 2000). A major underutilized by-product from cranberry juice production is cranberry press cake, containing seeds and skins. Cranberry press cake contains many phenolic compounds (Zheng & Shetty, 2000) and could be used as a potential source for preparing antioxidant extracts (Moure et al., 2001).

The potential of cranberry press cake and cranberry juice powder as antioxidants in meat and poultry products has been the interest of several researchers (Larrain et al., 2008; Raghavan & Richards, 2006, 2007). The cranberry juice powder extract (extracted with chloroform) was superior ($P < 0.05$) to cranberry press cake extract (extracted with either ethyl acetate or ethanol) in inhibiting lipid oxidation in vacuum-packaged mechanically separated turkey (MST) (Raghavan & Richards, 2006).

The ability of components of cranberry powder to inhibit lipid oxidation processes in MST and cooked ground pork was assessed by Lee et al. (2006). Fraction of extract enriched in flavonols showed the greatest inhibitory effect on lipid oxidation of cooked ground pork with 81% inhibition, in comparison to other fractions (phenolic acids, anthocyanin, and proanthocyanidin), over the entire storage period. Crude extract treated cooked ground pork exhibited up to 51% inhibition on TBARS formation. Concentrated cranberry juice powder (0.32%) was effective in retarding TBARS formation and rancidity development in MST during 14 days of storage at 2 °C. Quercetin, a non-glycosylated flavonol present in cranberry powder, inhibited lipid oxidation in MST at low concentrations. Ethanol was the most effective carrier solvent of polyphenolics compared to propylene glycol and water as carriers.

3.5.1.6 GRAPES

Grape seed extract has been reported to be one of the richest sources of natural polyphenols, comprising flavanols, phenolic acids, catechins, proanthocyanidins, and anthocyanins. Among these, catechins and proanthocyanidins are the major groups representing about 77.6% of total polyphenols (Silvan et al., 2013). The high amount of phenol groups in grape seed extract explains their strong lipid oxidation inhibition and antimicrobial activity in raw and cooked muscle foods (Ahn et al., 2007a; Brannan, 2008). Numerous authors have mentioned the potent antioxidant effect of grape polyphenols (*Vitis vinifera*) in pork (Carpenter et al., 2007; O'Grady et al., 2008), beef (Rojas & Brewer, 2007, 2008), and poultries (Brannan, 2009; Mielnik et al., 2006; Sayago-Ayerdi et al., 2009).

Kulkarni et al. (2011) compared grape seed extract (100, 300, 500 ppm) with ascorbic acid and PG (100 ppm of fat) in lean beef sausages cooked (70 °C), sliced and stored at −18 °C for four months and concluded that samples prepared with the grape seed extract and PG retained their freshness, had less rancid odor and had lower TBARS values compared to controls and ascorbic acid containing samples during the storage period. It was also demonstrated that frankfurters prepared with addition of different concentrations (0, 0.5, 1, 2, 3, 4, and 5%) of grape seed flour, had lower oxidation level and enhanced protein and total dietary fiber (TDF) content with increasing levels of grape seed flour (Ozvural & Vural, 2011). The addition of red grape pomace extract (0.06 g/100 g) to pork burgers resulted in color stability, lipid oxidation inhibition and yielded best overall acceptability after six days storage at 4 °C under aerobic conditions (Garrido et al., 2011).

Grape seed extract (ActiVin™) and pine bark extract (Pycnogenol®) significantly improved the oxidative stability of cooked beef at three days of refrigerated storage. TBARS values, hexanal content, and warmed over flavor were reduced during the storage period (Ahn et al., 2002). In another study, grape seed extract (ActiVin™), pine bark extract (Pycnogenol), oleoresin rosemary (Herbalox), and BHA/BHT were used in cooked ground beef. The control showed significantly higher TBARS and hexanal content over storage. BHA/BHT, ActiVin™, Pycnogenol, and Herbalox retarded the formation of TBARS by 75, 92, 94, and 92%, respectively, after nine days, and significantly lowered the hexanal content throughout the storage period. The color of cooked beef treated with ActiVin™ was less light ($L*$), more red ($a*$), and less yellow ($b*$) than those treated with BHA/BHT, Pycnogenols, and Herbaloxs. ActiVin™ and Pycnogenols effectively retained the redness in cooked beef during storage (Ahn et al., 2007a). The antioxidant effect of grape seed extract was determined in raw or cooked ground muscle during refrigerated or frozen storage (Brannan & Mah, 2007). It was found that grape seed extract was more effective than gallic acid in inhibiting oxidation. The formation of lipid hydroperoxides (LOOH) and TBARS was inhibited by grape seed extract (0.1 and 1.0%) compared to untreated controls. Furthermore, the results showed that grape seed extract at concentrations as low as 0.1% is a very effective inhibitor of primary and secondary oxidation products in various meat systems.

3.5.1.7 GUAVA

Guava (*Psidium guajava* L.), being recognized as *"super food"* is getting much attention in the agro-food industry due to the attractive characteristics of the fruit, its health-promoting bioactive components and functional elements. The fruit is considered as highly nutritious because it contains a high level of ascorbic acid (50–300 mg/100 g fresh weight) and has several carotenoids such as phytofluene, β-carotene, β-cryptoxanthin, lycopene, rubixanthin, cryptoflavin, lutein, and neochrome (Mercadante et al., 1999). Phenolic compounds such as myricetin and apigenin (Miean & Mohamed, 2001), ellagic acid, and anthocyanins (Misra & Seshadri, 1968) are also at high levels in guava fruits.

Reports regarding use of guava in the meat products either as antioxidant or as a source of dietary fiber are scarce. Guava powder (0.5, 1%) has been used as a source of antioxidant dietary fiber in sheep meat nuggets (Verma et al., 2013) and it was found that the powder was rich in dietary fiber (43.21%), phenolics (44.04 mg GAE/g), and possessed good radical scavenging activity as well as reducing power. Total phenolics, TDF, and product redness values were significantly increased ($P < 0.05$) in nuggets with added guava powder. Addition of powder retarded lipid peroxidation of cooked sheep meat nuggets as measured by TBARS number during refrigerated storage. Antioxidant potential of pink guava pulp (10%) was evaluated in raw pork emulsion during refrigerated storage for nine days under aerobic packaging (Joseph et al., 2014). The surface redness ($a*$ value) increased ($P < 0.05$) with the incorporation of pink guava pulp. Metmyoglobin formation and lipid oxidation were lower ($P < 0.05$) in guava-treated emulsions than in control. Overall, incorporation of pink guava pulp improved the visual color and odor scores of raw pork emulsion. Incorporation of guava fruits having several bioactive components in meat products would definitely enhance their physiological, functional, and nutritional values.

3.5.1.8   PLUM

Dried plums, known as prunes, have been extensively investigated for their potential human health benefits. The phenolic compounds in dried plum products have been shown to inhibit low-density lipoprotein cholesterol oxidation in humans (Stacewicz-Sapuntzakis et al., 2001) and have been shown to have better antioxidant ability than vitamins C and E *in vitro* (Vinson et al., 2005). Plum products exhibited antioxidant properties in a variety of meat products under several different processing and storage conditions (Lee & Ahn, 2005; Nuñez de Gonzalez et al., 2008a, 2008b; Yildiz-Turp & Serdaroglu, 2010).

Leheska et al. (2006) added 5 and 10% of dried plum puree (DPP) and dried blueberry purees in pork breakfast sausage patties, pre-cooked them prior to sensory evaluation and total phenolic level was measured. This research demonstrated that plum puree increased the phenolic content of the sausage more than the blueberry puree. Yıldız-Turp and Serdaroglu (2010) used different levels of plum puree (5, 10, and 15%) as an extender in low-salt beef patties. Addition of plum puree slightly increased redness and decreased yellowness and lightness both in cooked and uncooked samples. TBARS values of treated samples were lower than control at the end of the storage period. Lee and Ahn (2005) found that plum extract (California Dried Plum Board, Sunsweet Growers Inc., Yuba City, CA) used at 3% in irradiated (3 kGy) turkey breast rolls reduced ($P < 0.05$) lipid oxidation. TBARS value for the control product was 0.95 mg MDA/kg meat whereas the 3% plum extract sample had a reduced ($P < 0.05$) TBARS value of 0.84 mg MDA/kg meat after seven days of storage at 4 °C. Nuñez de Gonzalez et al. (2008a) added 3 and 6% each of DPP, dried plum and apple puree, or 0.02% BHA/BHT to sausage. For the raw sausage, there were no differences ($P < 0.05$) in TBARS values among treatments. The 3 and 6% DPP were about equal to the synthetic antioxidants BHA and BHT. In the pre-cooked/refrigerated sausage TBARS values were similar to the raw sausage, whereas the control without antioxidants exhibited higher ($P < 0.05$) TBARS values which indicated lower antioxidant capacity. Among the treatments for pre-cooked/frozen sausage only the 6% DPP inhibited lipid oxidation such that its TBARS values (0.46 mg MDA/kg) were similar ($P < 0.05$) to the raw, non-oxidized values. However, even 3% DPP was able to inhibit lipid oxidation to the same extent as BHA and BHT. These researchers concluded that plum products could be used as natural antioxidants and have the ability to replace BHA and BHT in inhibiting lipid oxidation. The work was continued by Nuñez de Gonzalez et al. (2008b) in roast beef brine formulations. They used 2.5 and 5% each of fresh plum puree (FPP), DPP, and spray dried plum powder (DPWD) in addition to sodium chloride, dextrose, alkaline phosphate, potassium lactate, and water. All treatments had reduced ($P < 0.05$) TBARS values compared to the control (0.62 mg MDA/kg), which further proved that dried plum ingredients were able to inhibit lipid oxidation.

Conversely, hams brined with the same amounts of plum products (2.5 and 5% FPP, DPP, and DPWD) and with sodium chloride, dextrose, alkaline phosphate, potassium lactate, sodium nitrite, and sodium erythorbate did not exhibit differences between treatments and control for TBARS values at 21 days post storage (Nuñez de Gonzalez et al., 2009). They suggested that the hams were not susceptible to lipid oxidation due to the inclusion of sodium nitrite and alkaline phosphates and therefore no differences were observed among treatments.

3.5.1.9   POMEGRANATE

Pomegranates (*Punica granatum*) have been used extensively in the folk medicine of many cultures (Longtin, 2003). Numerous studies on the anti-oxidant activity have shown that pomegranate juice contains high levels of antioxidants—higher than most other fruit juices and beverages (Gil et al., 2000; Seeram et al., 2008). The exceptionally high antioxidative capacity of the fruit juice might be the result of the remarkably high content and unique composition of soluble phenolic compounds (Gil et al., 2000; Poyrazoglu et al., 2002; Seeram et al., 2008). Phenolic concentration and composition in the pomegranate fruit are cultivar-dependent; the most abundant components are anthocyanins, catechins, ellagic tannins, and gallic and ellagic acids (El-Nemr et al., 1990; Gil et al., 2000; Poyrazoglu et al., 2002).

PRP was used at 10 mg tannic acid equivalent phenolics/100 g in fresh chicken, and then prepared as cooked chicken patties (Naveena et al., 2008a). Reduced TBARS values were observed ($P < 0.05$) in comparison to control. Chicken patties were treated with pomegranate, cooked to an internal temperature of 80 °C, and stored in low-density polyethylene pouches for 15 days at 4 °C. TBARS value for control was reported as 1.272 ± 0.13 mg MDA/kg meat, and the treatment with PRP had a value of 0.203 ± 0.04 mg MDA/kg. TBARS values also decreased 68% compared with samples treated with BHT (100 mg BHT/100 g meat) for the same product held under identical storage conditions. PRP and pomegranate juice powder have little effect on sensory or quality attributes when used at concentrations from 5 to 20 mg tannic acid equivalent phenolics/100 g meat (Naveena et al., 2008a; Naveena et al., 2008b). Naveena et al. (2008b) reported a decrease in L^* values (56.71 ± 0.74) compared with the control (63.8 ± 0.73) for cooked chicken patties with PRP at 20 mg equivalent phenolics/100 g meat. An 8–10-member trained sensory panel found no differences for off-odor, sweet flavor, and chicken flavor between the pomegranate samples at any of the concentrations compared with the control; however, chicken flavor was slightly reduced for the sample with 20 mg tannic acid equivalent phenolics/100 g meat. Vaithiyanathan et al. (2011) evaluated the effect of pomegranate fruit juice phenolics (PFJP) dipping solution on the shelf life of chicken meat held under refrigerated storage at 4 °C. TBARS were evaluated in two-day intervals for 28 days and it was reported that TBARS values were lower (0.35–0.75 mg MDA/kg of meat) in samples treated with PFJP. Sensory scores indicated that both samples treated with and without PFPJ performed well in all sensory attributes. However, on day 4, sensory attribute scores of samples without PFJP started to decline while scores of samples with PFJP remained high. Additionally, the acceptability scores of control samples decreased significantly ($P < 0.05$) on day 12 of storage. These studies demonstrated the potential of pomegranate components as antioxidants in refrigerated chicken and goat patties. Pomegranate was effective at inhibiting lipid oxidation and does not significantly affect the overall sensory attributes of the finished product.

3.5.2 HERBS- AND SPICES-BASED ANTIOXIDANTS

In addition to the application of herbs and spices as seasonings, meat industry has explored their potential as a natural antioxidants and antimicrobials too. Spices and herbs are excellent sources of antioxidants and have a long history of safe usage. They are rich sources of phytochemicals (Shan et al., 2005; Srinivasan, 2014; Surh, 2002; Zheng & Wang, 2001) (Table 3.2). Spices are derived from different parts of a plant other than the leaves while herbs from leaves of a plant. The basil, oregano, bay leaf, and thyme come from leaves, clove and saffron from flower or bud, clove, chilli, and black-pepper from fruits or berries, fennel and fenugreek from seeds, cinnamon and cassia from bark, onion and garlic from bulb, ginger and turmeric from root, and mace from aril. Antioxidant components of herbs and spices may be removed/concentrated as extracts, EOs, or resins.

TABLE 3.2 Major Active Constituents and Antioxidative Capacities of Herbs/Spices (Modified from Charles, 2013; Prior et al., 2003).

Herb/spice	Active constituents	Total ORAC value (μm TE/100) g
Basil	Eugenol, apigenin, limonene, ursolic acid, methyl cinnamate, 1,8-cineole, α-terpinene, anthocyanins, β-sitosterol, carvacrol, citronellol, farnesol, geraniol, kaempferol, menthol, p-coumaric acid, quercetin, rosmarinic acid, rutin, safrole, tannin, catechin	4805 (fresh)
Marjoram	Sinapic acid, ferulic acid, coumarinic acid, caffeic acid, syringic acid, vanillic acid, 4-hydroxybenzoic acid, limonene, ursolic acid, α-pinene, α-terpinene, p-cymene, rosmarinic acid, sterols, apigenin	27,297 (fresh)
Oregano	Apigenin, rosmarinic acid, luteolin, quercetin, myricetin, caffeic acid, p-coumaric acid, diosmetin, protocatechuic acid, eriodictyol, carvacrol, thymol	13,970 (fresh)
Sage	α-Pinene, β-pinene, geraniol, limonene, 1,8-cineole, perillyl alcohol, citral, β-sitosterol, farnesol, ferulic acid, gallic acid, β-carotene, catechin, apigenin, luteolin, saponin, ursolic acid, rosmarinic acid, carnosic acid, vanillic acid, caffeic acid, carnosol	32,004 (fresh)
Cinnamon	Cinnamic aldehyde, 2-hydroxycinnamaldehyde, eugenol, myristicin, cinnamate, phenolics	131,420 (ground)
Clove	Eugenol, isoeugenol, gallic acid, flavonoids, phenolic acids	290,283 (ground)
Ginger	Zingiberone, zingiberene, ar-curcumene, gingerol, paradol, shogaols, zingerone, curcumin, zerumbone	39,041 (ground)
Nutmeg	Caffeic acid, argenteane, myristicin, lignans, catechin	69,640 (ground)
Oregano	Apigenin, rosmarinic acid, luteolin, quercetin, myricetin, caffeic acid, p-coumaric acid, diosmetin, protocatechuic acid, eriodictyol, carvacrol, thymol	175,295 (dried)
Black pepper	Piperidine, piperine, limonene, α-pinene, β-pinene, sarmentine, guineesine, isoquercetin	34,053
Rosemary	Carnosol, carnosic acid, rosmanol, ursolic acid, 1,8-cineole, geraniol, α-pinene, limonene, β-carotene, apigenin, naringin, luteolin, caffeic acid, rosmarinic acid, rosmanol, vanillic acid, diosmetin	165,280 (dried)
Sage	α-Pinene, β-pinene, geraniol, limonene, 1,8-cineole, perillyl alcohol, citral, β-sitosterol, farnesol, ferulic acid, gallic acid, β-carotene, catechin, apigenin, luteolin, saponin, ursolic acid, rosmarinic acid, carnosic acid, vanillic acid, caffeic acid, carnosol	119,929 (ground)

TABLE 3.2 *(Continued)*

Herb/spice	Active constituents	Total ORAC value (μm TE/100) g
Thyme	Thymol, carvacrol, 1,8-cineole, α-pinene, limonene, apigenin, β-carotene, ursolic acid, luteolin, gallic acid, caffeic acid, rosmarinic acid, carnosic acid, hispidulin, cismaritin, diosmetin, naringenin, kaempferol, quercetin, hesperidin	27,426 (fresh)
Turmeric	Curcumin, curcuminoids, β-turmerin	127,068 (ground)
Garlic	Allicin, diallyl sulfide, diallyl disulfide, diallyl trisulfide allyl isothiocyanate, S-allylcysteine	5708 (raw)
Ginger	Zingiberone, zingiberene, ar-curcumene, gingerol, paradol, shogaols, zingerone, curcumin, zerumbone	14,840 (raw)

The herbs and spice extracts, including rosemary, oregano, clove, thyme, and so forth have been investigated for their antioxidant potential in several meat products. El-Alim et al. (1999) investigated the use of ground spices and spice extracts as antioxidants in raw ground chicken and ground pork. Ground chicken was treated with 1% of dried spices: marjoram, wild marjoram, caraway, clove, peppermint, nutmeg, curry, cinnamon, basil, sage, thyme, and ginger. TBARS formation was significantly inhibited in refrigerated and frozen samples that were treated with spices. During refrigerated storage (4 °C for seven days), cloves showed the largest reduction in TBARS values compared with the control. After six months of frozen storage at −18 °C, marjoram-treated samples showed the highest inhibition for TBARS formation. These researchers also examined use of spice extracts of basil, sage, thyme, and ginger @ 1 ml/10g as antioxidants in ground pork. After seven days of refrigerated storage, TBARS values for all treatments were significantly lower than control. Sage, thyme, and basil were more effective at inhibiting TBARS values than ginger. All treatments significantly reduced TBARS formation after six months frozen storage compared with the control. Efficacy of varying concentrations of dried holy basil powder (0.07, 0.18, and 0.35%) and its ethanolic extracts (0.02, 0.05, and 0.10%) in retarding oxidative rancidity was reported in cooked ground pork during refrigerated storage at 5 °C for 14 days (Juntachote et al., 2007). Ethanolic extracts of holy basil were less effective than dried holy basil powder in controlling oxidative stability. Dried holy basil powder at a concentration of 0.35% (w/w) was the most effective in retarding lipid oxidation in cooked ground pork during the storage period.

Mohamed et al. (2011) reported that addition of herbal extracts of marjoram, rosemary, and sage at concentration of 0.04% (v/w) to ground beef prior to irradiation (2 and 4.5 kGy) significantly lowered the TBARS values, off-odor scores, and increased color and acceptability scores. Similarly, Kanatt et al. (2007) found that radiation processed lamb meat treated with mint leaf extract (0.1 and 0.5%) showed greater antioxidant activity and decreased lipid oxidation during four weeks chilled storage compared with non-treated samples. Jayathilakan et al. (2007) showed that cinnamon and cloves (250 mg/100 g meat) were significantly effective in inhibiting TBARS formation in cooked ground beef, pork, and mutton stored at 5 °C for six days. No difference in TBARS values was observed between samples treated with cinnamon and samples treated with BHA, or PG at 0.02% in ground beef and pork. Cloves exhibited higher antioxidant activity than BHA and PG. However, TBHQ demonstrated the highest antioxidant activity of all tested antioxidants in all three types of meat.

According to Yu et al. (2002) aqueous rosemary extracts (0, 100, 250, and 500 ppm) improved the color stability of turkey rolls in addition to their inhibition of lipid oxidation. Sánchez-Escalante et al. (2001) reported that rosemary powder and rosemary along with ascorbic acid were most effective in inhibiting oxidation of both lipid and myoglobin, as revealed by the results of TBARS and the percentage of metmyoglobin, respectively, in beef patties stored at 2 ± 1 °C for 20 days. Both of these desirable effects contributed in maintaining desirable sensory characteristics of fresh beef patties in extending their shelf life. Antioxidant effectiveness of a commercial rosemary extract (FORTIUM™ R20) at concentrations of 1500 and 2500 ppm in frozen and pre-cooked-frozen pork sausage, and from 500 to 3000 ppm in refrigerated, fresh pork sausage was compared with BHA/BHT (Sebranek et al., 2005). Rosemary extract at 2500 ppm was as effective as the maximum permitted concentrations of BHA/BHT in refrigerated, fresh pork sausage and in cooked-frozen sausage, but was superior to BHA/BHT in raw-frozen pork sausage patties. Three kinds of *Rosmarinus officinalis* extract (powder-acetone, liquid-methanol, and liquid-acetone) were used by Rocío Teruel et al. (2015) to examine their effects on frozen chicken nuggets quality. The highest antioxidant activity was found for the powder-acetone extract followed by the liquid-methanol and liquid-acetone extracts. In a study with porcine liver pâté, Doolaege et al. (2012) found that addition of a rosemary extract had a positive effect on retarding lipid oxidation and maintaining higher concentrations of the antioxidants ascorbic acid, TM, and carnosic acid. It was also noticed that the sodium nitrite concentration in liver pâté, could be reduced from 120 to 80 ppm when rosemary extract was added, without negative effects on lipid oxidation, antioxidant level, and color stability.

Jinap et al. (2015) reported that local spices, that is, turmeric (4 g/100 g), lemongrass, torch ginger, and curry leaves (10 g/100 g) have effectively reduced the level of heterocyclic amines in grilled beef, thus spices could be exploited to avoid toxic and carcinogenic effects from these amines too.

3.5.3 ESSENTIAL OIL-BASED ANTIOXIDANTS

EOs are aromatic and volatile oily liquids that are extracted from plant materials, such as flowers, buds, roots, bark, leaves, seeds, peels, fruits, wood, and whole plants (Hyldgaard et al., 2012; Sanchez et al., 2010; Viuda-Martos et al., 2010) and are characterized by a strong odor (Burt, 2004). These have been widely used for centuries for their biological and flavoring characteristics. Some of the beneficial properties, for example, antiseptic, antioxidant, or, anti-inflammatory, have been supported by recent scientific investigations (Bakkali et al., 2008; Adorjan et al., 2010). The advantages associated with use of EOs in comparison to dried spice materials include better stability during storage, higher concentrations of flavor components, reduced need for storage space, ease of handling, microbial safety, and standardization (Tipsrisukond et al., 1998).

The composition of an EO is influenced by the extraction method, which ultimately influences its antioxidative properties. Among the different extraction methods that are used to obtain EOs, steam distillation is most commonly used on a commercial basis (Burt, 2004). Antioxidant properties play a pivotal role in some of EOs' biological activities. These attributes are due to the inherent ability of some of their components, particularly phenols, to stop or delay the aerobic oxidation of organic matter. However, there are some phenol-free EOs that express antioxidant behavior due to the radical chemistry of some terpenoids and other volatile constituents like sulfur-containing components of garlic (Valgimigli, 2012).

The effectiveness of a wide range of EOs against lipid oxidation has been demonstrated by many researchers over the years. The antioxidant activity of pork and beef (both raw and cooked) treated with oregano and sage EOs, during meat storage (4 °C for 12 days), was determined by Fasseas et al. (2007). Both meat samples showed significantly lower TBARS values and DPPH radical scavenging effects during refrigerated storage when treated EOs. Oregano EO was found more effective as an antioxidant than sage EO in both meat samples. Findings of Estevez et al. (2007) revealed that sage and rosemary EOs at the levels of 0.1% improve the oxidative stability of lipids in liver pates during refrigerated storage for 90 days, mainly by reducing the degradation of PUFAs and thereby preventing the formation of residual components such as malondialdehyde and lipid-derived volatiles. At 60 and 90 days, TBARS values and lipid-derived volatiles in pates with EOs were significantly lower than in the control samples. The effect of addition of rosemary and marjoram EOs (200 ppm) to beef patties formulated with mechanically deboned poultry meat (20%) was demonstrated by Mohamed and Mansour (2012). The findings showed that both marjoram and rosemary EOs reduced the lipid oxidation and improved the sensory attributes of beef patties during frozen storage for three months. The TBARS value of beef patties prepared with marjoram and rosemary EOs remained significantly lower compared with those of control beef patties during frozen storage. Additionally, these natural antioxidants were superior to BHT used in the study with regard to their role in sensory attributes besides lowering the lipid oxidation. Moreover, the addition of EOs to the beef patties significantly increased the flavor and overall acceptability scores of patty formulas processed with mechanically deboned poultry meat after processing and during the frozen storage period. Dzudie et al. (2004) prepared beef patties by incorporating ginger and basilica EOs together with maize oil. During 18 days of storage, the TBARS values of the beef patties with EOs were stable. Fratianni et al. (2010) investigated the effectiveness of balm and thyme EOs as natural antioxidants on fresh chicken breast meat that had been stored for three weeks at 4 °C. During their experiment, they found that both EOs reduced radical formation in meat compared with untreated control. Thyme EO, in particular, was the most effective out of the two EOs, with a DPPH radical inhibition percentage of 25–30%. Balm oil showed 15–20% DPPH radical inhibition and lower activity of ~10% was demonstrated in the control.

3.5.4   VEGETABLES-BASED ANTIOXIDANTS

Vegetables account for a small part of our daily caloric intake; however, their benefits to health surpass their caloric contribution mainly due to presence of dietary fiber, phenolic compounds, minerals, and vitamins. Epidemiological studies have indicated that the frequent consumption of fruits and vegetables significantly reduced the incidence of chronic diseases (WHO, 2003). In order to obtain maximum health benefits, intake of sufficient amounts of antioxidants from plant food (fruits, vegetables, etc.) is preferred. The antioxidants obtained in vegetables play the major role in maintenance of health and prevention of diseases (Paganga et al., 1999). It has been estimated that every serving increase in vegetable consumption reduces the risk of cancer by 15%, cardiovascular disease by 30%, and mortality by 20% (Steimez & Potter, 1996; Rimm et al., 1996), attributable to antioxidants such as ascorbic acid, vitamin E, carotenoids, lycopenes, polyphenols, and other phytochemicals. These antioxidants scavenge radicals and inhibit the chain initiation or break the chain propagation (the second defense line). Vitamin E and carotenoids also contribute to the first defense line against oxidative stress, because they quench singlet oxygen (Krinsky, 2001; Shi et al., 2001).

3.5.4.1   LEAFY GREEN VEGETABLES

Fresh leafy green vegetables contain important functional food components, such as β-carotene, ascorbic acid, riboflavin, folic acid, minerals (Grusak & DellaPenna, 1999), and a large amount of polyphenols (e.g., phenolic acids, flavonoids, and aromatic compounds). They are also known for their characteristic color, flavor, and therapeutic value (Gupta et al., 2005; Faller & Fialho, 2009). Ten common vegetables were screened for their antioxidant and anti-proliferative activities by Chu et al. (2002). Broccoli and spinach had the highest amount of free phenolics, followed by yellow onion, red sweet pepper, cabbage, carrot, potato, and lettuce. Cucumber had the lowest free phenolics of the 10 vegetables. The total antioxidant activity was determined by total oxyradical scavenging activity (TOSC) assay. Red pepper, broccoli, carrot, and spinach were in the group with higher antioxidant activities. The medium group comprised cabbage and yellow onion. The remaining four vegetables in the group with lower antioxidant activities included celery, potato, lettuce, and cucumber.

Antioxidant activities of 70% ethanolic extracts of 10 leafy green vegetables were determined and applied in raw beef patties (Kim et al., 2013a). The extracts and BHT were separately added to patties at 0.1 and 0.5% (w/w) concentrations and the patties were stored at 4 °C for 12 days. The addition of extracts and BHT resulted in concentration dependent decreases in TBARS values in the beef patties and also improved meat color stability. The fatsia (*Aralia elata*) extract had more effective antioxidant than the chamnamul (*Pimpinella brachycarpa*). In another study, the antioxidant efficacy of 70% ethanol and water extract of 10 leafy edible plants was evaluated in ground beef patties. Plant extracts (butterbur and broccoli extracts) and BHT were separately added to the patties at 0.1 and 0.5% (w/w) concentrations and stored at refrigerated conditions for 12 days. TBARS values were significantly lower in the samples containing plant extracts or BHT than the non-treated control. In addition, the beef patties formulated with the selected plant extracts showed significantly better color stability than those without antioxidants (Kim et al., 2013b).

3.5.4.2   CARROT

Carrot is one of the important widely consumed root vegetable with high nutritional value due to its enriched healthy composition, such as phytonutrients and minerals. It is a good source of natural antioxidants especially carotenoids and phenolic compounds, having the highest carotenoid content among foods (Arabshahi-D et al., 2007; Hsieh & Ko, 2008; Soria et al., 2009). Antioxidant activity of carrot juice (unconcentrated carrot juice, carrot juice concentrated by 35 and 60%) in gamma irradiated (0, 3, and 4.5 kGy) beef sausage was studied by Badr and Mahmoud (2011). Carrot juice exerted a significant antioxidant effect during the irradiation of sausages and the formation of hydroperoxides, and TBARS significantly decreased with increasing the concentration of the carrot juice. Formulation of sausages with carrot juice at the different concentrations decreased the formation of hydroperoxides by 24.68, 40.38, and 58.01%, respectively, in sausage samples exposed to the highest irradiation dose, while decreased the formation of TBARS in the samples by 28.86, 42.86, and 54.29%, respectively.

3.5.4.3   POTATO

Potato (*Solanum tuberosum* L.) tubers possess a wide range of carotenoid contents. Potato peel extract (PPE) was found to have the highest antioxidant activity owing to its high content of phenolic compounds and flavonoids (Mohdaly et al., 2010). Effectiveness of PPE in reducing lipid peroxidation of γ-irradiated lamb meat was examined by Kanatt et al. (2005). TBARS number and carbonyl content were reduced in irradiated meat containing PPE in comparison to the samples without PPE. These researchers found that the antioxidant activity of PPE was comparable to BHT and it did not affect flavour or aroma of the radiation processed meat.

3.5.4.4 CRUCIFEROUS VEGETABLES

The *Brassicaceae* (*Cruciferae*) family is composed of 350 genera and about 3500 species (Sasaki & Takahashi, 2002). *Brassica* is an inexpensive, though very nutritive, source of food, providing nutrients and health-promoting phytochemicals such as phenolic compounds, vitamins (Dekker et al., 2000; Vallejo et al., 2002, 2003; Vallejo et al., 2004), phytic acid, fiber, soluble sugars (Pedroche et al., 2004), glucosinolates (Fowke et al., 2003), minerals, polyphenols (Heimler et al., 2005), fat, and carotenoids (Zakaria-Rungkat et al., 2000). There is ever-increasing evidence that a higher consumption of *Brassica* vegetables, for example, broccoli, cabbage, kale, mustard greens, Brussels sprouts, and cauliflower, reduces the risk of several types of cancer (Kristal & Lampe, 2002; Wang et al., 2004). The anti-carcinogenic effect of these vegetables has been attributed to decomposition products of gluco-sinolates, indoles, and iso-thiocyanates (Zukalova & Vasak, 2002), phyto-alexins, and other antioxidants (Samaila et al., 2004; Hanf & Gonder, 2005). Extracts of the different species of the *Brassicaceae* family show antioxidant effects (Banerjee et al., 2012; Banerjee et al., 2015) and decrease oxidative damage (Ferguson, 1999). Phenolic compounds with vitamin C are the major antioxidants of *Brassica* vegetables, due to their high content and high antioxidant activity. On the contrary, lipid-soluble antioxidants such as carotenoids and vitamin E were responsible for up to 20% of the total antioxidant activity of *Brassica* vegetables (Podsedek, 2007). The order of the oxygen radical absorbance capacity (ORAC) values of the fresh weight extracts reported by Cao et al. (1996) was: kale > Brussels sprouts > broccoli > cauliflower > cabbage. Generally, among *Brassica* vegetables, Brussels sprouts, broccoli, and red cabbage have the highest antioxidant capacity. Common cabbage demonstrated rather low antioxidant activity. However, the antioxidant activities depend on the extraction method, and on the type of the reactive species in the reaction mixture (Azuma et al., 1999; Cao et al., 1996). Mustard leaf (*Brassica juncea*), a cruciferous vegetable originating from China has attracted a lot of attention as a functional food for maintenance of health and disease prevention (Kim et al., 2004). Lee et al. (2010) demonstrated the effectiveness of mustard leaf kimchi ethanolic extract (MK; 0.05, 0.1, and 0.2%) on microbial growth and lipid oxidation and in extending the shelf life of raw ground pork meats during storage at 4 °C for 14 days. The TBARS values indicated that at MK @ 0.1 or 0.2% was as effective as 0.02% L-ascorbic acid, and at the level of 0.2%, it suppressed lipid oxidation and reduced the formation of peroxides more than the ascorbic acid treatment, indicating the high protective effect of MK against oxidation in ground pork. The lowest free fatty acid value was reported in 0.2% MK treated pork. The antioxidant power of broccoli powder extract (1.0, 1.5, and 2.0%) was determined and evaluated in goat meat nuggets by Banerjee et al. (2012). Total phenolics, radical scavenging activity, and reducing power estimation indicated that broccoli powder has good antioxidant potential. Among treatments, TBARS number decreased with the higher levels of broccoli powder extract with significant effect at 2% level and its value was similar to the product with 100 ppm BHT. The antioxidant potential of cauliflower (*Brassica oleracea*) powder (CP) was evaluated in pork meatballs by Banerjee et al. (2015). The amount of total phenolics (mgGAE/g) was higher in aqueous extract (29.52) of CP as compared to the extract from acetone: water mix (24.22). Addition of CP in pork meatballs significantly increased the amount of total phenolics and TDF. It also reduced the lipid peroxidation and thus enhanced their oxidative stability of meatballs. Therefore, inclusion of CP in meat products makes them much healthier and stable without affecting their acceptability.

3.5.4.5   TOMATO

Tomatoes are an important dietary source of antioxidants—ascorbic acid, lycopene and carotenoids, phenolics, and vitamin E (Frusciante et al., 2007). Dietary intakes of tomatoes and tomato products containing lycopene have been shown to be associated with decreased risk of chronic diseases, such as cancer (van Breemen et al., 2011). Approximately one-third of the total weight of tomatoes in the form of skin and seeds is discarded during processing. However, majority of the flavonols in tomatoes are present in the skin (Stewart et al., 2000). George et al. (2004) reported that tomato skin had significant amounts of phenolics and ascorbic acid and on an average, it had 2.5 times higher lycopene levels than the pulp. Hence, adding tomato, tomato products, or lycopene to processed meat could lead to a significant increase in the amount of all the major antioxidants in the final products.

Selgas et al. (2009) analyzed the safety and shelf life of vaccum-packed, irradiated (2 or 4 kGy) raw hamburgers containing 30, 45, and 60 g/kg dry tomato peel as a source of lycopene. The lycopene concentration fell to 15% of the initial value after irradiation with 4 kGy on 17 days of storage period. Even with this decrease, hamburgers containing 6% dry tomato peel had a final lycopene concentration of 7.14 mg/100 g of hamburger, an amount very close to the recommended daily intake for a healthy diet. Dry tomato peel masked the brownish color characteristic of irradiated meat, and 6%

dry tomato peel imparts a characteristic redness ($a*$) to the hamburger, independent of the dose of radiation applied. The researchers also found that the higher lycopene concentration (6 g/kg) sufficiently masked the negative effects of irradiation on sensory characteristics to ensure an acceptable color and odor in the final product after the storage period. The nitrosomyoglobin (NOMb) content, lycopene content, oxidation level, and the sensory properties of frankfurters produced by both reducing the nitrite level and adding tomato powder were analyzed by Eyiler and Oztan (2011). The pH of the frankfurters produced with tomato powder was reduced, compared with samples which did not contain tomato powder. The addition of 2 g tomato powder/100 g decreased the level of oxidation; however, 4 g tomato powder/100 g caused a slight increase compared with the samples which did not contain tomato powder. According to sensorial evaluations, tomato powder also improved consumer acceptability.

Sanchez-Escalante et al. (2003) analyzed the stabilization of color and odor of beef patties using lycopene-rich tomato as a source of antioxidants, which they found exerted a significant antioxidative effect on the beef patties, depending on the lycopene concentration. These tomato products delayed meat deterioration and the shelf life of treated beef patties ranged between 8 and 12 days. Mercadante et al. (2010) analyzed the oxidative stability of sausages containing added natural pigments and stored under refrigeration. These authors reported that the addition of lycopene (10%) produced significant reductions in redness, although it did not exert any antioxidant effect.

3.5.5   SEAWEED-BASED ANTIOXIDANTS

Marine macroalgae or seaweeds have been used as food mainly in Asian countries and to a lesser extent in Europe and America. Seaweeds are also used as raw material for industrial production of some purified ingredients, for example, agar, carrageenan, alginates, or oils. Edible seaweeds contain good-quality protein, high concentrations of vitamins, high proportions of essential unsaturated fatty acids, particularly long chain n-3 PUFA, bioactive compounds with known antioxidant properties, and are an excellent source of most minerals and dietary fiber (Kolb et al., 2004; Sánchez-Machado et al., 2004). Among the marine organisms, marine macroalgae or seaweed represents one of the richest sources of natural antioxidants (Cox et al., 2010; Ruperez et al., 2002). Among the three algal groups, brown algae generally contains higher amount of polyphenols than red and green algae (Wang et al., 2009a). The bioactive compounds which have been isolated and identified from seaweeds include sulfated polysaccharides (laminarins and fucoidans), polyphenols such as phlorotannins (Zou et al., 2008), carotenoid pigments such as fucoxanthin (Airanthi et al., 2011) and astaxanthin, and sterols and mycosporine-like amino acids. The potential antioxidants identified in seaweeds include some pigments such as fucoxanthin and astaxanthin, polyphenols such as phlorotannins, chlorophyll related compounds, phospholipids, flavonoids, bromophenols, and polysaccharides.

Algal phlorotannins have up to eight interconnected rings and are thus more potent antioxidants than plant polyphenols (Wang et al., 2009a). These phlorotannins have been reported to scavenge free radicals, superoxide radicals (Kuda et al., 2005), peroxyl radical (Wang et al., 2009b), chelate ferrous ions (Chew et al., 2008), and nitric oxide (Valentão et al., 2010). Presence of polyphenols such as catechin, epicatechin, epigallocatechin gallate, and gallic acid are reported in the green seaweed *Halimada* (Yoshie et al., 2002). López et al. (2011) reported the presence of 14 polyphenols, namely gallic acid, catechin, epicatechin, rutin, *p*-coumaric acid, myricetin, quercetin and protocatechuic, vanillic, caffeic, ferulic, chlorogenic, syringic, and gentisic acids in the solvent extracts of *Stypocaulon scoparium*. Onofrejová et al. (2010) reported the extraction of bioactive phenolic acids (protocatechuic, p-hydroxybenzoic, 2,93-dihydroxybenzoic, chlorogenic, caffeic, *p*-coumaric, and salicylic acid), cinnamic acid, and hydroxybenzaldehydes (p-hydroxybenzaldehyde, 3,4-dihydroxybenzaldehyde) from food products of *Porphyra tenera* and *Undaria pinnatifida*. Fucoxanthin and phlorotannins have been identified as active antioxidant compounds from *Hijika fusiformis* (Yan et al., 1999) and *Sargassum kjellamanianum* (Yan et al., 1996), respectively.

Several researchers have investigated the antioxidant activities of the bioactive compounds purified from seaweeds and found that DPPH, alkyl, hydroxyl, superoxide radical scavenging, and metal chelating activities were comparable or even higher than most of the commercial antioxidants (Athukorala et al., 2003; Ahn et al., 2007b; Kim et al., 2007). Since seaweeds contain various bioactive compounds with potential health-beneficial properties, their use as functional ingredients paving the ways for their application in food processing, including meat products (Cofrades et al., 2008; Fleurence, 1999). However, limited attention has been paid to the use of edible seaweeds as a source of natural antioxidants in meat products.

López-López et al. (2009) reported that the addition of edible seaweeds, Sea Spaghetti (*Himanthalia elongata*), Wakame (*U. pinnatifida*), and Nori (*Porphyra umbilicalis*) to low-salt meat emulsion model systems supplied the meat samples with soluble polyphenolic compounds thereby enhancing the antioxidant capacity of the systems. The amount of soluble polyphenols varied from 820 mg GAE/100 g in samples containing Wakame to values as high as 2170 and 2570 mg GAE/100 g of sample in products with Nori and Sea Spaghetti, respectively. The highest antioxidant activity was observed in the samples containing Sea Spaghetti (3.69 µmol eq. Trolox/g of sample), while products containing Nori and Wakame had similar antioxidant potentials (1.18 and 1.09 µmol eq. Trolox/g, respectively). López-López et al. (2010) also studied the effect of adding Wakame seaweed on the characteristics of beef patties with low-salt and low-fat contents. However, no clear effect on lipid oxidation was found with the presence of seaweed.

Sasaki et al. (2008) observed the effects of major carotenoid pigment, fucoxanthin (from Wakame) @ 200 mg/kg on lipid peroxidation and meat color in ground chicken breast meat upon chilled storage before and after cooking. It was found that fucoxanthin could decrease the TBARS value on days 1 and 6 (63.1 and 58.5%, respectively) when stored after cooking. It also decreased the $L*$ value and increased $a*$ and $b*$ values thereby showing it to be a potent ingredient for the improvement of the appearance and shelf life of chicken meat and its products. However, the antioxidative activity of fucoxanthin during chilling storage after cooking was lower than that of TM.

Thus, the use of seaweeds can play key role in the development of functional foods and provide an opportunity to improve the nutritional profile of meat products and address consumer demands. Seaweeds can also supply meat products with valuable polyphenols to improve their oxidative stability during processing and storage besides enriching them with dietary fiber and minerals.

3.6   MARKET POTENTIAL

From tea bags to grape seeds, the term antioxidant is used as a marketing tool for food products especially processed foods. The increased demand for processed meat products will undoubtedly advance the use of antioxidants across the globe. The global market for processed meats is estimated to be USD 362 billion in 2012 and is projected to reach USD 799 billion by 2018 with a compound annual growth rate of 14.3%. Maximum growth of this sector is expected in China, India, Japan, and New Zealand. Similarly, the global natural antioxidants' market is expected to witness substantial growth and is forecasted to reach USD 4.14 billion by 2022, particularly due to increasing global meat consumption. Asia Pacific has experienced highest growth in 2014 followed by Europe and North America. Huge population and growing health awareness in developing nations such as India and China are influencing the growth of the antioxidants' market in the Asia Pacific region. Demand for vitamin C was highest accounting for over 80% of the global demand in 2014 and this trend is likely to continue in the coming years too. Growing demand for animal feed on account of increasing consumption of dairy products and meat products on a global scale is expected to augment vitamin E market growth over the forecast period. Demand-supply imbalance has escalated the price of natural antioxidants. In an effort to deal with the increasing raw material cost, food manufacturers are opting blending of natural antioxidants, for example, TM with rosemary extract or herbal extract with synthetic antioxidants, which produces a less expensive product without compromising the efficacy of antioxidants.

3.7 FUTURE PROSPECTS

Strong health awareness among the modern consumers is creating a genuine need for adopting a synthetic additive free diet, with increasing personalized value of convenience, cost, and taste. Plant materials have drawn enough attention as a source of natural antioxidants in meat system to maintain and improve quality and stability. However, still there is a long way ahead to make these herbal ingredients an appropriate substitute of synthetic antioxidants.

- Both synthetic and natural antioxidants have to face a number of considerations for their practical usage—natural availability, extraction efficiency and purification, and economical and practical aspect of the natural antioxidants as preservatives. The other factors considered should involve the incorporation of antioxidant into the final meat product; effect of the incorporation on their stability, solubility and sensory properties, and the interactions of the preservatives with other meat ingredients. Further, there are certain important regulatory considerations, such as generally recognized as safe (GRAS) status, the limit of the amount that can be incorporated, and so forth.
- Plant extracts are harvested both in aqueous and organic solvents and the extracts in organic solvents have been found to possess greater antioxidant potential. This could be mainly due to greater solubility and extraction of plant phenolics in organic solvents. However, for the application in food or meat system, these organic solvents need to be carefully removed from the extracts. It would be proper to extract

plant phenolics in aqueous medium to avoid any chances of residual organic solvents in foods. In order to enhance the yield of aqueous plant extracts there is need to device suitable extraction process and techniques.

- It is claimed that plant extracts have lower antioxidant potential than synthetic counterparts. The application of microencapsulation and nanoencapsulation should be considered to optimize delivery and release of natural antioxidants and enhance their efficacy. Active packaging technology can also be exploited to control the delivery and efficacy of natural antioxidants.
- Screening of plant extracts should be done with the help of mass spectrometry to track down any unwanted component.
- The interaction among antioxidant in different food system should be taken in account. On one hand, the synergistic action of TM and ascorbic acid can be utilized for preventing auto-oxidation or photo-oxidation of lipids, on the other hand, the strong antagonistic action between plant polyphenols and TM reported in pork fat should also be scrutinized.

3.8 CONCLUSION

Meat and meat products because of their chemical constituents and processing as well as post-processing conditions, become vulnerable to oxidation which subsequently lead to deterioration in color, texture, shelf life, and overall acceptability. Protection from such chemical and quality changes requires various managemental, processing, and post-processing steps. Application of natural antioxidants from plant sources in meat system to prevent quality deterioration is one of several steps. Plant extracts are being regularly screened for their antioxidant potential. These are applied in the meat and meat products to evaluate their efficacy and efficiency. Several plant extracts have been attempted in the meat system at different levels and the quality, acceptability, and oxidative stability of the products has been monitored through determination of protein and lipid peroxidation, color properties, and sensory acceptability. In the coming years there are need to carry out research work to enhance yield of plant phenolics, screen their active principles, and control their delivery and release in the meat system. It can be expected that more and more natural antioxidants with greater anti-oxidant potential and tailored delivery mechanisms would be explored in the near future to produce safer, fresh, acceptable, and stable meat products.

KEYWORDS

- **natural antioxidants**
- **plant material**
- **meat and meat products**
- **oxidation**

REFERENCES

Abascal, K.; Ganora, L.; Yarnell, E. The Effect of Freeze-Drying and Its Implications for Botanical Medicine: A Review. *Phytother. Res.* **2005**, *19*, 655–660.

Abdullakasim, P.; Songchitsomboon, S.; Techagumpuch, M.; Balee, N.; Swatsitang, P.; Sungpuag, P. Antioxidant Capacity, Total Phenolics and Sugar Content of Selected Thai Health Beverages. *Int. J. Food Sci. Nutr.* **2007**, *58*, 77–85.

Adorjan, B.; Buchbauer, G. Biological Properties of Essential Oils: An Updated Review. *Flavour Fragr. J.* **2010**, *25*, 407–426.

Ahn, G. N.; Kim, K. N.; Cha, S. H.; Song, C. B.; Lee, J. H.; Heo, M. S.; Yeo, I. K.; Lee, N. H.; Jee, Y. H.; Kim, J. S.; Heu, M. S.; Jeon, Y. J. Antioxidant Activities of Phlorotannins Purified from Ecklonia Cava on Free Radical Scavenging Using ESR and H_2O_2-Mediated DNA Damage. *Eur. Food Res. Technol.* **2007b**, *226*, 71–79.

Ahn, J.; Grun, I. U.; Mustapha, A. Effects of Plant Extracts on Microbial Growth, Color Change, and Lipid Oxidation in Cooked Beef. *Food Microbiol.* **2007a**, *24*, 7–14.

Ahn, J.; Grun, I. U.; Fernando, L. N. Antioxidant Properties of Natural Plant Extracts Containing Polyphenolic Compounds in Cooked Ground Beef. *J. Food Sci.* **2002**, *67*, 1364–1369.

Airanthi, M. K. W. A.; Hosokawa, M.; Miyashita, K. Comparative Antioxidant Activity of Edible Japanese Brown Seaweeds. *J. Food Sci.* **2011**, *76*, C104–C111.

Akarpat, A.; Turhan, S.; Ustun, N. S. Effects of Hot-Water Extracts from Myrtle, Rosemary, Nettle and Lemon Balm Leaves on Lipid Oxidation and Color of Beef Patties During Frozen Storage. *J. Food Process. Preserv.* **2008**, *32*, 117–132.

Amakura, Y.; Umino, Y.; Tsuji, S.; Tonogai, Y. Influence of Jam Processing on the Radical Scavenging Activity and Phenolic Content in Berries. *J. Agric. Food Chem.* **2000**, *48* (12), 6292–6297.

Anagnostopoulou, M. A.; Kefalas, P.; Papageorgiou, V. P.; Assimopoulou, A. N.; Boskou, D. Radical Scavenging Activity of Various Extracts and Fractions of Sweet Orange Peel (*Citrus sinensis*). *Food Chem.* **2006**, *94*, 19–25.

Anwar, F.; Naseer, R.; Bhanger, M. I.; Ashraf, S.; Talpur, F. N.; Aladededune, F. A. Physicochemical Characteristics of Citrus Seeds and Oils from Pakistan. *J. Am. Oil Chem. Soc.* **2008**, *85*, 321–330.

Arabshahi-D, S.; Devi, D. V.; Urooj, A. Evaluation of Antioxidant Activity of Some Plant Extracts and Their Heat, pH and Storage Stability. *Food Chem.* **2007**, *100*, 1100–1105.

Arts, I. C. W.; van de Putte, B.; Hollman, P. C. H. Catechin Contents of Foods Commonly Consumed in the Netherlands. 1. Fruits, Vegetables, Staple Foods, and Processed Foods. *J. Agric. Food Chem.* **2000,** *48,* 1746–1751.

Athukorala, Y.; Lee, K. W.; Song, C. B.; Ahn, C. B.; Shin, T. S.; Cha, Y. J. Potential Antioxidant Activity of Marine Red Alga Grateloupia Filicina Extracts. *J. Food Lipids.* **2003,** *10,* 251–265.

Avallone, R.; Plessi, M.; Baraldi, M.; Monzan, A. Determination of Chemical Composition of Carob (*Ceratonia Siliqua*): Protein, Fat Carbohydrates, and Tannins. *J. Food Compos. Anal.* **1997,** *10,* 166–172.

Azuma, K.; Ippoushi, K.; Ito, H.; Higashio, H.; Terao, J. Evaluation of Antioxidative Activity of Vegetable Extracts in Linoleic Acid Emulsion and Phospholipid Bilayers. *J. Sci. Food Agric.* **1999,** *79,* 2010–2016.

Badr, H. M.; Mahmoud, K. A. Antioxidant Activity of Carrot Juice in Gamma Irradiated Beef Sausage during Refrigerated and Frozen Storage. *Food Chem.* **2011,** *127,* 1119–1130.

Bailey, M. E. Inhibition of Warmed-Over Flavor, with Emphasis on Maillard Reaction Products. *Food Technol.* **1988,** *42* (6), 123–126.

Bakkali, F.; Averbeck, S.; Averbeck, D.; Idaomar, M. Biological Effects of Essential Oils – A Review. *Food Chem. Toxicol.* **2008,** *46,* 446–475.

Banerjee, R.; Verma, A. K.; Das, A. K.; Rajkumar, V.; Shewalkar, A. A.; Narkhede, H. P. Antioxidant Effects of Broccoli Powder Extract in Goat Meat Nuggets. *Meat Sci.* **2012,** *91,* 179–184.

Banerjee, R.; Verma, A. K.; Narkhede, H. P.; Kokare, P. G.; Manjhi A.; Bokde, P. M. Cauliflower Powder in Pork Meatballs: Effects on Quality Characteristics and Oxidative Stability. *Fleischwirtshaft Int.* **2015,** *30* (1), 97–102.

Bastida, S.; Sánchez-Muniz, F. J.; Olivero, R.; Pérez-Olleros, L.; Ruiz-Roso, B.; Jiménez-Colmenero, F. Antioxidant Activity of Carob Fruit Extracts in Cooked Pork Meat Systems During Chilled and Frozen Storage. *Food Chem.* **2009,** *116* (3), 748–754.

Batista, M. T.; Amaral, M. T.; Proença Da Cunha, A. In *Carob Fruits as a Source of Natural Antioxidants,* III International Carob Symposium, Cabanas-Tavira, Portugal (in Press). 1996.

Bertelsen, G.; Jakobsen, M.; Juncher, D.; Moller, J.; Kroger-Ohlsen, M.; Weber, C. In *Oxidation, Shelf-Life and Stability of Meat and Meat Products,* Proceedings of the 46th International Congress of Meat Science and Technology; Buenos Aires, Argentina, Aug 27– Sept 1, **2000;** pp 516–524.

Biswas, A. K.; Chatli, M. K.; Sahoo, J. Antioxidant Potential of Curry (*Murraya koenigii* L.) and Mint (*Mentha spicata*) Leaf Extracts and Their Effect on Colour and Oxidative Stability of Raw Ground Pork Meat During Refrigeration Storage. *Food Chem.* **2012,** *133,* 467–472.

Biswas, A. K.; Keshri, R. C.; Bisht, G. S. Effect of Enrobing and Antioxidants on Quality Characteristics of Precooked Pork Patties Under Chilled and Frozen Storage Conditions. *Meat Sci.* **2004,** *66,* 733–741.

Brannan, R. G. Effect of Grape Seed Extract on Physicochemical Properties of Ground, Salted, Chicken Thigh Meat during Refrigerated Storage at Different Relative Humidity Levels. *J. Food Sci.* **2008,** *73,* C36–C40.

Brannan, R. G. Effect of Grape Seed Extract on Descriptive Sensory Analysis of Ground Chicken during Refrigerated Storage. *Meat Sci.* **2009,** *81,* 589–595.

Brannan, R. G.; Mah, E. Grape Seed Extract Inhibits Lipid Oxidation in Muscle from Different Species during Refrigerated and Frozen Storage and Oxidation Catalyzed by Peroxynitrite and Iron/Ascorbate in a Pyrogallol Red Model System. *Meat Sci.* **2007,** *77,* 540–546.

Brown, C. E. Interactions among Carnosine, Anserine, Ophidine and Copper in Biochemical Adaptation. *J. Theor. Biol.* **1981,** *88,* 245–256.

Burt, S. Essential Oils: Their Antibacterial Properties and Potential Applications in Foods—A Review. *Int. J. Food Microbiol.* **2004,** *94,* 223–253.

Calabro, M. L.; Galtieri, V.; Cutroneo, P.; Tommasini, S.; Ficarra, P.; Ficarra, R. Study of the Extraction Procedure by Experimental Design and Validation of a LC Method for Determination of Flavonoids in Citrus Bergamia Juice. *J. Pharm. Biomed. Anal.* **2004,** *35,* 349–363.

Cao, G.; Sofic, E.; Prior, R. L. Antioxidant Capacity of Tea and Common Vegetables. *J. Agric. Food Chem.* **1996,** *44,* 3426–3431.

Carpenter, R.; O'grady, M. N.; O'callaghan, Y. C.; O'brien, N. M.; Kerry, J. P. Evaluation of the Antioxidant Potential of Grape Seed and Bearberry Extracts in Raw and Cooked Pork. *Meat Sci.* **2007,** *76,* 604–610.

Charles, D. J. *Antioxidant Properties of Spices, Herbs and Other Sources;* Springer: New York, 2013.

Chen, H.; Zuo, Y.; Deng, Y. Separation and Determination of Flavonoids and Other Phenolic Compounds in Cranberry Juice by High-Performance Liquid Chromatography. *J. Chromatogr. A.* **2001,** *913,* 387–395.

Chew, Y. L.; Lim, Y. Y.; Omar, M.; Khoo, K. S. Antioxidant Activity of Three Edible Seaweeds from Two Areas in South East Asia. *LWT.* **2008,** *41,* 1067–1072.

Choe, E.; Min, D. B. Chemistry and Reactions of Reactive Oxygen Species in Foods. *J. Food Sci.* **2005,** *70,* 142–159.

Chu, Y. F.; Sun, J.; Wu, X.; Liu, R. H. Antioxidant and Antiproliferative Activities of Common Vegetables. *J. Agric. Food Chem.* **2002,** *50* (23), 6910–6916.

Cofrades, S.; López-López, I.; Solas, M. T.; Bravo, L.; Jiménez-Colmenero, F. Influence of Different Types and Proportions of Added Edible Seaweeds on Characteristics of Low-Salt Gel/Emulsion Meat Systems. *Meat Sci.* **2008,** *79,* 767–776.

Corsi, L.; Avallone, R.; Cosenza, F.; Farina, F.; Baraldi, C.; Baraldi, M. Antiproliferative Effects of *Ceratonia siliqua* L. on Mouse Hepatocellular Carcinoma Cell Line. *Fitoterapia.* **2002,** *73,* 674–684.

Cox, S.; Abu-Ghannam, N.; Gupta, S. An Assessment of the Antioxidant and Antimicrobial Activity of Six Species of Edible Irish Seaweeds. *Int. Food Res. J.* **2010,** *17,* 205–220.

Das, A. K.; Rajkumar, V.; Verma, A. K. Bael Pulp Residue as a New Source of Antioxidant Dietary Fiber in Goat Meat Nuggets. *J. Food Process. Preserv.* **2014,** *39* (6), 1626–1635. DOI: 10.1111/jfpp.12392.

Das, A. K.; Rajkumar, V.; Verma, A. K.; Swarup, D. *Moringa oleifera* Leaves Extract: A Natural Antioxidant for Retarding Lipid Peroxidation in Cooked Goat Meat Patties. *Int. J. Food Sci. Technol.* **2012,** *47,* 585–591.

Das, K. C.; Das, C. K. Curcumin (*Diferuloylmethane*), a Singlet Oxygen ($^1O_2$) Quencher. *Biochem. Biophys. Res. Commun.* **2002,** *295,* 62–66.

Decker, E. A.; Mei, L. In *Antioxidant Mechanisms and Applications in Muscle Foods,* Proceedings of the 49th Reciprocal Meat Conference, Savoy, 1996; American Meat Science Association: Savoy, IL, 1996; pp 64–72.

Decker, E.; Livisay, S.; Zhou, S. Mechanisms of Endogenous Skeletal Muscle Antioxidants: Chemical and Physical Aspects. In *Antioxidants in Muscle Foods;* Decker, E., Faustman, C., Lopez-Bote, C., Eds.; Wiley-Interscience: New York, 2000; pp 25–60.

Decker, E. A.; Ivanov, V.; Zhu, B. Z.; Frei, B. Inhibition of Low-Density Lipoprotein Oxidation by Carnosine and Histidine. *J. Agric. Food Chem.* **2001,** *49,* 511–516.

Dekker, M.; Verkerk, R.; Jongen, W. M. F. Predictive Modelling of Health Aspects in the Food Production Chain: A Case Study on Glucosinolates in Cabbage. *Trends Food Sci. Technol.* **2000,** *11,* 174–181.

Devatkal, S. K.; Narsaiah, K.; Borah, A. Anti-Oxidant Effect of Extracts of Kinnow Rind, Pomegranate Rind and Seed Powders in Cooked Goat Meat Patties. *Meat Sci.* **2010,** *85* (1), 155–159.

Devatkal, S. K.; Naveena, B. M. Effect of Salt, Kinnow and Pomegranate Fruit By-product Powders on Color and Oxidative Stability of Raw Ground Goat Meat During Refrigerated Storage. *Meat Sci.* **2010,** *85* (2), 306–311.

Di Majo, D.; Giammanco, M.; La Guardia, M.; Tripoli, E.; Giammanco, S.; Finotti, E. Flavanones in Citrus Fruit: Structure-Antioxidant Activity Relationships. *Food Res. Int.* **2005,** *38,* 1161–1166.

Doolaege, E. H. A.; Vossen, E.; Raes, K.; De Meulenaer, B.; Verhé, R.; Paelinck, H.; DeSmet, S. Effect of Rosemary Extract Dose on Lipid Oxidation, Colour Stability and Antioxidant Concentrations in Reduced Nitrite Liver Pâtés. *Meat Sci.* **2012,** *90,* 925–31.

Dzudie, T.; Kouebou, C. P.; Essia-Ngang, J. J.; Mbofung, C. M. F. Lipid Sources and Essential Oils Effects on Quality and Stability of Beef Patties. *J. Food Eng.* **2004,** *65,* 67–72.

El-Alim, S. S. L. A.; Lugasi, A.; Hovari, J.; Dworschak, E. Culinary Herbs Inhibit Lipid Oxidation in Raw and Cooked Minced Meat Patties during Storage. *J. Sci. Food Agric.* **1999,** *79,* 277–285.

Elangovan, V.; Sekar, N.; Govindasamy, S. Chemoprotective Potential of Dietary Bipoflavonoids against 20-Methylcholanthrene- Induced Tumorigenesis. *Cancer Lett.* **1994,** *87,* 107–113.

El-Nemr, S. E.; Ismail, I. A.; Ragab, M., Chemical Composition of Juice and Seeds of Pomegranate Fruit. *Nahrung.* **1990,** *7,* 601–606.

Enser, M.; Hallett, K.; Hewett, B.; Fursey, G. A. J.; Wood, J. D. Fatty Acid Content and Composition of English Beef, Lamb and Pork at Retail. *Meat Sci.* **1996,** *44,* 443–458.

Estevez, M.; Ramirez, R.; Ventanas, S.; Cava, R. Sage and Rosemary Essential Oils Versus BHT for the Inhibition of Lipid Oxidative Reactions in Liver Pate. *LWT-Food Sci. Technol.* **2007,** *40,* 58–65.

Eyiler, E.; Oztan, A. Production of Frankfurters with Tomato Powder as a Natural Additive. *LWT-Food Sci. Technol.* **2011,** *44* (1), 307–311.

Faller, A. L. K.; Fialho, E. The Antioxidant Capacity and Polyphenol Content of Organic and Conventional Retail Vegetables after Domestic Cooking. *Food Res. Int.* **2009,** *42* (1), 210–215.

Fasseas, M. K.; Mountzouris, K. C.; Tarantilis, P. A.; Polissiou, M.; Zervas, G. Antioxidant Activity in Meat Treated with Oregano and Sage Essential Oils. *Food Chem.* **2007,** *106,* 1188–1194.

Ferguson, L. R. Prospects for Cancer Prevention. *Mutat. Res.* **1999,** *428,* 329–338.

Fernandez-Lopez, J.; Sevilla, L.; Sayas-Barbera, E.; Navarro, C.; Marin, F.; Perez-Alvarez, J. A. Evaluation of the Antioxidant Potential of Hyssop (*Hyssopus officinalis* L.) and Rosemary (*Rosmarinus officinalis* L.) Extracts in Cooked Pork Meat. *J. Food Sci.* **2003,** *68,* 660–664.

Fernandez-Lopez, J.; Zhi, N.; Aleson-Carbonell, L.; Perez-Alvarez, J. A.; Kuri, V. Antioxidant and Antibacterial Activities of Natural Extracts: Application in Beef Meatballs. *Meat Sci.* **2005,** *69,* 371–380.

Fleurence, J. Seaweed Proteins: Biochemical, Nutritional Aspects and Potential Uses. *Trends Food Sci. Technol.* **1999,** *10* (1), 25–28.

Foo, L. Y.; Lu, Y.; Howell, A. B.; Vorsa, N. The Structure of Cranberry Anthocyanidins Proanthocyanidins which Inhibit Adherence of Uropathogenic P-Fimbriated Escherichia Coli In Vitro. *Photochemistry.* **2000,** *54,* 173–181.

Formanek, Z.; Kerry, J. P.; Higgins, F. M.; Buckley, D. J.; Morrissey, P. A.; Farkas, J. Addition of Synthetic and Natural Antioxidants to Alpha-Tocopheryl Acetate Supplemented Beef Patties: Effects of Antioxidants and Packaging on Lipid Oxidation. *Meat Sci.* **2001,** *58* (4), 337–341.

Fowke, J. H.; Chung, F. L.; Jin, F.; Qi, D.; Cai, Q.; Conaway, C.; Cheng, J.; Shu, X.; Gao, Y.; Zheng, W. Urinary Isothiocyanate Levels, *Brassica,* and Human Breast. *Cancer Res.* **2003,** *63,* 3980–2986.

Fratianni, F.; Martino, L. D.; Melone, A.; Feo, V. D.; Coppola, R.; Nazzaro, F. Preservation of Chicken Breast Meat Treated with Thyme and Balm Essential Oils. *J. Food Sci.* **2010,** *75,* 528–535.

Frusciante, L.; Carli, P.; Ercolano, M. R.; Pernice, R.; Di Matteo, A.; Fogliano, V.; Pellegrini, N. Antioxidant Nutritional Quality of Tomato. *Mol. Nutr. Food Res.* **2007,** *51,* 609–617.

Garrido, M. D.; Auqui, M.; Marti, N.; Linares, M. B. Effect of Two Different Red Grape Pomace Extracts Obtained under Different Extraction Systems on Meat Quality of Pork Burgers. *LWT-Food Sci. Technol.* **2011,** *44,* 2238–2243.

George, B.; Kaur, C.; Khurdiya, D. S.; Kapoor, H. C. Antioxidants in Tomato *(Lycopersicon esculentum)* as a Function of Genotype. *Food Chem.* **2004,** *84,* 45–51.

Gil, M. I.; Tomas-Berberan, A.; Hess-Pierce, B.; Holcroft, D. M.; Kader, A. A. Antioxidant Activity of Pomegranate Juice and Its Relationship with Phenolic Composition and Processing. *J. Agric. Food Chem.* **2000,** *48,* 4581–4589.

Gohar, A.; Gedara, S. R.; Baraka, H. N. New Acylated Flavonol Glycoside from *Ceratonia siliqua* L. Seeds, *J. Med. Plants Res.* **2009,** *3,* 424–428.

Graf, E.; Eaton, J. W. Antioxidant Functions of Phytic Acid. *Free Radic. Biol Med.* **1990,** *8,* 61–69.

Gray, J. I.; Gomaa, E. A.; Buckley, D. J. Oxidative Quality and Shelf Life of Meats. *Meat Sci.* **1996,** *43,* S111–S123.

Grusak, M. A.; Della Penna, D. Improving the Nutrient Composition of Plants to Enhance Human Nutrition and Health. *Annu. Rev. Plant Biol.* **1999,** *50,* 133–161.

Gunes, G.; Liu, R. H.; Watkins, C. B. Controlled-Atmosphere Effects on Postharvest Quality and Antioxidant Activity of Cranberry Fruits. *J. Agric. Food Chem.* **2002,** *50,* 5932–5938.

Gupta, S.; Jyothi Lakshmi, A.; Manjunath, M. N.; Prakash, J. Analysis of Nutrient and Antinutrient Content of Underutilized Green Leafy Vegetables. *LWT-Food Sci. Technol.* **2005,** *38* (6), 339–345.

Guyot, S.; Marnet, N.; Drilleau, J. Thiolysis-HPLC Characterization of Apple Procyanidins Covering a Large Range of Polymerization States. *J. Agric. Food Chem.* **2001,** *49,* 14–20.

Hamid, A. A.; Aiyelaagbe, O. O.; Usman, L. A.; Ameen, O. M.; Lawal, A. Antioxidants: Its Medicinal and Pharmacological Applications. *Afr. J. Pure. Appl. Chem.* **2010,** *4* (1), 7–10.

Hanf, V.; Gonder U. Nutrition and Primary Prevention of Breast Cancer: Foods, Nutrients and Breast Cancer Risk. *Eur. J. Obstet. Gynaecol. Reprod. Biol.* **2005,** *123,* 139–149.

Hänsel, R.; Keller, K.; Rimpler, H.; Schneider, G. Uvae Ursi Folium *(Bärentraubenblätter).* In *Hagers Handbuch der Pharmazeutischen Praxis;* 5th ed.; Springer-Verlag: Berlin, 1992; pp 330–336.

Heimler, D.; Vignolini, P.; Dini, M. G.; Vincieri, F. F.; Romani, A. Antiradical Activity and Polyphenol Composition of Local *Brassicaceae* Edible Varieties. *Food Chem.* **2005,** *99,* 464–969.

Hsieh, C. W.; Ko, W. C. Effect of High-Voltage Electrostatic Field on Quality of Carrot Juice during Refrigeration. *LWT-Food Sci. Technol.* **2008,** *41,* 1752–1757.

Huang, B.; He, J.; Ban, X.; Zeng, H.; Yao, X.; Wang, Y. Antioxidant Activity of Bovine and Porcine Meat Treated with Extracts from Edible Lotus (*Nelumbo Nucifera*) Rhizome Knot and Leaf. *Meat Sci.* **2011,** *87,* 46–53.

Hyldgaard, M.; Mygind, T.; Meyer, R. L. Essential Oils in Food Preservation: Mode of Action, Synergies, and Interactions with Food Matrix Components. *Front. Microbiol.* **2012,** *3,* 1–24.

Jayaprakasha, G. K.; Patil, B. S. *In Vitro* Evaluation of the Antioxidant Activities in Fruit Extracts from Citron and Blood Orange. *Food Chem.* **2007,** *101,* 410–418.

Jayathilakan, K.; Sharma, G. K.; Radhakrishna, K.; Bawa, A. S. Antioxidant Potential of Synthetic and Natural Antioxidants and Its Effect on Warmed-Over-Flavour in Different Species of Meat. *Food Chem.* **2007,** *105,* 908–916.

Jinap, S.; Iqbal, S. Z.; Selvam, R. M. P. Effect of Selected Local Spices Marinades on the Reduction of Heterocyclic Amines in Grilled Beef (*Satay*). *LWT-Food Sci. Technol.* **2015,** *63,* 919–926.

Jo, S. C.; Nam, K. C.; Min, B. R.; Ahn, D. U.; Cho, S. H.; Park, W. P.; Lee, S. C. Antioxidant Activity of *Prunus mume* Extract in Cooked Chicken Breast Meat. *Int. J. Food Sci. Technol.* **2006,** *41,* 15–19.

Jones, D.; Coates, R.; Flagg, E. W.; Eley. J. W.; Block, G.; Greenberg, R. S.; Gunter, E. W.; Jackson B. Glutathione in Foods Listed in the National Cancer Institute's Health Habits and History Food Frequency Questionnaire. *Nutr. Cancer.* **1992,** *17,* 57–75.

Joseph, S.; Chatli, M. K.; Biswas, A. K.; Sahoo, J. Oxidative Stability of Pork Emulsion Containing Tomato Products and Pink Guava Pulp during Refrigerated Aerobic Storage. *J. Food Sci. Technol.* **2014,** *51* (11), 3208–3216.

Jung, C.; Maeder, V.; Funk, F.; Frey, B.; Sticher, H.; Frossard, E. Release of Phenols from *Lupinus albus* L. Roots Exposed to Cu and Their Possible Role in Cu Detoxification. *Plant Soil.* **2003,** *252,* 301–312.

Juntachote, T.; Berghofer, E.; Siebenhandl, S.; Bauer, F. The Effect of Dried Galangal Powder and Its Ethanolic Extracts on Oxidative Stability in Cooked Ground Pork. *LWT-Food Sci. Technol.* **2007,** *40,* 324–330.

Kamalakkannan, N.; Prince, S. M. Effect of *Aegle marmelos* Correa. (*bael*) Fruit Extract on Tissue Antioxidants in Streptozotocin Diabetic Rats. *Ind. J. Exp. Biol.* **2003b,** *41,* 1285–1288.

Kamalakkannan, N.; Prince, P. S. Hypoglycemic Effect of Water Extracts of *Aegle marmelos* Fruits in Streptozotocin Diabetic Rats. *J. Ethnopharmacol.* **2003a,** *87,* 207–210.

Kanatt, S. R.; Chander, R.; Radhakrishna, P.; Sharma, A. Potato Peel Extract- a Natural Antioxidant for Retarding Lipid Peroxidation in Radiation Processed Lamb Meat. *J. Agric. Food Chem.* **2005,** *53,* 1499–1504.

Kanatt, S. R.; Chander, R.; Sharma, A. Antioxidant Potential of Mint (*Mentha spicata* L.) in Radiation-Processed Lamb Meat. *Food Chem.* **2007,** *100,* 451–458.

Kanatt, S. R.; Chander, R.; Sharma, A. Chitosan and Mintmixture: A New Preservative for Meat and Meat Products. *Food Chem.* **2008,** *107,* 845–852.

Karre, L.; Lopez, K.; Getty, J. K. K. Natural Antioxidants in Meat and Poultry Products. *Meat Sci.* **2013,** *94,* 220–227.

Kim, J. I.; Choi, J. S.; Kim, W. S.; Woo, K. L.; Jeon, J. T.; Min, B. T.; Cheigh, H. S. Antioxidant Activity of Various Fractions Extracted from Mustard Leaf (*Brassica juncea*) and Their Kimchi. *J. Life Sci.* **2004,** *14* (2), 286–290.

Kim, S. H.; Choi, D. S.; Athukorala, Y.; Jeon, Y. J.; Senevirathne, M.; Rha, C. K. Antioxidant Activity of Sulfated Polysaccharides Isolated from *Sargassum fulvellum*. *J. Food Sci. Nutr.* **2007,** *12,* 65–73.

Kim, S. J.; Cho, A. R.; Han, J. Antioxidant and Antimicrobial Activities of Leafy Green Vegetable Extracts and Their Applications to Meat Product Preservation. *Food Control.* **2013a,** *29,* 112–120.

Kim, S. J.; Min, S. C.; Shin, H. J.; Lee, Y. J.; Cho, A. R.; Kim, S. Y.; Han, J. Evaluation of the Antioxidant Activities and Nutritional Properties of Ten Edible Plant Extracts and Their Application to Fresh Ground Beef. *Meat Sci.* **2013b,** *93,* 715–722.

Koga, T.; Moro, K.; Nakamori, K.; Yamakoshi, J.; Hosoyama, H.; Kataoka, S.; Ariga, T. Increase of Antioxidative Potential of Rat Plasma by Oral Administration of Proanthocyanidinrich Extract from Grape Seeds. *J. Agric. Food Chem.* **1999,** *47,* 1892–1897.

Kolb, N.; Vallorani, L.; Milanovic, N.; Stocchi, V. Evaluation of Marine Algae Wakame (*Undaria pinnatifida*) and Kombu (*Laminaria digitata japonica*) as Food Supplements. *Food Technol. Biotechnol.* **2004,** *42* (1), 57–61.

Krinsky, N. I. Carotenoids as Antioxidants. *Nutrition.* **2001,** *17,* 815–817.

Kristal, A. R.; Lampe, J. W. *Brassica* Vegetables and Prostate Cancer Risk: A Review of the Epidemiological Evidence. *Nutr. Cancer.* **2002,** *42,* 1–9.

Kuda, T.; Tsunekawa, M.; Hishi, T.; Araki, Y. Antioxidant Properties of Dried 'Kayamo-Nori', a Brown Alga Scytosiphon Lomentaria (*Scytosiphonales, Phaeophyceae*). *Food Chem.* **2005,** *89,* 617–622.

Kulkarni, S.; DeSantos, F. A.; Kattamuri, S.; Rossi, S. J.; Brewer, M. S. Effect of Grape Seed Extract on Oxidative, Color and Sensory Stability of Pre-Cooked, Frozen, Re-Heated Beef Sausage Model System. *Meat Sci.* **2011,** *88,* 139–144.

Kumazawa, S.; Taniguchi, M.; Susuki, Y.; Shimura, M.; Kwon M. S.; Nakayama, T. Antioxidant Activity of Polyphenols in Carob Pods, *J. Agric. Food. Chem.* **2002,** 50, 373–377.

Larrain, R. E.; Krueger, C. G.; Richards, M. P.; Reed, J. D. Color Changes and Lipid Oxidation in Pork Products Made from Pigs Fed with Cranberry Juice Powder. *J. Muscle Foods.* **2008,** *19,* 17–33.

Lee, C. H.; Reed, J. D.; Richards, M. P. Ability of Various Polyphenolic Classes from Cranberry to Inhibit Lipid Oxidation in Mechanically Separated Turkey and Cooked Ground Pork. *J. Muscle Foods.* **2006,** *17* (3), 248–266.

Lee, E. J.; Ahn, D. U. Quality Characteristics of Irradiated Turkey Breast Rolls Formulated with Plum Extract. *Meat Sci.* **2005,** *71,* 300–305.

Lee, M. A.; Choi, J. H.; Choi, Y. S.; Han, D. J.; Kim, H. Y.; Shim, S. Y.; Chung, H. K.; Kim, C. J. The Antioxidative Properties of Mustard Leaf (*Brassica juncea*) Kimchi Extracts on Refrigerated Raw Ground Pork Meat against Lipid Oxidation. *Meat Sci.* **2010,** *84* (3), 498–504.

Leheska, J. M.; Boyce, J.; Brooks, J. C.; Hoover, L. C.; Thompson, L. D.; Miller, M. F. Sensory Attributes and Phenolic Content of Precooked Pork Breakfast Sausage with Fruit Purees. *J. Food Sci.* **2006,** *71,* S249–S252.

Longtin, R. The Pomegranate: Nature's Power Fruit? *J. Natl. Cancer Inst.* **2003,** *95,* 346–348.

López, A.; Rico, M.; Rivero, A.; de Tangil, M. S. The Effects of Solvents on the Phenolic Contents and Antioxidant Activity of *Stypocaulon scoparium* Algae Extracts. *Food Chem.* **2011,** *125,* 1104–1109.

López-López, I.; Bastida, S.; Ruiz-Capillas, C.; Bravo, L.; Larrea, M. T.; Sánchez-Muniz, F.; Cofrades, S.; Jiménez-Colmenero, F. Composition and Antioxidant Capacity of Low-Salt Meat Emulsion Model Systems Containing Edible Seaweeds. *Meat Sci.* **2009,** *83,* 492–498.

López-López, I.; Cofrades, S.; Yakan, A.; Solas, M. T.; Jiménez-Colmenero, F. Frozen Storage Characteristics of Low-Salt and Low-Fat Beef Patties as Affected by Wakame Addition and Replacing Pork Back Fat with Olive Oil-in-Water Emulsion. *Food Res. Int.* **2010,** *43,* 1244–1254.

Makris, D.; Kefalas P. Carob Pods (*Ceratonia siliqua* L.) as a Source of Polyphenolic Antioxidants. *Food Technol. Biotechnol.* **2004,** *42,* 105–108.

Mansour, E. H.; Khalil, A. H. Evaluation of Antioxidant Activity of Some Plant Extracts and Their Application to Ground Beef Patties. *Food Chem.* **2000,** *69,* 135–141.

Manthley, J. A.; Grohmann, K. Phenols in Citrus Peel Byproducts. Concentrations of Hydroxycinnamates and Polymethoxylated Flavones in Citrus Peel Molasses. *J. Agric. Food Chem.* **2001,** *49,* 3268–3273.

Martın, F. R.; Frutos, M. J.; Perez-Alvarez, J. A.; MartınezSanchez, F.; Del Rıo, J. A. Flavonoids as Nutraceuticals: Structural Related Antioxidant Properties and Their Role on Ascorbic Acid Preservation. In *Studies in Natural Products Chemistry;* AttaUr-Rahman., Ed.; Elsevier Science: Amsterdam, 2002; pp 324–389.

Masuda, T.; Inaba, Y.; Maekawa, T.; Takeda, Y.; Tamura, H.; Yamaguchi, H. Recovery Mechanism of the Antioxidant Activity from Carnosic Acid Quinone, an Oxidized Sage and Rosemary Antioxidant. *J. Agric. Food Chem.* **2002,** *50* (21), 5863–5869.

McCarthy, T. L.; Kerry, J. P.; Kerry, J. F.; Lynch, P. B.; Buckley, D. J. Assessment of the Antioxidant Potential of Natural Food and Plant Extracts in Fresh and Previously Frozen Pork Patties. *Meat Sci.* **2001a,** *57,* 177–184.

McCarthy, T. L.; Kerry, J. P.; Kerry, J. F.; Lynch, P. B.; Buckley, D. J. Evaluation of the Antioxidant Potential of Natural Food/Plant Extracts as Compared with Synthetic Antioxidants and Vitamin E in Raw and Cooked Pork Patties. *Meat Sci.* **2001b,** *57,* 45–52.

Mercadante, A. Z.; Capitani, C. D.; Decker, E. A.; Castro, I. A. Effect of Natural Pigments on the Oxidative Stability of Sausages Stored under Refrigeration. *Meat Sci.* **2010,** *84,* 718–726.

Mercadante, A. Z.; Steck, A.; Pfander, H. Carotenoids from Guava (*Psidium guajava* L): Isolation and Structure Elucidation. *J. Agric. Food Chem.* **1999,** *47,* 145–151.

Miean, K. H.; Mohamed S. Flavonoid (*Myricetin, Quercetin, Kaempferol, Luteolin, and Apigenin)* Content of Edible Tropical Plants. *J. Agric. Food Chem.* **2001,** *49,* 3106–3112.

Mielnik, M. B.; Olsen, E.; Vogt, G.; Adeline, D.; Skrede, G. Grape Seed Extract as Antioxidant in Cooked, Cold Stored Turkey Meat. *LWT-Food Sci. Technol.* **2006,** *39,* 191–198.

Misra, K.; Seshadri T. Chemical Components of the Fruits of *Psidium guava. Phytochemical.* **1968,** *7,* 641–645.

Mohamed, H. M.; Mansour, H. A.; Farag, M. D. The Use of Natural Herbal Extracts for Improving the Lipid Stability and Sensory Characteristics of Irradiated Ground Beef. *Meat Sci.* **2011,** *87,* 33–39.

Mohamed, H. M. H.; Mansour, H. A. Incorporating Essential Oils of Marjoram and Rosemary in the Formulation of Beef Patties Manufactured with Mechanically Deboned Poultry Meat to Improve the Lipid Stability and Sensory Attributes. *LWT-Food Sci. Technol.* **2012,** *45,* 79–87.

Mohdaly, A. A.; Sarhan, M. A.; Smetanska, I.; Mahmoud, A. Antioxidant Properties of Various Solvent Extracts of Potato Peel, Sugar Beet Pulp and Sesame Cake. *J. Sci. Food Agric.* **2010,** *90* (2), 218–226.

Moure, A.; Cruz, J. M.; Franco, D.; Dominguez, J. M.; Sineiro, J.; Dominguez, H.; Jose Nunez, M.; Parajo, J. C. Natural Antioxidants from Residual Sources. *Food Chem.* **2001,** *72* (2), 145–171.

Muthukumar, M.; Naveena, B. M.; Vaithiyanathan, S.; Sen, A. R.; Sureshkumar, K. Effect of Incorporation of *Moringa oleifera* Leaves Extract on Quality of Ground Pork Patties. *J. Food Sci. Technol.* **2014,** *51* (11), 3172–3180.

National Agricultural Statistics Service (NASS); *Agricultural Statistics Board, Cranberries*; Annual Reports, Fr Nt 4; USDA: Washington, DC, 2001.

Naveena, B. M.; Sen, A. R.; Kingsly, R. P.; Singh, D. B.; Kondaiah, N. Antioxidant Activity of Pomegranate Rind Powder Extract in Cooked Chicken Patties. *Int. J. Food Sci. Technol.* **2008a,** *43,* 1807–1812.

Naveena, B. M.; Sen, A. R.; Vaithiyanathan, S.; Babji, Y.; Kondaiah, N. Comparative Efficacy of Pomegranate Juice, Pomegranate Rind Powder Extract and BHT as Antioxidants in Cooked Chicken Patties. *Meat Sci.* **2008b,** *80,* 304–308.

Naveena, B. M.; Vaithiyanathan, S.; Muthukumar, M.; Sen, A. R.; Kumar, Y. P.; Kiran, M.; Shaju, V. A.; Chandran, K. M. Relationship between the Solubility, Dosage and Antioxidant Capacity of Carnosic Acid in Raw and Cooked Ground Buffalo Meat Patties and Chicken Patties. *Meat Sci.* **2013,** *95,* 195–202.

Nuñez de Gonzalez, M. T.; Hafley, B. S.; Boleman, R. M.; Miller, R. M.; Rhee, K. S.; Keeton, J. T. Qualitative Effects of Fresh and Dried Plum Ingredients on Vacuum-Packaged, Sliced Hams. *Meat Sci.* **2009,** *83,* 74–81.

Nuñez de Gonzalez, M. T.; Boleman, R. M.; Miller, R. K.; Keeton, J. T.; Rhee, K. S. Antioxidant Properties of Dried Plum Ingredients in Raw and Precooked Pork Sausage. *J. Food Sci.* **2008a,** *73,* H63–H71.

Nuñez de Gonzalez, M. T.; Hafley, B. S.; Boleman, R. M.; Miller, R. K.; Rhee, K. S.; Keeton, J. T. Antioxidant Properties of Plum Concentrates and Powder in Precooked Roast Beef to Reduce Lipid Oxidation. *Meat Sci.* **2008b,** *80,* 997–1004.

O'Grady, M. N.; Carpenter, R.; Lynch, P. B.; O'Brien, N. M.; Kerry, J. P. Addition of Grape Seed Extract and Bearberry to Porcine Diets: Influence on Quality Attributes of Raw and Cooked Pork. *Meat Sci.* **2008,** *78,* 438–446.

O'Brien, N. M.; Carpenter, R.; O'Callaghan, Y. C.; O'Grady, M. N.; Kerry, J. P. Modulatory Effects of Resveratrol, Citroflavan-3-Ol, and Plant-Derived Extracts on Oxidative Stress in U937 Cells. *J. Med. Food.* **2006,** *9* (2), 187–195.

Onofrejová, L.; Vašíčková, J.; Klejdus, B.; Stratil, P.; Mišurcová, L.; Kráčmar, S.; Kopeckýb, J.; Vaceka, J. Bioactive Phenols in Algae: The Application of Pressurized-Liquid and Solid-Phase Extraction Techniques. *J. Pharm. Biomed. Anal.* **2010,** *51,* 464–470.

Owen, R. W.; Haubner, R.; Hull, W. E.; Erben, G.; Spiegelhalder, B.; Bartsch, H.; Haber, B. Isolation and Structure Elucidation of the Major Individual Polyphenols in Carob Fibre. *Food Chem. Toxicol.* **2003,** *41,* 1727–1738.

Özvural, E. B.; Vural, H. Grape Seed Flour is a Viable Ingredient to Improve the Nutritional Profile and Reduce Lipid Oxidation of Frankfurters. *Meat Sci.* **2011,** *88* (1), 179–183.

Paganga, G.; Miller, N.; Rice-Evans, C. A. The Polyphenolic Content of Fruits and Vegetables and Their Antioxidant Activities. What does a Serving Constitute? *Free Radic. Res.* **1999,** *30,* 153–162.

Papagiannopoulos, M.; Wollseifen, H. R.; Mellenthin, A.; Haber, B.; Galensa, R. Identification and Quantification of Polyphenols in Carob Fruits (*Ceratonia siliqua* L.) and Derived Products by HPLC-UVESI/MSn. *J. Agric. Food Chem.* **2004,** *52,* 3784–3791.

Pedroche, J.; Yust, M. M.; Lqari, H.; Giron-Calle, J.; Alaiz, M.; Vioque, J.; Millan, F. *Brassica carinata* Protein Isolates: Chemical Composition, Protein Characterization and Improvement of Functional Properties by Protein Hydrolysis. *Food Chem.* **2004,** *88* (3), 337–346.

Pegg, R. B.; Amarowicz, R.; Naczk, M. Antioxidant Activity of Polyphenolics from a Bear-berry Leaf *(Arctostaphylos uvaursi* L. *Sprengel)* Extract in Meat Systems. In *Phenolic Compounds in Foods and Natural Health Products;* Shahidi, F., Ho, C. T., Eds.; American Chemical Society: Washington, DC, 2005; pp 67–82.

Podsedek, A. Natural Antioxidants and Antioxidant Capacity of *Brassica* Vegetables: A Review. *LWT-Food Sci. Technol.* **2007,** *40* (1), 1–11.

Poyrazoglu, E.; Gokmen, V.; Artik, N. Organic Acids and Phenolic Compounds in Pomegran-ates (*Punica granatum* L.) Grown in Turkey. *J. Food Comp. Anal.* **2002,** *15,* 567–575.

Prior, R. L.; Hoang, H.; Gu, L.; Wu, X.; Bacchiocca, M.; Howard, L.; Hampsch-Woodill, M.; Huang, D.; Ou, B.; Jacob, R. Assays for Hydrophilic And Lipophilic Antioxidant Capacity (Oxygen Radical Absorbance Capacity [ORACFL]) of Plasma and other Biological and Food Samples. *J. Agric. Food Chem.* **2003,** *51,* 3273–3279.

Prior, R. L.; Lazarus, S. A.; Cao, G.; Muccitelli, H.; Hammerstone, J. F. Identification of Procyanidins and Anthocyanins in Blueberries and Cranberries (*Vaccinium spp.*) Using High Performance Liquid Chromatography/Mass Spectrometry. *J. Agric. Food Chem.* **2001,** *49,* 1270–1276.

Purchas, R.; Busboom, J. The Effect of Production System and Age on Levels of Iron, Taurine, Carnosine, Coenzyme Q10, and Creatine in Beef Muscles and Liver. *Meat Sci.* **2005,** *70,* 589–596.

Purchas, R.; Rutherfurd, S.; Pearce, P. D.; Vathera, R.; Wilkinson, B. H. P. Concentrations in Beef And Lamb of Taurine, Carnosine, Coenzyme Q10, and Creatine. *Meat Sci.* **2004,** *66,* 629–637.

Rababah, T. M.; Ereifej, K. I.; Alhamad, M. N.; Al-Qudah, K. M.; Rousan, L. M.; Al-Mahasneh, M. A.; Al-u'datt, M. H.; Yang, W. Effects of Green Tea and Grape Seed and TBHQ on Physicochemical Properties of Baladi Goat Meats. *Int. J. Food Prop.* **2011,** *14,* 1208–1216.

Raghavan, S.; Richards, M. P. Partitioning and Inhibition of Lipid Oxidation in Mechanically Separated Turkey by Components of Cranberry Press Cake. *J. Agric. Food Chem.* **2006,** *54,* 6403–6408.

Raghavan, S.; Richards, M. P. Comparison of Solvent and Microwave Extracts of Cranberry Press Cake on the Inhibition of Lipid Oxidation in Mechanically Separated Turkey. *Food Chem.* **2007,** *102,* 818–826.

Rahman, S.; Pravin, R. Therapeutic Potential of *Aegle marmelos* (L) – An Overview. *Asian Pac. J. Trop. Dis.* **2014,** *4* (1), 71–77.

Ramulu, P.; Rao, P. U. Total, Insoluble and Soluble Dietary fiber Contents of Indian Fruits. *J. Food Compos. Anal.* **2003,** *16,* 677–685.

Rimm, E. B.; Ascherio, A.; Grovannucci, E.; Spielgelman, D.; Stampfer M. J.; Willett, W. C. Vegetable, Fruit and Cereal Fiber Intake and Risk of Coronary Heart Disease among Men. *JAMA.* **1996,** *275,* 447–451.

Rocío Teruel, M.; Garrido, M. D.; Espinosa, M. C.; Linares, M. B. Effect of Different Format-Solvent Rosemary Extracts (*Rosmarinus officinalis*) on Frozen Chicken Nuggets Quality. *Food Chem.* **2015,** *172,* 40–46.

Rojas, M. C.; Brewer, S. Effect of Natural Antioxidants on Oxidative Stability of Frozen, Vacuum-Packaged Beef and Pork. *J. Food Qual.* **2008,** *31,* 173–188.

Rojas, M. C.; Brewer, S. Effect of Natural Antioxidants on Oxidative Stability of Cooked, Refrigerated Beef and Pork. *J. Food Sci.* **2007,** *72,* S282–S288.

Roy, S. K.; Khurdiya, D. S. Other Subtropical Fruit. In *Handbook of Fruit Science and Tech-nology: Production, Composition, Storage and Processing;* Salunkhe D. K., Kadam, S. S., Eds.; CRC Press: New York, 1995; pp 539–562.

Rupérez, P.; Ahrazem, O.; Leal, J. A. Potential Antioxidant Capacity of Sulphated Polysaccharides from Edible Brown Seaweed Fucus Vesiculosus. *J. Agric. Food Chem.* **2002,** *50,* 840–845.

Samaila, D.; Ezekwudo, D. E.; Yimam, K. K.; Elegbede, J. A. Bioactive Plant Compounds Inhibited the Proliferation and Induced Apoptosis in Human Cancer Cell Lines, *In Vitro. Trans. Int. Biomed. Inform. Enabling Tech. Symp. J.* **2004,** *1,* 34–42.

Sanchez, E.; Garcia, S.; Heredia, N. Extracts of Edible and Medicinal Plants Damage Membranes of *Vibrio cholerae. Appl. Environ. Microbiol.* **2010,** *76,* 6888–6894.

Sánchez-Escalante, A.; Djenane, D.; Torrescan, G.; Beltrán, J.A.; Roncalés, P. The Effects of Ascorbic Acid, Taurine, Carnosine and Rosemary Powder on Colour and Lipid Stability of Beef Patties Packaged in Modified Atmosphere. *Meat Sci.* **2001,** *58,* 421–429.

Sanchez-Escalante, A.; Torrescano, G.; Djenane, D.; Beltrán, J. A.; Roncalés, P. Stabilisation of Colour and Odour of Beef Patties by Using Lycopene Rich Tomato and Peppers as a Source of Antioxidants. *J. Sci. Food Agric.* **2003,** *83,* 187–194.

Sánchez-Machado, D. I.; López-Cervantes, J.; López-Hernández, J.; Paseiro-Losada, P. Fatty Acids, Total Lipid, Protein and Ash Contents of Processed Edible Seaweeds. *Food Chem.* **2004,** *85* (3), 439–444.

Sasaki, K.; Ishihara, K.; Oyamada, C.; Sato, A.; Fukushi, A.; Arakane, T., Motoyama, M.; Yamazaki, M.; Mitsumoto, M. Effects of Fucoxanthin Addition to Ground Chicken Breast Meat on Lipid and Colour Stability during Chilled Storage, before and after Cooking. *Asian-Australas. J. Anim. Sci.* **2008,** *21,* 1067–1072.

Sasaki, K.; Takahashi T. A Flavonoid from *Brassica rapa* Flower as UV-Absorbing Nectar Guide. *Phytochemical.* **2002,** *61,* 339–343.

Sayago- Ayerdi, S. G.; Brines, A.; Goñi, I. Effect of Grape Antioxidant Dietary Fiber on the Lipid Oxidation of Raw and Cooked Chicken Hamburger. *LWT-Food Sci. Technol.* **2009,** *42,* 971–976.

Sayari, N.; Sila, A.; Balti, R.; Abid, E.; Hajlaoui, K.; Nasri, M.; Bougatef, A. Antioxidant and Antibacterial Properties of *Citrus Paradisi* Barks Extracts during Turkey Sausage Formulation and Storage. *Biocatal. Agric. Biotechnol.* **2015,** doi:http://dx.doi.org/10.1016/j.bcab.2015.10.004

Scalbert, A.; Williamson, G. Dietary Intake and Bioavailability of Polyphenols. *J. Nutr.* **2000,** *130,* 2073–2085.

Schieber, A.; Stintzing, F. C.; Carle, R. Byproducts of Plant Food Processing as a Source of Functional Compounds – Recent Developments. *Trends Food Sci.Technol.* **2001,** *12,* 401–413.

Sebranek, J. G.; Sewalt, V. J. H.; Robbins, K. L.; Houser, T. A. Comparison of a Natural Rosemary Extract and BHA/BHT for Relative Antioxidant Effectiveness in Pork Sausage. *Meat Sci.* **2005,** *69,* 289–296.

Seeram, N. P.; Aviram, M.; Zhang, Y.; Henning, S. M.; Feng, L.; Dreher, M.; Heber, D. Comparison of Antioxidant Potency of Commonly Consumed Polyphenol-Rich Beverages in the United States. *J. Agric. Food Chem.* **2008,** *56,* 1415–1422.

Selgas, M. D.; García, M. L.; Calvo, M. M. Effects of Irradiation and Storage on the Physico-Chemical and Sensory Properties of Hamburgers Enriched with Lycopene. *Int. J. Food Sci. Technol.* **2009,** *44,* 1983–1989.

Shan, B.; Cai, Y. Z.; Brooks, J. D.; Corke, H. Antibacterial and Antioxidant Effects of Five Spice and Herb Extracts as Natural Preservatives of Raw Pork. *J. Sci. Food Agric.* **2009,** *89,* 1879–1885.

Shan, B.; Cai, Y. Z.; Sun, M.; Corke, H. Antioxidant Capacity of 26 Spice Extracts and Characterization of Their Phenolic Constituents. *J. Agric. Food Chem.* **2005,** 53, 7749–7759.

Shi, H.; Noguchi, N.; Niki, E. Introducing Natural antioxidants. In *Antioxidants in Food Practical Application;* Pokorny J., Yanishlieva, N., Gordon, M., Eds.; CRC Press Woodhead Publishing Ltd.: Cambridge, UK, 2001; pp 147–158.

Shi, J.; Nawaz, H.; Pohorly, J.; Mittal, G.; Kakuda, Y.; Jiang, Y. Extraction of Polyphenolics From Plant Material for Functional Foods-Engineering and Technology. *Food Rev. Int.* **2005,** *21,* 139–166.

Silvan, J. M.; Mingo, E.; Hidalgo, M.; de Pascual-Teresa, S.; Carrascosa, A. V.; Martinez-Rodriguez, A. J. Antibacterial Activity of a Grape Seed Extract and Its Fractions against *Campylobacter* spp. *Food Control.* **2013,** *29,* 25–31.

Soria, A. C.; Sanz, M. L.; Villamiel, M. Determination of Minorcarbohydrates in Carrot (*Daucus carota* L.) by GC–MS. *Food Chem.* **2009,** *114,* 758–762.

Srinivasan, K. Antioxidant Potential of Spices and Their Active Constituents. *Crit. Rev. Food Sci. Nutr.* **2014,** *54,* 352–372.

Stacewicz-Sapuntzakis, M.; Bowen, P. E.; Hussain, E. A.; Damayanti-Wood, B. I.; Farnsworth, N. R. Chemical Composition and Potential Health Effects of Prunes: A Functional Food? *Crit. Rev. Food Sci. Nutr.* **2001,** *41,* 251–286.

Stalikas, C. D. Extraction, Separation, and Detection Methods for Phenolic Acids and Flavonoids. *J. Sep. Sci.* **2007,** *30,* 3268–3295.

Stavric, B. Antimutagens and Anticarcinogens in Foods. *Food Chem. Toxicol.* **1993,** *32,* 79–90.

Steimez, K. A.; Potter, J. D. Vegetables, Fruits and Cancer Prevention: A Review. *J. Am. Diet. Assoc.* **1996,** *96,* 1027–1039.

Stewart, A. J.; Bozonnet, S.; Mullen, W.; Jenkins, G. I.; Lean, M. E. J.; Crozier, A. Occurrence of Flavonols in Tomatoes and Tomato-Based Products. *J. Agric. Food Chem.* **2000,** *48,* 2663–2669.

Sun, J.; Chu, Y. F.; Wu, X.; Liu, R. H. Antioxidant and Antiproliferative Activities of Common Fruits. *J. Agric. Food Chem.* **2002,** *50,* 7449–7454.

Surh, Y. J. Anti-Tumor Promoting Potential of Selected Spice Ingredients with Antioxidative and Anti-Inflammatory Activities: A Short Review. *Food Chem. Toxicol.* **2002,** *40,* 1091–1097.

Suvimol, C.; Anprung, P. Bioactive Compounds and Volatile Compounds of Thai Bael Fruit (*Aegle Marmelos (L.) Correa*) as a Valuable Source for Functional Food Ingredients. *Int. Food Res. J.* **2008,** *15,* 287–295.

Tayel, A. A.; El-Tras, W. F. Plant Extracts as Potent Biopreservatives for Salmonella Typhimurium Control and Quality Enhancement in Ground Beef. *J. Food Saf.* **2012,** *32,* 115–121.

Tichivangana, J. Z.; Morrissey, P. A. Metmyoglobin and Inorganic Metals as Pro-Oxidants in Raw and Cooked Muscle Systems. *Meat Sci.* **1985,** *15,* 107–116.

Tipsrisukond, N.; Fernando, L. N.; Clarke, A. D. Antioxidant Effects of Essential Oil and Oleoresin of Black Pepper from Supercritical Carbon Dioxide Extractions in Ground Pork. *J. Agric. Food Chem.* **1998,** *46,* 4329–4333.

Turkmen, N.; Sari, F.; Velioglu, Y. S. Effects of Extraction Solvents on Concentration and Antioxidant Activity of Black and Black Mate Tea Polyphenols Determined by Ferrous Tartrate and Folin–Ciocalteu Methods. *Food Chem.* **2006,** *99,* 835–841.

Vaithiyanathan, S.; Naveena, B. M.; Muthukumar, M.; Girish, P. S.; Kondaiah, N. Effect of Dipping in Promegranate (*Punica granatum*) Fruit Juice Phenolic Solution on the Shelf Life of Chicken Meat Under Refrigerated Storage (4 °C). *Meat Sci.* **2011,** *88,* 409–414.

Valentão, P.; Trindade, P.; Gomes, D.; de Pinho, P. G.; Mouga, T.; Andrade, P. B. *Codium tomentosum* and *Plocamium cartilagineum*: Chemistry and Antioxidant Potential. *Food Chem.* **2010,** *119,* 1359–1368.

Valgimigli, L. Essential Oils: An Overview on Origins, Chemistry, Properties and Uses. In *Essential Oils as Natural Food Additives;* Valgimigli, L., Ed.; Nova Science Publishers: New York, 2012; pp 1–24.

Vallejo, F.; Gil-Izquierdo, A.; Perez-Vicente, A.; Garcia-Viguera, C. *In Vitro* Gastrointestinal Digestion Study of Broccoli Inflorescence Phenolic Compounds, Glucosinolates, and Vitamin C. *J. Agric. Food Chem.* **2004,** *52,* 135–138.

Vallejo, F.; Tomas-Barberan, F.; Garcia-Viguera, C. Health-Promoting Compounds in Broccoli as Influenced by Refrigerated Transport and Retail Sale Period. *J. Agric. Food Chem.* **2003,** *51,* 3029–3034.

Vallejo, F.; Tomas-Barberan, F. A.; Garcia-Viguera, C. Potential Bioactive Compounds in Health Promotion from Broccoli Cultivars Grown in Spain. *J. Sci. Food Agric.* **2002,** *82,* 1293–1297.

van Breemen, R. B.; Sharifi, R.; Viana, M.; Pajkovic, N.; Zhu, D.; Yuan, L.; Yang, Y.; Bowen, P. E.; Stacewicz-Sapuntzakis, M. Antioxidant Effects of Lycopene in African American Men with Prostate Cancer or Benign Prostate Hyperplasia: A Randomized, Controlled Trial. *Cancer Prev. Res. (Phila).* **2011,** *4* (5), 711–718.

Vaya, J.; Mahmood, S. Flavonoid Content in Leaf Extracts of the Fig (*Ficus carica* L.), Carob (*Ceratonia siliqua* L.) and Pistachio (*Pistacia lentiscus* L.). *Biofactors.* **2006,** *28,* 169–175.

Verma, A. K.; Rajkumar, V.; Banerjee, R.; Biswas, S.; Das, A. K. Guava (*Psidium guajava* L.) Powder as an Antioxidant Dietary Fibre in Sheep Meat Nuggets. *Asian-Australas. J. Anim. Sci.* **2013,** *26* (6), 886–895.

Vinson, J. A.; Zubik, L.; Bose, P.; Samman, N.; Proch, J. Dried Fruits: Excellent *In Vitro* and *In Vivo* Antioxidants. *J. Am. Coll. Nutr.* **2005,** *24,* 44–50.

Viuda-Martos, M.; Ruiz-Navajas, Y.; Fernandez-Lopez, J.; Perez-Alvarez, J.A. Effect of Added Citrus Fibre and Spice Essential Oils on Quality Characteristics and Shelf-Life of Mortadella. *Meat Sci.* **2010,** *85,* 568–576.

Viuda-Martos, M.; Ruiz-Navajas, Y.; Fernández-López, J.; Pérez-Álvarez, J. A. Effect of Adding Citrus Waste Water, Thyme and Oregano Essential Oil on the Chemical, Physical and Sensory Characteristics of a Bologna Sausage. *Innov. Food Sci. Emerg. Tech.* **2009,** *10* (4), 655–660.

Wang, L. I.; Giovannucci, E. L.; Hunter, D.; Neuberg, D.; Su, L.; Christiani, D. C. Dietary Intake of Cruciferous Vegetables, Glutathione S-Transferase (GST) Polymorphisms and Lung Cancer Risk in a Caucasian Population. *Cancer Cause. Control.* **2004,** *15,* 977–985.

Wang, S. J.; Stretch, A. W. Antioxidant Capacity in Cranberry is Influenced by Cultivar and Storage Temperature. *J. Agric. Food Chem.* **2001,** *49,* 969–974.

Wang, S. Y.; Jiao, H. J. Scavenging Capacity of Berry Crops on Superoxide Radicals, Hydrogen Peroxide, Hydroxyl Radicals, and Singlet Oxygen. *J. Agric. Food Chem.* **2000,** *48,* 5677–5684.

Wang, T.; Jónsdóttir, R.; Ólafsdóttir, G. Total Phenolic Compounds, Radical Scavenging and Metal Chelation of Extracts from Icelandic Seaweeds. *Food Chem.* **2009a,** *116,* 240–248.

Wang, Y.; Xu, Z.; Bach, S. J.; McAllister, T. A. Sensitivity of Escherichia Coli to Seaweed (Ascophyllum nodosum) Phlorotannins and Terrestrial Tannins. *Asian-Australas. J. Anim. Sci.* **2009b,** *22,* 238–245.

Wood, J. D.; Enser, M.; Fisher, A. V.; Gute, G. R.; Sheard, P. R.; Richardson, R. I.; Hughes, S. I.; Whittington, F. M. Fat Deposition, Fatty Acid Composition and Meat Quality: A Review. *Meat Sci.* **2008,** *78,* 343–358.

Wood, J.; Nute, G.; Richardson, R.; Whittington, F.; Southwood, O.; Plastow, G.; Mansbridge, R.; da Costa, N.; Chang, K. C. Effects of Breed, Diet and Muscle on Fat Deposition and Eating Quality in Pigs. *Meat Sci.* **2004,** *67,* 651–667.

World Health Orgnization (WHO); *Diet, Nutrition and the Prevention of Chronic Diseases,* WHO Technical Report Series No 916; WHO: Geneva, Switzerland, 2003.

Yamaguchi, F.; Yoshimura, Y.; Nakazawa, H.; Ariga, T. Free Radical Scavenging Activity of Grape Seed Extract and Antioxidants by Electron Spin Resonance Spectrometry in an H_2O_2/NaOH/DMSO System. *J. Agric. Food Chem.* **1999,** *47,* 2544–2548.

Yan, X. J.; Li, X. C.; Zhou, C. X.; Fan, X. Prevention of Fish Oil Rancidity by Phlorotannins from *Sargassum kjellmanianum. J. Appl. Phycol.* **1996,** *8,* 201–203.

Yan, X.; Chuda, Y.; Suzuki, M.; Nagata, T. Fucoxanthin as the Major Antioxidantin *Hijikia Fusiformis,* A Common Edible Seaweed. *Biosci. Biotechnol. Biochem.* **1999,** *63,* 605–607.

Yan, X.; Murphy, B.; Hammond, G. B.; Vinson, J. A.; Neto C. C. Antioxidant Activities and Antitumor Screening of Extracts from Cranberry Fruit (*Vaccinium macrocarpon*). *J. Agric. Food Chem.* **2002,** *20,* 5844–5849.

Yildiz-Turp, G.; Serdaroglu, M. Effects of Using Plum Puree on Some Properties of Low Fat Beef Patties. *Meat Sci.* **2010,** *86,* 896–900.

Yoshie, Y.; Wand, W.; Hsieh, Y. P.; Suzuki, T. Compositional Difference of Phenolic Compounds between Two Seaweeds, *Halimeda* spp. *J. Tokyo Univ. Fisher.* **2002,** *88,* 21–24.

Yu, L.; L. Scanlin, J.; Wilson, G. Schmidt, Rosemary Extract as Inhibitors of Lipid Oxidation and Color Change in Cooked Turkey Products during Refrigerated Storage. *J. Food Sci.* **2002,** *76,* 582–585.

Zakaria-Rungkat, F.; Djaelani, M.; Setiana, Rumondang, E.; Nurrochmah. Carotenoid Bioavailability of Vegetables and Carbohydrate-Containing Foods Measured by Retinol Accumulation in Rat Livers. *J. Food Comp. Anal.* **2000,** *13,* 297–310.

Zheng, W.; Wang, S. Y. Oxygen Radical Absorbing Capacity of Phenolics in Blueberries, Cranberries, Chokeberries, and Lingonberries. *J. Agric. Food Chem.* **2003,** *51,* 502–509.

Zheng, Z.; Shetty, K. Solid-State Bioconversion of Phenolics from Cranberry Pomace and Role of *Lentinus edodes* Beta-Glucosidase. *J. Agric. Food Chem.* **2000,** *48,* 895–900.

Zheng, W.; Wang, S. Y. Antioxidant Activity and Phenolic Compounds in Selected Herbs. *J. Agric. Food Chem.* **2001,** *49,* 5165–5170.

Zou, Y.; Qian, Z. J.; Li, Y.; Kim, M. M.; Lee, S. H.; Kim, S. K. Antioxidant Effects of Phlorotannins Isolated from Ishige Okamurae in Free Radical Mediated Oxidative Systems. *J. Agric. Food Chem.* **2008,** *56,* 7001–7009.

Zukalova, H.; Vasak, J. The Role and Effect of Glucosinolates of *Brassica* Species—A Review. *Rostl. Vyroba.* **2002,** *48,* 175–180.

Zuo, Y.; Wang, C.; Zhan, J. Separation, Characterization, and Quantitation of Benzoic and Phenolic Antioxidants in American Cranberry Fruit by GC–MS. *J. Agric. Food Chem.* **2002,** *50,* 3789–3794.

CHAPTER 4

NATURAL ANTIOXIDANTS: CONTROL OF OXIDATION IN FISH AND FISH PRODUCTS

UGOCHUKWU ANYANWU and REZA TAHERGORABI*

Department of Family and Consumer Sciences, Food and Nutritional Sciences, North Carolina Agricultural and Technical State University, Greensboro, NC, USA

Corresponding author. E-mail: rtahergo@ncat.edu

CONTENTS

ABSTRACT

Seafood is known to contain high amounts of polyunsaturated fatty acids (PUFAs); thereby lipid oxidation is the main cause of quality loss in seafood. Lipid oxidation may cause off-flavor as well as lowering of nutritive value, which can be retarded by incorporation of additives having antioxidative properties. The use of synthetic antioxidants has long been practiced in retarding lipid oxidation. However, due to the potential health concerns of synthetic antioxidants, natural antioxidants such as polyphenolic compounds, essential oils, peptides, and microbial antioxidants, have been used as an alternative natural antioxidants to retard lipid oxidation in different seafood systems. This chapter reviews in detail the lipid oxidation in fish products, antioxidants, and their natural sources as well as focuses on the application of these compounds on the prevention of lipid oxidation in different seafood systems.

4.1   INTRODUCTION

The demand for fish and fishery products in the global market has been increasing with increase in the world population. Fish is an important part of a healthy diet (Mozaffarian & Rimm, 2006). It is an important source of a number of nutrients, particularly protein, vitamin D, retinol, iodine, vitamin E, selenium, and the essential long-chain polyunsaturated fatty acids (PUFAs), that is, eicosapentaenoic acid (EPA) and docosahexaenoic acid (DHA) (Welch et al., 2002). It is recommended that fish and seafood products take a prominent position in the human diet due to their beneficial effect on chronic degenerative diseases. The consumption of fish may protect against cancers and cardiovascular diseases (Nestel, 2000), hence the food industry and health authorities have a joint interest in increasing the consumption of fish.

Oxidation as a chemical reaction that involves the transfer of electrons from one compound to another has long been reported to have negative effects especially in physiological context. Oxidation of molecules such as DNA or lipids has resulted in many problems such as cancer and heart diseases, both of which are important for proper life function. Oxidation reactions can occur in food items mainly as a result of prolonged exposure to the atmosphere. These reactions can cause rancidity, browning, and development of unpleasant flavor. Oxidation is one of the main factors that determine food quality loss and shelf-life reduction. Hence, delaying oxidation is highly relevant to food processors. Oxidative processes in fish products can lead to the degradation of lipids and proteins which, in turn, contribute to the deterioration in texture and color (Decker et al., 1995).

To combat oxidation in food and physiological context, compounds known as antioxidants are utilized. Antioxidants are generally defined as compounds that prevent oxidation. They vary greatly in size, molecular weight, and composition. While some are small in size others are enormous and can even be macromolecules such as proteins. They provide electron density to compounds likely to undergo oxidation, hence preventing them from losing electrons. Since there are different types and sources of antioxidants, they can be grouped into two major categories: natural and synthetic. Natural antioxidants are harvested directly from organic sources or compounds found in foods consumed without much processing whereas synthetic antioxidants are artificial compounds created in laboratories and generally added to processed or pre-packaged foods acting as preservatives. Although, synthetic antioxidants have been widely used to inhibit lipid oxidation, the trend is to decrease their use because of the growing concern among consumers about such synthetic materials (Chastain et al., 1982). Therefore, the search for natural additives, especially of plant origin, has increased in recent years. Compounds obtained from natural sources such as grains, fruit, oilseeds, spices, and vegetables have been investigated (Chen & Ho, 1997). The development and application of natural products with antioxidant activities in fish products may be necessary and useful to prolong their shelf life and potential for preventing fish spoilage.

Natural antioxidants have shown to have significant health benefits especially in prevention of cancer and heart diseases. Due to this, many food manufacturers have begun publicizing this fact and have achieved prominence on many food labels on various food products. Although natural antioxidants are known as key supplements by vitamin and health food manufacturers, consumers must remain aware of the sources from which these compounds are procured as well as their concentrations. Therefore, the aim of this chapter is to elucidate the natural antioxidants commonly applied to fish, their mode of chemical reaction, and consumer awareness of these compounds.

4.2   OXIDATION IN FISH PRODUCTS

Oxidation in fish and fish products possess a high risk of quality loss leading to rancid taste, off-flavor, and development of many different compounds which have adverse effect to human health (Ames et al., 1993). Oxidation limits storage time and this affects marketing and distribution of fish and fish products. Oxidation is high in fish because of presence of omega-3 PUFAs susceptible to peroxidation and results in restriction to storage and processing possibilities (Gray et al., 1996). Products of peroxidation-aldehydes, react with specific amino acids to form carbonyls and protein aggregates which causes additional nutritional loss (Uchida & Stadtman, 1993); for instance, in red fish such as salmon, oxidation not only deteriorates the lipids but also affects the color, thus affecting visual consumer acceptability of fish products (Scaife et al., 2000). Two forms of oxidation occur in fish products—lipid and protein oxidation and they are discussed in the following subsections. While lipid oxidation leads to formation of unhealthy compounds and off-flavors such as rancid, protein oxidation affects the functional properties, including texture, and may potentially affect the taste of fish products. Lipid and protein oxidation occur as a result of the presence of reactive oxygen species (ROS) which include oxygen radicals such as superoxide anion (O_2^-), hydroxyl (HO^-), peroxy (ROO^-), alkoxy (RO^-), and hydroperoxy (HOO) radicals. Non-radical derivates of oxygen such as hydrogen peroxide (H_2O_2), ozone (O_3), and singlet oxygen ($^1O_2^-$) are also ROS (Choe & Min, 2009).

The thiobarbituric acid reactive substances (TBARS) as reported by Botsoglou et al. (1994) are naturally present in biological specimens and composed of lipid hydroperoxides (HPOs) and aldehydes which increase in concentration as a response to oxidative stress. The sensitivity of measuring TBARS has made this assay the method of choice for screening and monitoring lipid peroxidation which is a major indicator of oxidative stress. TBARS assay values are usually reported in malonaldehyde (MDA) equivalents, which is a compound that results from the decomposition of PUFA lipid peroxides. This assay is well recognized and an established method for quantifying lipid peroxides.

4.2.1   LIPID OXIDATION

Lipids are one of the important functional and structural components of foods. They provide energy and essential nutrients such as EPA, DHA, and fat-soluble vitamins including, vitamins A, D, E, and K. Christie (1998) defined lipids as "fatty acids, their derivatives and substances related biosynthetically or functionally to these components." Although they constitute a minor component of food, contribute to the feeling of satiety, and help in making food products palatable, they have been known to significantly affect food quality. This effect on food quality is as a result of constant exposure of lipids, particularly unsaturated fatty acids to air. The susceptibility of lipids to oxidation is one of the main causes of quality deterioration in various types of fresh food products as well as in processed foods.

Lipid oxidation is believed to be one of the factors limiting the shelf life of fish as well as many other complex products (Jacobsen, 1999). Lipid oxidation is most evident in products with a high amount of unsaturated fatty acids leading to rancidity, off-flavor, taste, color, and nutritional value such as reductions in omega-3 fatty acids, some vitamins, and formation of potentially toxic substances (Medina et al., 2009). Lipid oxidation to some extent affects the safety of fish products for human consumption; hence, a notable determinant in consumers' preference for fish and fish products. Within the food industry, a great deal of research and attention is spent on the on-going oxidative processes with the aim of protecting raw materials and products from oxidation during production process and storage. Lipid oxidation, initiated by hemoglobin (Hb) (Christensen et al., 2011), occurs by a reaction between free radicals and oxygen in the presence of other initiators (metal, light, and heat) that results in the formation of HPOs and peroxy radicals (Andersen et al., 2007).

The primary products undergo further reactions to form more stable compounds such as hydroxy acids (that can contribute to bitter taste) or epoxides (Grosch et al., 1992). The interaction of lipid HPOs and secondary oxidation products with proteins and other components in complex food systems, significantly impact oxidative, flavor stability and texture during processing, cooking, and storage (Erickson, 1992). Oxidized lipids can further react with amines, amino acids, and proteins to form brown macromolecular products (Frankel, 1998). According to Pan et al. (2004), color formation is known to be primarily influenced by the degree of fatty acid unsaturation, water activity, oxygen pressure, and the presence of phenolic compounds. Metal, metalloproteins, and enzymes are important factors affecting lipid oxidation in fish products. Water activity, lipid interactions, proteins, and sugars are important elements affecting the quality of processed fish.

4.2.2 PROTEIN OXIDATION

Like lipid oxidation, ROS also take part in protein oxidation. Their mechanisms are very similar to that of lipid oxidation and caused by metal catalysis, irradiation, light exposure, and peroxidation. The negative effect of protein oxidation is seen in the oxidation of sulfhydryl groups, reactions with aldehydes, establishment of cross-links between proteins, formation of oxidative adducts on amino acids, and protein fragmentations (Stadtman & Oliver, 1991). The outcome of oxidation of the majority of amino acids leads to the generation of several products and formation of protein carbonyls viz., glutamic semialdehyde from alanine, hydroperoxides and carbonyl compounds from arginine, sulfenyle chloride from cysteine, hydroperoxides from glutamic acids, asparatate and asparagines form histidine, alcohols and carbonyl compounds from isoleucine, hydroxyleucine and alcohols carbonyls from leucine, etc. (Stadtman and Oliver, 1991).

According to Stadtman (2006), the mechanisms that can lead to protein–protein cross-linking by reactive species are described as:

- Formation of disulfide cross-links (RSSR) as a result of oxidation of cysteine sulfhydryl groups on the proteins.
- Interaction between active groups on different proteins such as the reaction of an aldehyde group of a 4-Hydroxynonenal (HNE) protein adduct and the $-NH_2$ group of a lysine residue in two separate proteins.
- Reaction of MDA with the reactive $-NH_2$ groups of lysine residues in two different protein molecules.
- The formation of covalent linkages between carbon-centered radicals in two different protein molecules.
- The interaction between a carbonyl group of a glycated protein and an $-NH_2$ group of lysine residue in another protein.

Oxidation of amino acids as reported by Orrenius et al. (2007) can be used as a marker for protein oxidation and modification. In protein oxidation, ROS abstract a hydrogen atom which leads to the creation of a protein with a free radical (P•). The protein carbon-centered radical, in the presence of oxygen, reacts with a peroxyl radical (POO•) and as a result of its reactivity the peroxyl radical abstracts a hydrogen atom from another molecule leading to the formation of alkyl peroxide (POOH•) (Lund et al., 2011).

Protein oxidation leads to changes in protein hydrophobicity in processed fish thereby reducing the water holding capacity (WHC) and altering the texture of the product. Metal catalyzed reactions oxidize amino acids such as arginine, proline, and lysine into carbonyl residues (Lund et al., 2011). Amino acids such as cysteine or methionine react in cross-linking or sulfur derivatives. Furthermore, H_2O_2 reacts with metmyoglobin (Mb (Fe^{3+})) which accumulates in the muscle after slaughter, leading to formation of perferrylmyoglobin and ferrylmyoglobin. Lund et al. (2011) went further to state that protein oxidation can also be catalyzed by non-heme iron and other transition metals in the presence of H_2O_2. Myofibrillar proteins are oxidized in the presence of ferric iron (Fe^{3+}) and H_2O_2, resulting in formation of a semialdehyde and ferrous iron (Fe^{2+}). Ferrous iron in the presence of H_2O_2 continues the oxidative reaction degrading amino acids into hydroxyl radicals. Also, reaction between H_2O_2 and the thio groups of amino acids, such as cysteine leads to the formation of sulfhydryl groups. The results of the oxidation of thio groups are sulfenic acid (RSOH), sulfinic acid (RSOOH), and RSSR (Lund et al., 2011).

The formation of carbonyl during protein oxidation can alter the tertiary structure of protein and lead to various degrees of irreversible and irreparable protein unfolding (Aldini et al., 2007). Protein oxidation causes the loss of normal functions, such as enzymatic activity, channel-forming properties, and the proteins become more susceptible to proteolytic degradation. However, with alteration in the tertiary structure and increased hydrophobicity by oxidation, the rate of protein degradation is reduced (Matsuishi & Okitani, 1997).

4.3 OXIDATION MECHANISM

Fish and fish products are made up of several compounds especially lipids that can easily undergo oxidation due to its high tendency to lose electrons. The centerpiece of this reaction is the molecular species known as free radicals. Free radicals are molecules or atoms that have unpaired electrons and can vary greatly in their energy. Auto-oxidation of lipids triggered by exposure to light, ionizing radiations, metalloprotein catalyst, or heat can have a deteriorating effect on color, texture, flavor, quality, and safety of fish products (Urquiaga & Leighton, 2000). Two major components are involved in lipid oxidation; unsaturated fatty acids and oxygen. In the oxidation process, oxygen from the atmosphere is added to fatty acids particularly oleate, linoleate, and linolenate, creating unstable intermediates which break down to form unpleasant compounds. This process involves primary auto-oxidative reactions which are further accompanied by various secondary reactions having oxidative and non-oxidative characteristics (Scott, 2012). Fats contained in food are chemically composed of triglycerides and oxidation leading to the rancidity of foods occurs at the unsaturated sites of triglycerides. Oxidation in fish and fish products occurs in both free fatty acids and fatty acyl groups. The oxidation mechanism consists of three steps:

Initiation (Formation of Fatty Acid Radicals): This is the first step in the oxidative mechanism of lipids. It involves the abstraction of hydrogen from fatty acid to form a fatty acid radical known as the alkyl radical. Initiators react by binding a hydrogen atom from an unsaturated lipid leading to formation of a free radical. Stabilization of the free radical by delocalization over the double bonds results in double bond shifting or formation of conjugated double bonds. This leads to production of *cis* or *trans* configuration (Damodaran et al., 2007). Pro-oxidants such as ionizing radiation or light, transition metals, and temperature are responsible for initiation of oxidation reactions (Fennema et al., 2008).

$$RH \rightarrow R \cdot + H$$

$$RH \rightarrow ROO \cdot + H \cdot$$

where $ROO \cdot$ is a lipid peroxy radical, $R \cdot$ is a lipid radical, and RH is an unsaturated lipid. The peroxides may break down to carbonyls, form polymers or react with vitamins, proteins, pigments, and so forth. The ease of formation of fatty acid radicals increases with an increase in unsaturation. Hydrogen abstraction becomes easier and lipid oxidation is faster due to decrease in carbon–hydrogen bond energy as a result of bond dissociation. For instance, linolenic acid has been estimated to be 10–40 times more susceptible to oxidation than oleic acid due to its rate of bond dissociation (Damodaran et al., 2007).

Propagation (Fatty Acids Radical Reaction): The initial reaction in this step involves the addition of an oxygen molecule which binds to the fatty acid radical (alkyl radical) leading to the formation of peroxyl-fatty acid radical and a weak covalent bond. As a result of the weak covalent bonds of unsaturated fatty acids, they are susceptible to react with peroxyl radicals. Furthermore, the high energy of peroxyl radicals allows them to promote the abstraction of hydrogen from another molecule. This reaction leads to formation of fatty acid hydroperoxyl and fatty acid radical. Further hydrogen addition to the peroxyl radical results in formation of a fatty acid HPO and a new alkyl radical on another fatty acid. Thus, the reaction moves from one fatty acid to another.

$$R\cdot + O_2 \rightarrow ROO\cdot$$

$$ROO\cdot + RH \rightarrow R\cdot + ROOH$$

Termination: This reaction describes the combination of two fatty acid radicals leading to the formation of non-radical products in the presence of oxygen. The reactions between peroxyl and alkoxyl radicals take place under atmospheric conditions while reactions between alkyl and radicals lead to the formation of fatty acid dimers under low oxygen levels (Abidi & Rennick, 2003).

$$2RO_2\cdot \rightarrow O_2 + RO_2R$$

$$RO_2\cdot \rightarrow RO_2R$$

4.4 ANTIOXIDANTS

Antioxidants are regarded as key agents for improving oxidative stability of fish products and act as preservatives which are added to food items at various stages of processing (McClain & Bausch, 2003). Antioxidant effectiveness is related to activation energy, oxidation–reduction potential, rate constants, ease with which the antioxidant is lost or destroyed (volatility and heat susceptibility), and antioxidant solubility (O'Connor & O'Brien, 2006). They are compounds that delay auto-oxidation by inhibiting formation of free radicals by one (or more) of several mechanisms:

1) Scavenging species that initiate peroxidation.
2) Chelating metal ions such that they are unable to decompose lipid peroxides or generate reactive species.
3) Breaking the auto-oxidative chain reaction, and/or reducing localized O_2 concentrations.
4) Quenching O_2^- preventing formation of peroxides.

Antioxidants have been reported to play an important role in the body by protection against oxidative damage (especially damage to the DNA), and delay chronic health problems like cataracts (Morton et al., 2000). Although the Food and Drug Administration (FDA) defines antioxidants only as dietary supplements to be taken with normal food consumption, they also serve as preservatives for packaged foods. They are added to fish products to maintain freshness, prevent rancidity and are particularly important in food containing large amount of lipids. Therefore, addition of antioxidants to fish products greatly extends shelf life and maintains flavor and aroma for as long as possible. The antioxidant activity of a particular compound, mixture of compounds, or a natural source, is generally related to its ability to scavenge free radicals, decompose them, or quench singlet oxygen (Shahidi, 1997).

Antioxidants can be derived naturally as well as chemically synthesized although their performance levels differ. Both antioxidants function by donating electron density to fats, thus preventing their breakdown. Generally, natural antioxidants are known to have higher beneficial health effects such as their ability to prevent disease, for example, cancer and heart disease as compared to the synthetic ones (Morton et al., 2000).

4.4.1 SYNTHETIC ANTIOXIDANTS

Synthetic antioxidants do not occur in nature hence, are chemically synthesized to help prevent lipid oxidation in food. The use of synthetic antioxidants dates back to the 1940s when butylated hydroxyanisole (BHA) was found to retard oxidation and the effectiveness of several alkyl esters of gallic acid was unraveled. Furthermore, it was also evident that the harmful effects of transition metals such as iron and copper had to be counteracted; hence, certain acids, such as citric acid (CA), and their derivatives, were found to act as metal deactivators in combination with phenolic antioxidants (Barlow & Hudson, 1990). In 1954, butylated hydroxytoluene (BHT) was also approved for food use in the United States and tert-butylhydroquinone (TBHQ) was commercialized in 1972. Synthetic antioxidants are divided into primary and secondary antioxidants. Primary antioxidants which prevent the formation of free radicals are further divided into:

- Radical terminators: These constitute the bulk of synthetic antioxidants which prevent lipid oxidation by terminating the free radical chains. Examples include: BHA, BHT, propyl gallate (PG), TBHQ, dodecyl gallate (DG), and octyl gallate (OG).
- Oxygen scavengers: These function as reducing agents. Example: sulphites, ascorbyl palmitate, and glucose oxidase.
- Chelating agents: These prevent lipid oxidation by binding the catalysts such as heavy metals (copper, iron, etc.), either by precipitating

the metal or by occupying all its coordination sites. Examples include: ethylene diaminetetraacetic acid (EDTA) and polyphosphatases.

- Secondary antioxidants function by breaking down HPOs formed during lipid oxidation into stable products. Examples include: dilauryl theodipropionate and thiodipropionic acid (Naidu, 2010).

BHT and BHA are the most prevalent synthetic antioxidants in food. Chemically, they are monohydric phenol with BHA consisting of two isomers 3-tertiary butyl 4-hydroxyanisole and 2-tertiary butyl 4-hydroxy-anisol in the ratio 9:1. It is available as white waxy flakes, while BHT is a white crystalline solid and both are extremely soluble in fats but not in water as a result of their phenolic structure with bulky hydrocarbon side chains. Because of their carry through properties, both compounds can withstand various processing steps such as baking and frying as well as maintaining their functionality (Devlieghere et al., 2004). They are effective in protecting the flavor and color of foods.

The FDA stated that the presence of synthetic antioxidants used in foods be mentioned on food labels with an explanation of their intended usage. Their permissible levels in food is decided on the basis of the fat content of the food and usually limited to 0.02% total antioxidants (Naidu, 2010). When used within recommended levels, they have shown to prevent lipid deterioration in food thereby extending the shelf life of foods. Even though at current levels of intake, synthetic antioxidants seem to pose no reasonable threat to health, but long-term ingestion may aid in modifying the acute toxicity of several carcinogenic and mutagenic chemicals and lead to chronic side effects. Therefore, in recent time, there has been growing concern over possible carcinogenic effects of synthetic antioxidants in foods. BHA, TBHQ, as well as other synthetic antioxidants are no longer allowed for food application in Japan and a number of other countries although still in use at recommended levels in certain countries; there is a general desire to replace synthetic antioxidants with natural ingredients (Venkatesh, 2011).

4.4.2   NATURAL ANTIOXIDANTS

The mention of natural antioxidants brings about an association with spices and herbs, in that they are utilized by-product developers as replacements for synthetic antioxidants. However, other natural products such as nuts, cereals, oilseeds, legumes, animal products, and microbial products can serve as sources of natural antioxidants. Phenols, polyphenolics, and phenolic acid derivatives are antioxidants common to all plant sources of natural antioxidants. Furthermore, modified proteins and amino acids are antioxidants derived from animal and microbial products. The various sources of natural antioxidants are cereals (whole wheat products, oat, rice, bran), vegetables (leaf vegetables, potatoes), fruits (apples, bananas, berries, olives), oilseeds (sesame seeds, hazelnuts, almonds), legumes (beans, peanuts, soybeans), cocoa products (chocolate), beverages (tea, coffee, red wine, beer, fruit juices), and herbs and spices (rosemary, sage, oregano, savory).

4.4.2.1 ANIMAL ORIGIN

Amino acids, peptides, and carotenoids are three animal products that could serve as natural antioxidants. Glutathione peroxidase, superoxide dismutase, and catalase are antioxidant enzymes present in muscle systems. Anserine, carnosine, and ophidine are histidine-containing dipeptides reported to chelate metals and scavenge radicals (Chan et al., 1994). L-Histidine as part of a small peptide/protein or in the free form can scavenge hydroxyl radicals and quench singlet oxygen, which can react with the double bond of L-histidine to form a peroxyl radical (Wade & Tucker, 1998). L-Histidine has the ability to quench singlet oxygen three-fold higher than tryptophan and five-fold higher than methionine.

Carotenoids typically associated with the color of fruits and vegetables are also found in many animals. Crustacea demonstrate a multitude of carotenoid pigments found in nature. According to Zagalsky et al. (1990), carotenoids of invertebrates are associated with protein in a complex defined as carotenoprotein. The carotenoids present in the exoskeleton of crustaceans may provide the best opportunity to develop natural antioxidants from animal sources. Red crabs contain β-carotene and astaxanthin while blue crabs contain 4-hydroxyechinenone, canthaxanthin, 3-hydroxycanthaxanthin, echinenone, isocryptoxanthin, β-carotene, and astaxanthin. These compounds are the most common and important pigments from animal sources that serve as natural antioxidants although limited research have been completed on the antioxidant activity of carotenoids in crustacean (Ramírez et al., 2001). However, their activity would be expected to be similar to plant carotenoids due to the structural similarities between plant and animal carotenoids. Development of extraction or concentration processes is required for the production of adequate amount of natural antioxidants from animal sources.

4.4.2.2 PLANT ORIGIN

Plant-derived additives offer natural alternatives to synthetic antioxidants. In the modern world changes in lifestyle have triggered a growing awareness that particular ingredients in food may favorably modify diet-related problems. The interest in using naturally occurring nutritive and non-nutrient antioxidants for food preservative purpose is due to their possible prevention of a number of diseases, in the etiology of which oxidation mechanisms are involved. Antioxidants can be sourced from selected herbs, spices, fruits, nuts, and other plants (Boskou, 2006). The classes of compounds that act as antioxidants from plant sources include: tocochromanols (lipophilic plant-derived antioxidants) and the more polar phenolic compounds, including phenolic acids, simple phenolics, flavonoids, anthocyanins, tannins, hydroxytyrosol and derivatives, and constituents of essential oils (Pokorný, 2007). Frequently encountered natural antioxidants in plants are phenolic acids and hydroxybenzoic acid (vanillic acid), hydroxycinnamic acid series (ferulic acids, chlorogenic acid), flavonoids (quercetin, catechin, rutin), anthocyanins (delphidin), tannins (procyanidin, ellagic acid, tannic acid), lignans (sesamol), stibenes (resveratrol), coumarines (ortho-coumarine), and essential oils (S-carvone).

Phenolic compounds are plant secondary metabolites and are commonly found in herbs, vegetables, fruits, grains and cereals, coffee, red and white wines, and green and black tea. Phenolic acids are phenols that possess carboxylic acid functionality and they are made up of two distinguishing constitutive carbon frameworks. The flavonoids consist of a group of low-molecular weight polyphenolic substances. According to the degree of oxidation of the C-ring, flavonoids can be categorized into the subclasses flavones, isoflavones, flavanones, flavanonols (dihydroflavonols), flavanols (catechins), flavonols, anthocyanins, and proanthocyanidins. The antioxidant activity of these compounds arises from their direct reaction with free radicals (acting as primary antioxidants) and via their chelation of free metals, which prevents further involvement of these metals in reactions that finally generate radicals.

Tocochromanols are natural compounds known as tocopherols and tocotrienols. They are found mainly in plant oils, nuts, and seeds. Experimental data indicate that they have a radical chain-breaking activity and reducing ability (Kamal-Eldin & Appelqvist, 1996).

4.4.2.3 MICROBIAL SOURCES

Microorganisms are one of the most abundant and diverse species found on earth and their exploitation to produce food ingredients has been going on for the past decade. However, the isolation of microbial antioxidants became a focus of research in the early 1980s, Forbes et al. (1958) established a relationship between antioxidants and microorganisms and since this early work, a vast number of compounds and microorganisms have been characterized. Several studies have demonstrated the antioxidant activity of microorganisms. Using the thiocyanate method, the antioxidant activity of ethyl acetate extracts of several *Penicillium* and *Aspergillus* species was evaluated (Yen & Lee, 1996). Extracts of these species protected linoleic acid better than the control. In a study by Yen and Chang, it was reported that sucrose or lactose and ammonium sulphate in culture media enhanced the *Aspergillus candidus* production of antioxidants (Yen & Chang, 1999). Extracts with similar activity were produced from ethyl acetate extraction of the broth and mycelium.

In a study conducted by Aoyama et al. (1982), 750 filamentous fungi isolated from soil were screened. Two antioxidants were identified as citrinin and protocatechuic. A third compound, curvulic acid, isolated from an unidentified *Penicillium* was also evaluated for antioxidant activity in linoleic acid. All three compounds were reported to have good antioxidant activity. The curvulic acid had the largest antioxidant activity followed by the curvulic acid methyl ester, protocatechuic acid, and citrinin.

According to Esaki et al. (1997) *Aspergillus* species are effective producers of antioxidant activity compounds. In their study, 30 strains of *Aspergillus* were evaluated and it was found that methanol extracts of fermented soybeans (MEFS) prevented oxidation of methyl linoleate. The MEFS of 28 strains had better antioxidant activity than the non-fermented soybean while all strains were better than the control. Separation of the MEFS revealed 2,3-dihydroxybenzoic acid as a component of the most active fraction. Hayashi et al. (1995) also identified this compound in *Penicillium roquefortii* IFO 5956 cultures.

In another study, Esaki et al. (1997) evaluated the antioxidant activity of methanol extracts (MEs) of miso, natto, and tempeh and found that tempeh was the most effective followed by miso. They further stated that fermentation by mold cultures are more active than bacterial (*Bacillus natto*) ones in producing antioxidants. This was evident as a result of the antioxidant activity of the natto ME being less than that of other fermented products but was equivalent to that of unfermented soybeans. Hoppe et al. (1997)

identified 5-(6-tocopheroxy)-d-tocopherol as an antioxidant obtained from tempeh fermented by *Rhizopus oligosporus*. Gallic acid has been isolated from cultures of *Penicillium* and *Aspergillus* and is known as a phenolic acid found in many natural sources including microbial products. Methylenebis (5-methyl-6-tert-butyl-phenol) has been identified as an antioxidant from *Penicillium janthinellum*. *Eurotium* species have also been found to produce several antioxidants. Three of the seven metabolites were found to have antioxidant activity and were identified as dihydroauroglaucin, auroglaucin, and flavoglaucin. Furthermore, Atroventin was isolated from *Penicillium paraherquei* and found to have good antioxidant activity. Demethylnaphterpin and Carazostatin are free radical scavengers isolated from *Streptomyces chromofucus* and *Streptomyces prunicolor*, respectively (Shin-Ya et al., 1992).

Carotenoids are also group of antioxidants that can be synthesized by microorganisms. Nelis and Leenheer reported that lycopene from *Blakeslea trispora* and *Streptomyces chrestomyceticus*, subsp. *rubescens* and β-carotene from *B. trispora* and *Duniella salina* were approved for human foods as colorants (Johnson & Schroeder, 1996). Also, astaxanthin from microbial sources, for example, *Xanthophyllomyces dendrorhous*, found to have excellent singlet oxygen quenching activity has been approved for use in fish foods. The antioxidant activity of carotenoids including lutein, β-carotene, and astaxanthin was confirmed using a fluorometric assay (Naguib, 2000). The use of microbial fermentation as a method for producing natural antioxidants has promise; therefore, more work is needed to optimize production conditions.

4.5 APPLICATION OF NATURAL ANTIOXIDANTS IN FISH PRODUCTS

The necessity to stabilize food against oxidation was realized before World War II and, surprisingly, natural antioxidants have been in use because synthetic antioxidants for edible uses were not yet available at that time (Musher, 1944). However, composition and efficiency of the natural preparations were found extremely variable and their activity was considered insufficient. This led to the invention of synthetic antioxidants which were chemically pure, possessed antioxidant activity, tested for safety, then made readily available in the market. However, natural antioxidants isolated from herbs, tea, grapes, and seeds have gained interest as replacement for synthetic antioxidants (Samaranayaka & Li-Chan, 2011). They are readily acceptable by consumers as they are considered safe. Natural extracts or pure compounds have been used for supplementing food products made of minced fish muscle or surimi. Rosemary, olive oil, ginger, vegetable extracts especially from tea, grape seeds composed of flavonoids, terpenoids, and so forth, have successfully inhibited the rancidity of seafood products such as fish patties, canned fish, fermented fish, and emulsified fish (Tang et al., 2001).

Some other natural extracts obtained from materials such as light fish muscle have been also utilized in fish systems (Sannaveerappa et al., 2007). Furthermore, procyanidins, catechins and their gallate esters, flavonoids, and hydroxytyrosol have also been used in fish muscle (Pazos et al., 2008). They have exhibited a high ability to inhibit oxidation in fish and fish products.

Fish muscle is known to be highly susceptible to oxidation primarily because of the high level of unsaturation found in its lipids. Antioxidant compounds have been studied as a means of increasing the oxidative stability of fish and fish products. Several natural antioxidants have been used in fish oils and fish products to retard lipid oxidation. Cuppett (2001) reported that using a surface application of a rosemary oleoresin on muscle from fish supplemented with tocopherol enhanced the stability of rainbow trout muscle. Dry oregano was reported to be effective in preventing oxidation in mackerel oil (Tsimidou et al., 1995). Zheng and Wang (2001) reported that phenolic antioxidants from rosemary leaves were successfully used in sardine and cod liver oil. Another study found that green tea polyphenols protected silver carp from oxidation and phenolic extracts from grape by-products were successfully applied in fish muscle (Sánchez-Alonso et al., 2007). Frankel et al. (1996) reported that rosemary extracts (carnosol and carnosic acid) were effective antioxidants in fish oils tested in bulk systems. Tea catechins were found to be more efficient than tocopherol in inhibiting minced muscle lipid oxidation in fish patties (Tang et al., 2001). Medina et al. (1998) reported that polyphenols extracted from virgin olive oil retarded oxidation of canned tuna, fish oils, and horse mackerel. A combination of polyphenols and other antioxidant have been reported to be more effective than synthetic antioxidants in preventing lipid oxidation in marine oils and frozen fish (Pazos et al., 2005). Medina et al. (2007) reported that common phenolic acids, such as caffeic acid, when added at the relatively low concentrations of 10–50 mg/kg, are very active in inhibiting increases in TBARS and peroxide value (PV) in horse mackerel.

Many studies have shown the antioxidative effectiveness of natural plant polyphenolic extracts in fish model systems. In a recent study, lipid oxidation was retarded in washed seabass mince with added Hb and menhaden oil (MHO) using ethanolic kiam wood extract (EKWE) (0.1%, w/w). The effectiveness was demonstrated by decreased PV and TBARS values in the samples with added EKWE (Maqsood & Benjakul, 2013). They also found that the formation of volatile lipid oxidation compounds in the washed seabass mince with added MHO and Hb during iced storage could be retarded using 0.1% EKWE. Brown lead (*Leucaena leucocephala*) seed extracts (100 and 200 ppm) without prior chlorophyll removal showed a protective impact on lipid oxidation in a dose-dependent manner in minced mackerel. Brown lead seed extract exhibited scavenging action toward reactants and radicals causing lipid oxidation. The extract was also capable of delaying lipid oxidation in emulsion and liposome model systems in a dose-dependent way (Benjakul et al., 2013). In another study, quince extracts (8.9 ± 0.4 mg phenolics/mL) containing procyanidin B dimer (50.8%) and hydroxycinnamic acids (36.62%) lowered the PV and restrained the development of TBARS in the mackerel (*Scomber scombrus*) fillet fat fraction during refrigerated storage (4 °C) (Fattouch et al., 2008). The bioactive constituents of this extract possessed antioxidant activities as evaluated by antiradical test (Fattouch et al., 2008).

In a recent study by Anyanwu (2015), bay essential oil (BEO) which has been reported to have high antioxidant activity was used to inhibit lipid oxidation in surimi seafood. Alaska pollock surimi seafood was nutrified with omega-3 PUFA-rich oils from flaxseed and salmon and stabilized with BEO during storage time. Omega-3 oils were added at 5% by replacing ice at 1:1 along with 0 (control), 0.5, 1% BEO, followed by cooking (90 °C for 30 min) in hotdog casings, vacuum packed and stored at 4 °C for six days. Whiteness of surimi gels increased significantly with the addition of BEO between treatments and storage time. Lipid oxidation significantly decreased over storage time for treatments with 1% BEO. Addition of BEO and omega-3 rich oils had no detrimental effect on the texture of surimi gels. These studies indicate and promote the efficacy of natural antioxidants in the inhibition of lipid oxidation in fish and fish products as well as other food products.

4.6   MARKET AND CONSUMER ACCEPTABILITY OF NATURAL ANTIOXIDANTS

Due to ability of natural antioxidants to prevent formation of free radicals, they have been found to be useful in preventing certain diseases such as cancer and heart diseases by decreasing the amount of plaque buildup in the blood vessels (McClain & Bausch, 2003). Additionally, they are reported to increase the amount of high density lipoproteins (HDL), commonly known as "good cholesterol" in the blood, thus as preventing heart disease (Yao et al., 2004). These beneficial properties have put natural antioxidant on the forefront of recent advertising and public awareness concerning natural antioxidants and their positive effect has increased greatly. Asahara (1987) studied the antioxidant effect of natural tocopherol mixture on marinated sardine during cold storage which was compared with the effect of BHA. Thiobarbituric acid (TBA) value was determined on lipids of samples during storage time (200 days). Sensory evaluation revealed no negative effect on the organoleptic properties on the samples. Lemon balm and oregano have been reported to have a safe history of use as herbal food ingredients, and their natural extractives are listed as generally recognized as safe (GRAS) in the United States. Furthermore, herbs and flower tips of *Origanum vulgare* and *Melissa officinalis* have been allocated the status N2 by the Council of Europe; N2 comprises admissible natural sources of flavorings (Boskou, 2006).

4.7 CONCLUSION

Antioxidants are compounds that are present either naturally or added to food items to prevent oxidation which always leads to rancidity, browning, and general lack of freshness. Fish and fish products contain unsaturated fatty acids which are especially susceptible to oxidation because of their electron deficient double bonds. The breakdown products of oxidation can produce off-odors, loss of nutrient content, new flavors, and color deterioration. To manufacture high-quality, stable fish products, the most effective solution is the addition of antioxidants, especially natural, which can serve as "chain breakers," by intercepting generation of free radicals during various stages of oxidation or to chelate metals. A common feature of these compounds is that they have one or more aromatic rings (often phenolic) with one or more –OH groups capable of donating H to the oxidizing lipid. The facts that they are natural, and have antioxidative activity that is as good as or even better than the synthetic antioxidants, make them particularly attractive for commercial food processors. It is clear that consumers are becoming increasingly aware of and selective against foods that are perceived by them to be unnatural and containing additives. This means that controlling oxidation in fish with applied antioxidants has to be carefully considered. This would account for the trend toward investigating the use of more natural antioxidants such as herb extracts. Hence, there is a need to screen for new and perhaps more efficacious natural extracts because of consumer demand for natural ingredients.

KEYWORDS

- **seafood**
- **lipid oxidation**
- **natural antioxidants**
- **synthetic antioxidants**

REFERENCES

Abidi, S.; Rennick, K. Determination of Nonvolatile Components in Polar Fractions of Rice Bran Oils. *J. Am. Oil Chem. Soci.* **2003,** *80* (11), 1057–1062.

Aldini, G.; Dalle-Donne, I.; Facino, R. M.; Milzani, A.; Carini, M. Intervention Strategies to Inhibit Protein Carbonylation by Lipoxidation-Derived Reactive Carbonyls. *Med. Res.Rev.* **2007,** *27* (6), 817–868.

Ames, B. N.; Shigenaga, M. K.; Hagen, T. M. Oxidants, Antioxidants, and the Degenerative Diseases of Aging. *Proc. Natl. Acad. Sci.* **1993,** *90* (17), 7915–7922.

Andersen, E.; Andersen, M. L.; Baron, C. P. Characterization of Oxidative Changes in Salted Herring (*Clupea harengus*) during Ripening. *J. Agri. Food Chem.* **2007,** *55* (23), 9545–9553.

Anyanwu, U. C. Effect of Bay (*Laurus nobilis*) Essential Oil on Physicochemical Properties of Alaska Pollock (*Theragra chalgoramma*) Surimi Nutrified with Salmon and Flaxseed Oils under Refrigerated Storage. Master Thesis, North Carolina Agricultural and Technical State University, Greensboro, NC, 2015.

Aoyama, T.; Nakakita, Y.; Nakagawa, M.; Sakai, H. Screening for Antioxidants of Microbial Origin. *Agri. Biol. Chem.* **1982,** *46* (9), 2369–2371.

Asahara, M. Antioxidant Effects of Natural Tocopherol Mixture, Calcium Phytate and Inositol on Marinated Sardine [*Sardinops melanostictus*]. *Bull. Jpn. Soc. Sci. Fish.* **1987,** *53,* 1617–1621.

Barlow, S.; Hudson, B. *Food Antioxidants;* Springer: Netherlands, 1990; 253–307.

Benjakul, S.; Kittiphattanabawon, P.; Shahidi, F.; Maqsood, S. Antioxidant Activity and Inhibitory Effects of Lead (*Leucaena leucocephala)* Seed Extracts Against Lipid Oxidation in Model Systems. *Food Sci. Technol. Int.* **2013,** *19* (4), 365–376.

Boskou, D. *Culinary Applications;* AOCS Publishing: Urbana, IL, 2006; Vol. 1.

Botsoglou, N. A.; Fletouris, D. J.; Papageorgiou, G. E.; Vassilopoulos, V. N.; Mantis, A. J.; Trakatellis, A. G. Rapid, Sensitive, and Specific Thiobarbituric Acid Method for Measuring Lipid Peroxidation in Animal Tissue, Food, and Feedstuff Samples. *J. Agri. Food Chem.* **1994,** *42* (9), 1931–1937.

Chan, K. M.; Decker, E. A.; Feustman, C. Endogenous Skeletal Muscle Antioxidants. *Crit. Rev. Food Sci. Nutri.* **1994,** *34* (4), 403–426.

Chastain, M.; Huffman, D.; Hsieh, W.; Cordray, J. Antioxidants in Restructured Beef/Pork Steaks. *J. Food Sci.* **1982,** *47* (6), 1779–1782.

Chen, J. H.; Ho, C. T. Antioxidant Activities of Caffeic Acid and Its Related Hydroxycinnamic Acid Compounds. *J. Agri. Food Chem.* **1997,** *45* (7), 2374–2378.

Choe, E.; Min, D. B. Mechanisms of Antioxidants in the Oxidation of Foods. *Com. Rev. Food Sci. Food Saf.* **2009,** *8* (4), 345–358.

Christensen, M.; Andersen, E.; Christensen, L.; Andersen, M. L.; Baron, C. P. Textural and Biochemical Changes During Ripening of Old-Fashioned Salted Herrings. *J. Sci. Food Agri.* **2011,** *91* (2), 330–336.

Christie, W. W. Gas Chromatography-Mass Spectrometry Methods for Structural Analysis of Fatty Acids. *Lipids.* **1998,** *33* (4), 343–353.

Cuppett, S. L. *The Use of Natural Antioxidants in Food Products of Animal Origin;* Woodhead Publishing Ltd: Cambridge, UK, 2001; pp 285–310.

Damodaran, S.; Parkin, K. L.; Fennema, O. R.. *Fennema's Food Chemistry;* CRC Press: Boca Raton, FL, 2007.

Decker, E. A.; Chan, W. K.; Livisay, S. A.; Butterfield, D. A.; Faustman, C. Interactions between Carnosine and the Different Redox States of Myoglobin. *J. Food Sci.* **1995,** *60* (6), 1201–1204.

Devlieghere, F.; Vermeiren, L.; Debevere, J. New Preservation Technologies: Possibilities and Limitations. *Int. Dairy J.* **2004,** *14* (4), 273–285.

Erickson, M. C. Changes in Lipid Oxidation during Cooking of Refrigerated Minced Channel Catfish Muscle. In *Lipid Oxidation in Food;* ACS Symposium Series; American Chemical Society: Washington, DC, 1992; pp 344–350.

Esaki, H.; Onozaki, H.; Kawakishi, S.; Osawa, T. Antioxidant Activity and Isolation from Soybeans Fermented with *Aspergillus* spp. *J. Agri. Food Chem.* **1997,** *45* (6), 2020–2024.

Fennema, O.; Damodaran, S.; Parkin, K. *Fennema's Food Chemistry;* CRC Press: Boca Raton, FL, 2008.

Forbes, M.; Zilliken, F.; Roberts, G.; György, P. A New Antioxidant from Yeast. Isolation and Chemical Studies1. *J. Am. Chem. Soci.* **1958,** *80* (2), 385–389.

Fattouch, S.; Sadok, S.; Raboudi-Fattouch, F.; Ben, S. M. Damage Inhibition during Refrigerated Storage of Mackerel (*Scomber scombrus*) Fillets by a Presoaking in Quince (*Cydonia oblonga*) Polyphenolic Extract. *Int. J. Food Sci. Technol.* **2008,** *143* (11), 2056–2064.

Frankel, E. Antioxidants. *Lipid Oxid.* **1998,** *2,* 209–258.

Frankel, E. N.; Huang, S. W.; Prior, E.; Aeschbach, R. Evaluation of Antioxidant Activity of Rosemary Extracts, Carnosol and Carnosic Acid in Bulk Vegetable Oils and Fish Oil and Their Emulsions. *J. Sci. Food Agri.* **1996,** *72* (2), 201–208.

Gray, J.; Gomaa, E.; Buckley, D. Oxidative Quality and Shelf-Life of Meats. *Meat sci.* **1996,** *43,* 111–123.

Grosch, W.; Konopka, U. C.; Guth, H. In *Characterization of Off-Flavors by Aroma Extract Dilution Analysis;* ACS Symposium Series; American Chemical Society: Washington, DC, 1992.

Hayashi, K. I.; Suzuki, K.; Kawaguchi, M.; Nakajima, T.; Suzuki, T.; Numata, M.; Nakamura, T. Isolation of an Antioxidant from *Penicillium roquefortii* IFO 5956. *Biosci. Biotechnol. Biochem.* **1995**, *59* (2), 319–320.

Hoppe, M. B.; Jha, H. C.; Egge, H. Structure of an Antioxidant from Fermented Soybeans (Tempeh). *J. Am. Oil Chem. Soci.* **1997**, *74* (4), 477–479.

Jacobsen, C. Sensory Impact of Lipid Oxidation in Complex Food Systems. *Eur. J. Lipid. Sci. Technol.* **1999**, *101* (12), 484–492.

Johnson, E. A.; Schroeder, W. A. Microbial Carotenoids. In *Downstream Processing Biosurfactants Carotenoids;* Springer: New York, 1996; pp 119–178.

Kamal-Eldin, A.; Appelqvist, L. Å. The Chemistry and Antioxidant Properties of Tocopherols and Tocotrienols. *Lipids.* **1996**, *31* (7), 671–701.

Lund, M. N.; Heinonen, M.; Baron, C. P.; Estevez, M. Protein Oxidation in Muscle Foods: A Review. *Mol. Nutri. Food Res.* **2011**, *55* (1), 83–5.

McClain, R. M.; Bausch, J. Summary of Safety Studies Conducted with Synthetic Lycopene. *Reg. Toxicol. Pharmacol.* **2003**, *37* (2), 274–285.

Matsuishi, M.; Okitani, A. Proteasome from Rabbit Skeletal Muscle: Some Properties and Effects on Muscle Proteins. *Meat Sci.* **1997**, *45* (4), 451–462.

Maqsood, S.; Benjakul, S. Effect of Kiam (*Cotylelobium lanceolatum* Craib) Wood Extract on the Haemoglobin-Mediated Lipid Oxidation of Washed Asian Sea Bass Mince. *Food Bioprocess Tech.* **2013**, *6,* 61–72.

Medina, I.; Gallardo, J. M.; Aubourg, S. P. Quality Preservation in Chilled and Frozen Fish Products by Employment of Slurry Ice and Natural Antioxidants. *Int. J. Food Sci. Tech.* **2009**, *44* (8), 1467–1479.

Medina, I.; Gallardo, J.; Gonzalez, M.; Lois, S.; Hedges, N. Effect of Molecular Structure of Phenolic Families as Hydroxycinnamic Acids and Catechins on Their Antioxidant Effectiveness in Minced Fish Muscle. *J. Agri. Food Chem.* **2007**, *55* (10), 3889–3895.

Medina, I.; Sacchi, R.; Biondi, L.; Aubourg, S. P.; Paolillo, L. Effect of Packing Media on the Oxidation of Canned Tuna Lipids. Antioxidant Effectiveness of Extra Virgin Olive Oil. *J. Agri. Food Chem.* **1998**, *46* (3), 1150–1157.

Morton, L. W.; Caccetta, R. A. A.; Puddey, I. B.; Croft, K. D. Chemistry and Biological Effects of Dietary Phenolic Compounds: Relevance to Cardiovascular DFisease. *Clin. Exp. Pharmacol. Physiol.* **2000**, *27* (3), 152–159.

Mozaffarian, D.; Rimm, E. B. Fish Intake, Contaminants, and Human Health: Evaluating the Risks and the Benefits. *JAMA.* **2006**, *296* (15), 1885–1899.

Musher, S. Oat Product. Google Patents, 1944.

Naguib, Y. M. Antioxidant Activities of Astaxanthin and Related Carotenoids. *J. Agri. Food Chem.* **2000**, *48* (4), 1150–1154.

Naidu, A., *Natural Food Antimicrobial Systems;* CRC Press: Boca Raton, FL, 2010.

Nestel, P. J. Fish Oil and Cardiovascular Disease: Lipids and Arterial Function. *Am. J. Clin. Nutri.* **2000**, *71* (1), 228S–231S.

O'Connor, T.; O'Brien, N. Lipid Oxidation. In *Advanced Dairy Chemistry Volume 2 Lipids;* Springer: New York, 2006; pp 557–600.

Orrenius, S.; Gogvadze, V.; Zhivotovsky, B., Mitochondrial Oxidative Stress: Implications for Cell Death. *Annu. Rev. Pharmacol. Toxicol.* **2007**, *47*, 143–183.

Pan, X.; Ushio, H.; Ohshima, T. Photo-Oxidation of Lipids Impregnated on the Surface of Dried Seaweed (*Porphyra yezoensis Ueda*). Hydroperoxide Distribution. *J. Am. Oil Chem. Soc.* **2004**, *81* (8), 765–771.

Pokorný, J. Are Natural Antioxidants Better–and Safer–Than Synthetic Antioxidants? *Eur. J. Lipid Sci. Technol.* **2007,** *109* (6), 629–642.

Pazos, M.; Alonso, A.; Sanchez, I.; Medina, I. Hydroxytyrosol Prevents Oxidative Deterioration in Foodstuffs Rich in Fish Lipids. *J. Agri. Food Chem.* **2008,** *56* (9), 3334–3340.

Pazos, M.; Gallardo, J. M.; Torres, J. L.; Medina, I. Activity of Grape Polyphenols as Inhibitors of the Oxidation of Fish Lipids and Frozen Fish Muscle. *Food Chem.* **2005,** *92* (3), 547–557.

Ramírez, J.; Gutierrez, H.; Gschaedler, A. Optimization of Astaxanthin Production by *Phaffia rhodozyma* through Factorial Design and Response Surface Methodology. *J. Biotechnol.* **2001,** *88* (3), 259–268.

Samaranayaka, A. G.; Li-Chan, E. C. Food-Derived Peptidic Antioxidants: A Review of Their Production, Assessment, and Potential Applications. *J. Funct. Foods.* **2011,** *3* (4), 229–254.

Sannaveerappa, T.; Carlsson, N. G.; Sandberg, A. S.; Undeland, I. Antioxidative Properties of Press Juice from Herring (*Clupea harengus*) against Hemoglobin (Hb) Mediated Oxidation of Washed Cod Mince. *J. Agri. Food Chem.* **2007,** *55* (23), 9581–9591.

Sánchez-Alonso, I.; Solas, M. T.; Borderías, A. J. Physical Study of Minced Fish Muscle with a White-Grape By-Product Added as an Ingredient. *J. Food Sci.* **2007,** *72* (2), E94–E101.

Scaife, J.; Onibi, G.; Murray, I.; Fletcher, T.; Houlihan, D. Influence of a-Tocopherol Acetate on the Short-and Long-Term Storage Properties of Fillets from Atlantic Salmon Salmo Salar Fed a High Lipid Diet. *Aquacult. Nutr.* **2000,** *6,* 65–71.

Scott, G. Antioxidants *in vitro* and *in vivo. Atm. Oxid. Antiox.* **2012,** *3,* 205.

Shahidi, F. *Natural Antioxidants: Chemistry, Health Effects, and Applications;* The American Oil Chemists Society: Urbana, IL, 1997.

Shin-Ya, K.; Shimazu, A.; Hayakawa, Y.; Seto, H. 7-Demethylnaphterpin, a New Free Radical Scavenger from *Streptomyces prunicolor. J. Antibiot.* **1992,** *45* (1), 124–125.

Stadtman, E.; Oliver, C. Metal-Catalyzed Oxidation of Proteins. Physiological Consequences. *J. Biol. Chem.* **1991,** *266* (4), 2005–2008.

Stadtman, E. R.. Protein Oxidation and Aging. *Free Rad. Res.* **2006,** *40* (12), 1250–1258.

Tang, S.; Kerry, J. P.; Sheehan, D.; Buckley, D. J.; Morrissey, P. A. Antioxidative Effect of Added Tea Catechins on Susceptibility of Cooked Red Meat, Poultry and Fish Patties to Lipid Oxidation. *Food Res. Int.* **2001,** *34* (8), 651–657.

Tsimidou, M.; Papavergou, E.; Boskou, D. Evaluation of Oregano Antioxidant Activity in Mackerel Oil. *Food Res. Int.* **1995,** *28* (4), 431–433.

Uchida, K.; Stadtman, E. Covalent Attachment of 4-Hydroxynonenal to Glyceraldehyde-3-Phosphate Dehydrogenase. A Possible Involvement of Intra-and Intermolecular Cross-Linking Reaction. *J. Bio. Chem.* **1993,** *268* (9), 6388–6393.

Urquiaga, I.; Leighton, F. Plant Polyphenol Antioxidants and Oxidative Stress. *Biol. Res.* **2000,** *33* (2), 55–64.

Venkatesh, R. *A Review of the Physiological Implications of Antioxidants in Food;* Worcester Polytechnic Institute: Worcester, MA, 2011.

Wade, A. M.; Tucker, H. N., Antioxidant Characteristics of L-Histidine. *J. Nutr. Biochem.* **1998,** *9* (6), 308–315.

Welch, A.; Lund, E.; Amiano, P.; Dorronsoro, M.; Brustad, M.; Kumle, M.; Rodriguez, M.; Lasheras, C.; Janzon, L.; Jansson, J. Variability of Fish Consumption within the 10 European Countries Participating in the European Investigation into Cancer and Nutrition (EPIC) Study. *Public Health Nutr.* **2002,** *5* (6b), 1273–1285.

Yao, L. H.; Jiang, Y.; Shi, J.; Tomas-Barberan, F.; Datta, N.; Singanusong, R.; Chen, S. Flavonoids in Food and Their Health Benefits. *Plant Foods Hum. Nutr.* **2004,** *59* (3), 113–122.

Yen, G. C.; Lee, C. A. Antioxidant Activity of Extracts from Molds. *J. Food Prot.* **1996,** *59* (12), 1327–1330.

Yen, G. C.; Chang, Y. C. Medium Optimization for the Production of Antioxidants from *Aspergillus candidus*. *J. Food Prot.* **1999,** *62* (6), 657–661.

Zagalsky, P.; Eliopoulos, E.; Findlay, J. The Architecture of Invertebrate Carotenoproteins. *Comp. Biochem. Physiol. B, Com. Biochem.* **1990,** *97* (1), 1–18.

Zheng, W.; Wang, S. Y. Antioxidant Activity and Phenolic Compounds in Selected Herbs. *J. Agri. Food Chem.* **2001,** *49* (11), 5165–5170.

CHAPTER 5

NATURAL ANTIOXIDANTS IN POULTRY PRODUCTS

A. K. BISWAS[1,*], M. K. CHATLI[2], and GAURI JAIRATH[3]

[1]*Division of Post-Harvest Technology, ICAR-Central Avian Research Institute, Izatnagar, Bareilly 243 122, Uttar Pradesh, India*

[2]*Department of Livestock Products Technology, GADVASU, Ludhiana 141 004, Punjab, India*

[3]*Department of Livestock Products Technology, LUVAS, Hisar 125 001, Haryana, India*

*Corresponding author. E-mail: biswaslpt@gmail.com

CONTENTS

ABSTRACT

Arising consumer interests in "natural" and "fresh" muscle foods with minimum chemical preservatives have motivated researchers to explore antioxidant efficacy of many naturally occurring compounds. Many plants or their parts, fruits, vegetables, herbs, and spices contain diverse bioactive compounds namely phenolics, flavonoids, β-carotene, α-tocopherol, glucosides, dietary fibres, and macro-minerals especially potassium. These bioactive compounds, especially phenolic compounds, glucosides, and some important vitamins like vit. C and vit. E possess significant ABTS$^+$ radical cation, 1, 1 diphenyl-2-picrylhydrazyl (DPPH) radical, and super oxide anion scavenging activity. This antioxidant potential of leaves, seeds, vegetables, fruits can be utilized in different form such as paste, powder, solvent extract etc. to augment the shelf life of poultry products and could replace the synthetic antioxidants effectively. Addition of naturally derived compounds found to have substantial desirable effects on the physico-chemical, microbiological, and sensory attributes of poultry products. It has been reported that poultry products are rich in diverse bioactive compounds which exert chemomodulatory effects through a variety of physiological processes. Dietary intake of fruits and vegetables rich in phenolic/flavonoid compounds prevents excessive free radical formation in the cell, thereby, reduces risk of coronary heart diseases including stroke and cancers. Hence, utilization of these naturally occurring compounds in variety of poultry products will not only increase the storage stability, but also provide adequate health promoting effect to the consumer. In view of these facts, large scope exists for the use of naturally occurring antioxidants in processing of animal food products including poultry. Research studies were carried out on the beneficial aspects of natural antioxidants in improvement of poultry products quality. It is assumed that this chapter will be of great value to the researchers, academicians, entrepreneurs, students, and also some poultry producers/processors who are directly or indirectly linked with the processing of various poultry products.

5.1   INTRODUCTION

Poultry products are rich source of proteins, lipids, vitamins, and minerals, out of which lipids are considered to be the integral components of these food products. Being an integral component, the fatty acid compositions of cell membranes' phospholipid fractions are especially important in determining the stability of foods and act as a site for oxidative process initiation. Further, the fatty acids mostly exist in unsaturated state which makes them more prone to oxidation. Lipid oxidation is influenced by the composition of phospholipids, polyunsaturated fatty acids (PUFA), metal ions, oxygen, heme pigments, and addition of salt and processing approaches. It is one of the major factors affecting the quality of animal foods; however, in cured meat, due to formation of stable pink color by ferrous form of pigment, there is less oxidation as compared to cooked uncured meat. This oxidation is initiated when PUFA react with molecular oxygen via free radical chain mechanism form peroxides and accomplished with discoloration and production of malodorous compounds. These compounds not only give rancid odors but significantly affect human health. Color and flavor are the first stimuli for the consumers to purchase meat and meat products, and in this regard, lipid oxidation is the main limiting factor (Gray et al., 1996). Further, the low oxidative stability of fresh meat, precooked and restructured meat products is a problem for all those involved in the meat production chain, including the primary producers, processors, distributors, and retailers, and a major challenge for meat scientists. The scientific literature pertaining to propensity of lipid oxidation in animal foods including poultry products and other biological systems is vast and suggests various remedies to combat this problem; however, before adopting the solution there is a need to understand the concept of lipid oxidation.

5.2 LIPID OXIDATION

Lipid oxidation is often the decisive factor in determining the useful storage life of food products, even when their fat content is as low as 0.5–1%. Oxidative rancidity is initiated by the so called "reactive oxygen species" (ROS) (Evans & Halliwell, 2001). The ROS in-turn combines various free radicals which not only include oxygen-centered free radicals but also non-radical derivatives of oxygen. Free radicals contain one or more unpaired electrons and are capable of independent existence. Lipid oxidation in muscle foods is initiated in the highly unsaturated phospholipid fraction of subcellular bio-membranes; unsaturated portions of fatty acid esters react with molecular oxygen to form peroxides, hydroperoxides, and carbonyl compounds. The lipid hydroperoxides formed during the propagation of the per-oxidation processes are unstable and are reductively cleaved in the presence of trace elements to give a range of new free radicals and non-radical compounds including aldehydes, ketones, alcohols, and acids that cause the off-odor, off-color, change in nutritive value, and safety of muscle foods. Besides all, lipid oxidation products also affect protein solubility, emulsification, water-binding capacity, texture, and the other rheological properties through the interactions between lipid and protein oxidation products (Hall, 1987). Cholesterol may be oxidized to form oxysterols, which are also as toxic as fatty acid-derived hydroperoxides (Smith & Johnson, 1989). Consumption of oxysterol-containing foods may be potentially harmful to cellular physiology (Morel & Lin, 1996). Oxidation of meat lipids is a complex process and its dynamics depend on numerous factors including chemical composition of the meat, light, and oxygen access as well as storage temperature. The products of lipid oxidation also interfere with the absorption of protein or folic acid and have been found to cause pathological changes in the mucous membrane of the digestive tract.

5.2.1 LIPID OXIDATION MECHANISM

The first step in lipid oxidation is the removal of hydrogen ion from a methylene carbon of a fatty acid (RH). It becomes easier as the number of double bonds in the fatty acid increases, which is why PUFA are particularly susceptible to oxidation. The initiation step can be catalyzed by OH^- or by certain iron–oxygen complexes (e.g., ferryl or perferryl radicals).

$$RH + OH^- \longrightarrow R^- + H_2O \tag{5.1}$$

The fatty acyl radical (R^-) reacts rapidly with O_2 to form a peroxyl radical (ROO^-):

$$R^- + O_2 \longrightarrow ROO^- \tag{5.2}$$

The rate-constant (K_2) for this reaction is 3×10^8 M^{-1} S^{-1}. Because ROO^- is more highly oxidized than the fatty acyl radical or the fatty acid itself, it will preferentially oxidize other unsaturated fatty acids and propagate the chain reaction:

$$ROO^- + RH \longrightarrow ROOH + R^- \tag{5.3}$$

The rate-constant (K_3) for this step is relatively low (1×10^1 $M^{-1}S^{-1}$). Lipid hydroperoxides (ROOH) formed in the propagation reaction are both products of oxidation and substrates for further reaction with Fe^{++} and Cu^+ to yield ROO^- and alkoxyl radicals (RO^-). The ferrous (Fe^{++}) reductively cleaves ROOH (3) as follows:

$$Fe^{++} + ROOH \longrightarrow Fe^{+++} + RO^- + OH^- \qquad (5.4)$$

and Fe^{++} can be regenerated as follows:

$$Fe^{+++} + ROOH \longrightarrow ROO^- + Fe^{++} + H^+ \qquad (5.5)$$

Oxygen (O_2) also reduces ferric iron to ferrous and cupric copper to cuprous *in vivo,* allowing a redox cycle in which the transition metal ion is used several times:

$$O_2 + Fe^{+++} \longrightarrow Fe^{++} + O_2$$

$$O_2 + Cu^{++} \longrightarrow Cu^+ + O_2$$

Other strong reductants such as ascorbic acid (AA) and parquet also reduce Fe^{+++} to Fe^{++}. Both ROO^- and RO^- can initiate further reactions (3) and the following:

$$RO^- + RH \longrightarrow ROH + R \qquad (5.6)$$

The RO^- can also undergo β-scission and degrade to alkyl radicals ($R-CH^{2-}$) and a range of aldehydes (R–CHO) depending on the particular hydroperoxide present.

$$RO^- \longrightarrow R-CH^- + R-CHO \qquad (5.7)$$

($R-CH^{2-}$) can initiate further chain reactions resulting in the formation of ethane and pentane, while the aldehydes, including hexanal, malondialdehyde, and 4-hydroxynonenal, can react readily with e-amino groups of proteins to yield Maillard-type complexes.

5.2.2 FACTORS PREDISPOSING LIPID OXIDATION

The propensity of meats and meat products to undergo oxidation depends on several factors including pre-slaughter events such as stress and post-slaughter events such as early post-mortem pH, carcass temperature, cold shortening, and techniques such as electrical stimulation. Oxidation can also be accelerated due to disruption of the integrity of muscle membranes by mechanical deboning, grinding, restructuring, or cooking which alters cellular compartmentalization and releases the catalytic metal ions.

Post-slaughter changes which predispose muscle foods to oxidation are:

a) Stunning and bleeding—circulation of blood ceases.
b) Anaerobic metabolism—lactic acid accumulates and pH declines to approximately 5.5.
c) Circulation of nutrients rapidly ceases.
d) Preventative antioxidant enzyme system namely superoxide dismutase, catalase, glutathione peroxidase, glutathione reductase etc., unlikely to function.
e) Acute phase proteins which scavenge Fe^+ aeruloplasmin, transferrin, haptoglobin—unlikely to be activated.
f) Sarcoplasmic reticulum loses its calcium accumulating ability.
g) Calcium dependent proteinases degrade muscle proteins.
h) Some destruction of cell compartmentalization.
i) Low molecular weight chelatable iron is released.
j) Iron-catalyzed chain reactions.
k) Membranal lipid oxidation initiated.

5.2.3   CATALYSTS OF LIPID OXIDATION

Some confusion prevails about the nature of the initiation process in lipid oxidation, but spontaneous lipid radical formation or direct reaction of unsaturated fatty acids with molecular oxygen is thermodynamically unfavorable. Most researchers believe that the presence of transition metals, notably iron, is pivotal in the generation of species capable of abstracting a proton from an unsaturated fatty acid. During handling, processing, cooking, and storage, iron is released from high molecular weight sources (e.g., hemoglobin, myoglobin, ferritin, hemosiderin) may directly cause oxidation or made available to low molecular weight compounds such as amino acids, nucleotides, and phosphates with which it is believed to form chelates. These chelates are thought to be responsible for the catalysis of lipid oxidation in biological tissues. However, the relative contributions of the different forms of iron have not been clearly defined. Much of the information pertaining to lipid oxidation in meat deals with hydroperoxide dependent lipid oxidation. Pure lipid hydroperoxides are fairly stable at physiological temperatures, but in the presence of transition metal complexes, especially iron salts, their decomposition is greatly accelerated.

5.2.4 ANTIOXIDANTS IN PREVENTING LIPID OXIDATION

Antioxidants are organic molecules of either synthetic origin or natural origin, which can avoid or delay the progress of oxidative rancidity. It is mainly imparted to their phenol-derived structure. Antioxidants work by donating hydrogen to the lipid free radical to reform the fat molecule or by donating one hydrogen to a peroxide free radical to form a hydroperoxide and a stable antioxidant free radical. They "sacrifice themselves" by giving up a hydrogen atom, then rearrange to a stable conformation.

$$A: H + RO\bullet \rightarrow A\bullet + ROH \tag{5.8}$$

Antioxidants can be classified according to their protective properties at different stages of the oxidation process and since they act by different mechanisms, they are divided into two main types: primary and secondary antioxidants.

a) Primary antioxidants can inhibit or retard oxidation by scavenging free radicals by donation of hydrogen atoms or electrons, which converts them to more stable products, for example vit. C and E, phytochemicals like flavonoids.

b) Secondary antioxidants function by many mechanisms, including binding of metal ions, scavenging oxygen, converting hydroperoxides to non-radical species, absorbing UV radiation, or deactivating singlet oxygen, for example butylated hydroxyl anisole (BHA), butylated hydroxyrotoluene (BHT), propyl gallate (PG), and metal chelating agent (EDTA).

5.3 SOURCES OF NATURAL ANTIOXIDANTS

Natural antioxidants are presumed to be safe and are available in variable amounts in plant and animal kingdoms. Plant phenolics, "phytochemicals," are multifunctional and can act as reducing agents, free radical terminators, metal chelators, and singlet oxygen quenchers. Flavonoids and other classes of phenolic compounds are important phytochemicals. Extracts from plants which contribute health benefits to consumers, arising from protection from free radical-mediated deteriorations, and which cause retardation of lipid oxidation have stronger antioxidant activity (AOA) than that of synthetic antioxidants (Table 5.1).

TABLE 5.1   Some Important Natural Antioxidant and Their Effective Concentration in Chicken Products.

Name	Part	Active component	Effective concentration (products)	References
Plant sources				
Bamboo	Leaves	Flavones, lactones, phenolic acids	0.5% in chicken wings	(Zhang et al., 2007)
Beetroot	Ethanol extract	–	0.5 ml/kg in chicken meat	(Packer et al., 2015)
Canola	Seed extract	–	500 and 1000 ppm in chicken meat	(Wanasundara and Shahidi, 2007)
Cinnamon	Cinnamon powder	Cinnamaldehyde, Eugenol, cinnamicaldehyde	2% in minced chicken meat	(Yadav et al., 2002)
Cocoa	Leaves/extract	Polyphenols	800 mg/kg in MDCM	(Osman et al., 2004)
Dark green leafy vegetables	Lutein	Phenolic compounds	300 µg extract/100g	
Garlic	Fresh garlic	Allicin, diallyl sulfide and diallyl trisulfide	50 g/kg in chicken sausages	(Sallam et al., 2004)
	Garlic powder	–	9 g/kg in chicken sausages	
	Garlic oil	–	0.06 g/kg in chicken sausages	
Grapes	Peed extract (GSE)	Proanthocyaninidins	0.1% in turkey	(Brannan & Mah, 2007)
	GADF	Flavonoids	2% in chicken hamburger	(Ayerdi et al., 2009)
	Pomace (dietary)	–	60 parts GPC in chicken patties	(Ayerdi et al., 2009)
Green tea	Leaves/Extract	Catechins- epicatechins,-epigallo-catechin, epigallocatechin gallate, Epicatechin gallate	200–400 mg/kg in raw poultry meat	(Tang et al., 2001)
	(Dietary)	–	300 mg/kg in poultry feed	(Tang et al., 2000)
Honey	–	–	15% in turkey slices	(Antony et al., 2006)

TABLE 5.1 *(Continued)*

Name	Part	Active component	Effective concentration (products)	References
Onion	Freeze-dried onion powder	Quercetin	1.6% in cooked chicken meat	(Karastogiannidou, 1999)
	Juices	Quercetin	50% strength onion juice brine in turkey breast rolls	(Tang & Cronin, 2007)
Oregano	Oil and extracts	Rosmarinic acid	200 mg/kg in turkey meat	(Govaris et al., 2004)
Plums	Plum extract	Chlorogenic acid neoclorogenic acid, cryptochlorogenic acid	> 2% in turkey breast rolls	(Lee & Ahn, 2005)
Potato	Peel extract	Catechin, chlorogenic acid	0.04% in irradiated meat	(Kanatt et al., 2005)
Pomegranate	Juice rind powder	Tannins, anthocyanins and flavonoids	10 m.equ/100 g in chicken patties	(Naveena et al., 2008)
Rice	Hull extract	Phenolic compounds	0.1%, w/w in turkey breast	(Lee & Ahn, 2005)
Sage	Oil	Carnosol, rosmanol, rosemadiol, carnosic acid	3% in chicken meat	(Mariutti et al., 2008)
Sesame oil	Sesamol	Phenolic compounds	500–2000 µg/ml	
Dry soya sprouts			30 g/kg in chicken patties	(Romero et al., 2008)
Animal sources				
Bone protein hydrolysates	–	–	2%	
Milk proteins	Casein phosphopeptides	–	2 % in muscle foods	(Sakanaka et al., 2005)

– Data not recorded.

5.3.1 PLANT SOURCES

5.3.1.1 α-TOCOPHEROL (VIT. E)

It is a widely studied antioxidant and has the ability to function in biological systems. Free radicals are neutralized by tocopherol before lipid oxidation propagates among highly unsaturated fatty acids in cellular and subcellular membranes. The chromanol ring of α-tocopherol is located among the polar head groups of the phospholipids, and the phytol side chain interacts with the unsaturated fatty acyl chains of the phospholipids through van-der-Waals interactions in the interior of the membrane. This specific localization of α-tocopherol in the membrane and the molecule's lateral mobility allow it to function very efficiently to protect highly oxidizable PUFA from per-oxidation by ROS produced by adjacent membrane bound enzymes. A logical hypothesis is that tocopherol quenches free radicals originating from lipid oxidation and in-turn protects oxymyoglobin oxidation. It inhibits free radical oxidation by reacting with peroxyl radicals to stop chain propagation and with alkoxyl radicals to inhibit the decomposition of the hydroperoxides and decrease the formation of aldehydes.

Various attempts have been made to reduce pigment and lipid oxidation in meats by endogenous and exogenous vit. E treatments. Dietary supplementation of vit. E at 2500 mg for 40 days resulted in a 7–10 day extension of color, shelf life of meat steaks (Taylor et al., 1994). It has also been reported that animal fed a diet containing 75 or 1000 mg vit. E had higher concentration of vit. E in meat and lower discoloration. High concentrations of vit. E have been shown to exert a pro-oxidant effect (Mahoney & Graf, 1986). Post-mortem addition of vit. E was less effective in retarding the oxidation of pigment and lipid than endogenous one. Thus, dietary vit. E supplementation would be a safer and more effective method for retarding pigment and lipid oxidation.

Vit. E had been reported to be effective in improving shelf life of poultry meat and meat products through delay in oxidation. But the efficacy of oxidative stability of poultry meat is related to tissue concentration of tocopherol. As the level of vit. E in diet increases there is concomitant increase in its concentration in tissues. High level of vit. E in poultry meat not only improves the oxidative stability but also prevents excessive drip loss, as it is believed that α-tocopherol reduces leakage of sarcoplasmic components from muscle cells by maintaining the integrity of cellular membranes and thereby reduces drip (Goni et al., 2007). The sensory quality of meat assessed from scoring of aroma, flavor, taste, etc. is reported to be improved by adding vit. E in diets of broilers. Vit. E was reported to improve phagocytic ability of the immune system in broilers (Konjufca et al., 2004). It has also been reported that under heat stress condition dietary vit. E supplementation improved the immune response of broilers. Even supplementation of vit. E decreased the production of fishy color because of fishmeal oil in diet and warmed over color in refrigerated cooked meat and raw frozen meat (Niu et al., 2009). The stability of the meat has been correlated with the tissue concentration of α-tocopherol. For poultry, levels of 7 µg/kg α-tocopherol in muscle are recommended to prolong the keeping quality of meat. The supplemented tissues were more stable in cold store. Feeding of vitamin supplemented diets (10–20 ppm) showed significantly lower levels of thiobarbituric acid reactive substances (TBARS) in the meat measured after 5 and 10 days of storage (Lohakare et al., 2005).

The effects of vit. E in ducks are different. The effect of feeding of α-tocopheryle acetate at different levels to day old white Pekin ducklings revealed significantly increased concentrations of vit. E in breast, thigh, liver, and heart tissues in dose dependant manner. Supplementation though enhances oxidative stability, was more effective for thigh muscle than breast.

α-tocopherol, which easily gets deposited in the egg yolk, is used as antioxidant in animal nutrition (Galobart et al., 2001). But, other existing substances in yolks may also prohibit oxidation, as selenium or carotenoids (Yaroshenko et al., 2004). Antioxidants are deposited in yolks according to dietary levels. Enriching egg yolks with α-tocopherol or carotenoids can be done in a wide range and does not affect egg quality, except that with increasing levels of carotenoids intensity of yolk color increases. While supplementing carotenoids to diets, it has to be considered that yellow and red pigments should be added in ratios given by the manufacturer to avoid off-colors (Galobart et al., 2001). The artificial carotenoids (canthaxanthin) usually supplemented to the layer's diet cover more than 90% of total carotenoids in the egg. Enriching eggs with selenium is more complicated as high levels of selenium in the food are toxic for humans. Nevertheless, it is easy to enrich eggs with 35 µg Se which amounts to 50% of the recommended daily intake (RDI) for humans (Yaroshenko et al., 2004). No negative impacts of antioxidants on any egg quality criteria are known.

Although, the eggs provide considerable amounts of vit. E and selenium, but, enriching eggs with n-3 PUFA will result in a higher liability of eggs to oxidation and to off-colors (Tserveni-Gousi et al., 2004). Off-colors may further occur by the use of fish oil as a dietary source. Therefore, for the production of omega-3 enriched eggs only high quality fish oil should be used and distinct amounts of antioxidants (α- tocopherol) should be supplemented, as well (Galobart et al., 1999). Studies have shown no clear negative impact of omega-3 enrichment on other quality criteria of eggs, including their functional properties.

5.3.1.2   ALOE VERA

Aloe is a genus containing about four hundred species of flowering succulent plants belonging to Lileaceae family (Mohammad, 2003). True aloe vera plant is called *Aloe barbadensis* Miller (Shahzad et al., 2009). The mucilaginous jelly from the parenchyma cells of the peeled, spineless leaves of the plant is referred as aloe vera gel. The gel is a watery-thin, viscous, colorless liquid which contains 99% water and 1–0.5% solid matter at pH 4.5 (Shahzad et al., 2009). The gel exhibits antioxidant, antibacterial, antifungal, as well as antiviral activity as it contains aloin, aloe-emodin, barbloin, emodin, anthraquinone glycosides, glycoprotein, gamma-lanoline acid, prostaglandins, and mucopolysaccharides (Shafi et al., 2000; Singh et al., 2010). Ethanolic extract of fresh aloe vera juice possessed stronger radical scavenging activity than that of aloe vera powder, and the antioxidative effect of aloe vera extracts was correlated to its development stage (Zhang et al., 2001).

5.3.1.3   ASCORBIC ACID

AA is a chelating agent that binds metal ions; it also scavenges free radicals and acts as a reducing agent. At high levels (> 1000 mg/kg), AA inhibits oxidation; however, at low levels (< 100 mg/kg) it can catalyze oxidation and warmed over flavor (WOF) development (Ahn & Nam, 2004). In the presence of AA, iron stimulates oxidation in muscle membranes, presumably through the involvement of hydroxyl radicals. Sepe et al. (2005) found that sodium ascorbate and sodium erythorbate more effectively maintained red color and maintained myoglobin in the reduced state in cooked ground meat patties than AA and ascorbyl palmitate. The solubility of ascorbate affects its ability to prevent discoloration (Mancini et al., 2006). The lack of effectiveness of the hydrophobic antioxidant may be a result of localization of components responsible for bone discoloration within the aqueous phase. AA and phosphates appear to work synergistically to inhibit lipid oxidation. AA and tocopherol reduction of lipid oxidation in meat can be enhanced by adding sesamol, especially as storage time increases (Ismail et al., 2008).

It has been theorized that AA functions to maintain a portion of the iron in the reduced state. The amount of AA permitted in meat products varies depending on the route of introduction: brine, incorporation, or surface spray. A 10% AA solution can be applied as a spray to cure carcass surfaces. Sodium ascorbate or erythorbate is also used as cure accelerator.

5.3.1.4 BAMBOO LEAVES

Dried bamboo leaves, yellow or brown colored powder, are commonly used as antioxidant in various food systems. The antioxidative property is mainly due to flavones, lactones, and phenolic acids. It can either inhibit lipid auto-oxidation chain reaction, or chelate transition metal ions, and can be used as primary or secondary antioxidant. Moreover, bamboo leaf powder can help in eliminating the nitrites in cured meat. It inhibits the synthesis of *N*-nitro-samine, and has anti-bacterial, bacteriostatic, deodorizing, aroma enhancing, etc. functions (Zhang et al., 2007). It is commonly used in oil-containing food, meat products, fishery products, expanded foods, etc.

5.3.1.5 BEETROOT EXTRACTS, CLOUDBERRY, WILLOW HERB

The AOA of these plant extracts in meat system has been investigated parallel to pure quercetin, rutin, and caffeic acid. It is found that cloudberry extract and quercetin are the most potent; caffeic acid intermediate and pure reutin have the lowest AOA (Reya et al., 2005). Hexanal production was inhibited by the high level of beetroot, but TBARS production was not, perhaps because the red color of beetroot extract interfered with the determination of the pink thiobarbituric acid chromogen.

5.3.1.6 CAMELINA MEAL

The phenolic composition in camelina meal was predominated by flavonols (quercetin and glycosides), hydroxycinnamic acid derivatives (sinapine and sinapic acids), flavanols, and tocopherols inhibiting hexanal formation ($\geq 80\%$ inhibition). The lower efficacy of camelina meal than that of rapeseed meal is due to lower sinapine and α-tocopherol (24 μg/g) content. Camelina meal phenolics have potential effects in inhibiting the lipids oxidation of broiler meat when incorporated in ration (Aziza et al., 2010).

5.3.1.7   CANOLA EXTRACT

The antioxidative properties of canola extract were compared with various synthetic antioxidants. Canola extracts at 500 and 1000 ppm were more active than BHA, BHT, and BHA/BHT/MGC8 (methyl 3,4,5-tris(*n*-octyloxy) benzoate) and less effective than tert-butylhydroquinone (TBHQ) at a level of 200 ppm (Wanasundara & Shahidi, 1994).

5.3.1.8   CARAWAY

It is rich source of carvone and limonene. Caraway has been used as antioxidant in chicken meat stored under frozen conditions (El-Alim et al., 1999).

5.3.1.9   CAROTENOIDS

Carotenoids are yellow, orange, and red lipid-soluble pigments that occur widely in plants, fruits, and vegetables. They are 40-carbon isoprenoids with varying structures, and can be classified as carotenes and xanthophylls. Certain carotenoids are also referred to as pro-vitamins such as β-carotene, α-carotene, and β-cryptoxanthin. Carotenoids are antioxidant nutrients that act mainly as secondary antioxidants in foods by quenching singlet oxygen. They may also prevent oxidation by trapping free radicals in the absence of singlet oxygen. Carotenoids are a good synergist with tocopherols. Beta-carotene, lutein, lycopene, and isozeaxanthin are typical carotenoids that effectively retard oxidation in foods. Astaxanthin has AOA that is ten times greater than that of β-carotene, lutein, zeaxanthin, and canthaxanthin, and is often used in fish products. In a high-oxygen concentration, β-carotene may exhibit a pro-oxidant, rather than an antioxidant effect in food products. Carotenoids are natural constituents of foods and have generally recognized as safe (GRAS) status. No permissible limits on their addition level have been stipulated.

5.3.1.10   CHERRY

Cherry fractions contain phenolic compounds such as flavones, isoflavones, anthocyanins, anthocyanidins, and phenolic glycosides. The effects of tart cherry tissue added at an 11.5% level on the oxidation of lipids in raw and cooked ground meat patties and on the formation of heterocyclic aromatic amines (HAAs) in the fried patties were investigated. TBARS values and cholesterol oxidation for raw and cooked ground patties containing cherry tissue were significantly lower than those for the control samples. The formation of mutagenic/carcinogenic HAAs during frying of the patties was inhibited by components in the cherry tissue. The concentrations of 2-amino-1-methyl-6-phenylimidazo [4, 5-b] pyridine (PhIP), the principal HAA in cooked muscle foods, were reduced 93 and 87 % by cherry tissue, respectively.

5.3.1.11   CLOVE

It has been observed that 1 and 2% clove oil have very good antioxidant and antimicrobial effects in chicken frankfurters (Mytle et al., 2006); 0.2 and 0.5 % clove oleoresin in chicken meat marination (Carlos & Harrison, 1999); clove powder as phyto-preservative and antimicrobial in chicken nuggets (Kumar & Tanwar, 2011). Hao et al. (1998) applied eugenol to meat slices or cooked chicken and it was proved that eugenol inhibited the growth of *Aeromonas hydrophila* and *Listeria monocytogenes*.

5.3.1.12   COCOA LEAVES

These are effective antioxidant as green tea polyphenols (Osman et al., 2004). These can be used as extract and it has lower astringency and bitterness. Therefore extracts of cocoa leaves can be used in higher concentrations, for higher effectiveness as antioxidant.

5.3.1.13   CURRY LEAF

Curry leaf (*Murray koenigii*) is native from East-Asian countries and mostly used as a color ingredient in variety of products. The extract contains monoterpene hydrocarbons and monoterpene-derived alcohols which have recently been recognized for their efficacy in providing significant AOA to the human foods (Ningappa et al., 2008). However, AOA of curry leaf extracts may vary depending on extraction methods, purity, types, and quantity of active compounds present according to climate, soil composition, plant organ, age, and stage in the vegetative cycle. Antioxidant effects of curry leaf powder and extract have been investigated for their use in chicken mince and patties (Biswas et al., 2006). It has been reported that use of curry leaf powder could reduce production of malondialdehyde content in raw and precooked chicken meat patties.

5.3.1.14 DARK GREEN LEAFY VEGETABLES

Lutein is the main active compound from extracts of dark green leafy vegetables. It is an oxygenated carotenoid (xanthophylls) abundantly present and is one of the most important dietary antioxidants for eye health. Lutein significantly reduces the risk of age-related macular degeneration, atherosclerosis, and UV damage (O'Connell et al., 2008). AOA of lutein is based on its singlet oxygen quenching ability.

5.3.1.15 DRIED PLUMS

Plum derived food ingredients, due to their flavonoid and polyphenol content, are reported to function as antioxidants, antimicrobials, fat replacents, and flavorants. The principal phytochemicals in dried plums are phenolic acid derivatives, flavonoids, and coumarins. In addition, dried plums also contain large amounts of neochlorogenic, chlorogenic, and cryptochlorogenic acid. Chlorogenic acid isomers have a high AOA and inhibit low-density lipoprotein oxidation. Nunez et al. (2008) concluded that 2.5% dried plum or fresh plum juice concentrates or spray dried plum powder can be used in roasts without any detrimental effects and with potential benefit of reducing lipid oxidation and warmed over flavor. Similarly, it has been observed that plum extract used at > 2% in turkey breast rolls and irradiated at 3.0 kGy was effective in retarding oxidation and enhancing juiciness (Lee & Ahn, 2005).

5.3.1.16 DRY SOYA SPROUTS

It possesses high total phenolics and flavonoids content, and also a little amount of vit. C. Although reduction power is not good, 1, 1 diphenyl-2-picrylhydrazyl (DPPH) radical scavenging activity was comparable to BHA, and could be used as a cheap natural antioxidant source for meat and meat products (Romero et al., 2008).

5.3.1.17   DRUMSTICK LEAF

Drumstick leaf (*Moringa oleofera*) is characterized by high antioxidant potential as they are rich in phenolic compounds. Mature leaves of this plant contain 0.68 g total phenolic per 100 g portion, high amount of β-carotene, iron (Fe), potassium (K), calcium (Ca), and vit. C. Drumstick leaf has the highest amount of essential amino acids and significant quantities of minerals (Moyo et al., 2012). AOA of drumstick leaves has been reported in meat and meat products.

5.3.1.18   EVENING PRIMROSE

Evening primrose extract contains phenolics which inhibit the formation of conjugated dienes, hexanal, and total hydrogen peroxide (H_2O_2), hydroxyl radical ($^{.}OH$), and superoxide radical ($O_2^{.-}$) by 43.6 and 72.6% when used at 1 and 2%; w/w, respectively, in cooked comminuted meat products.

5.3.1.19   GARLIC

It is used as a flavor enhancer and known to have medicinal properties due to wide spectrum of actions such as antibacterial, antiviral, antifungal, and anti-protozoal. It also has beneficial effects on the cardiovascular and immune systems (Leuschner & Ielsch, 2003). Besides its medicinal effects, garlic showed effective AOA in-vivo and in-vitro due to organic sulfur compounds and their precursors, allicin, diallyl sulfide, and diallyl trisulfide. Garlic is used as fresh, powder, and oil forms in food systems. Sallam et al. (2004) graded the AOA of the various materials in the following order fresh garlic > garlic powder > BHA > garlic oil in raw chicken sausage during cold storage (3 °C); however, their activity is concentration dependent. In higher concentration it has very strong flavor and cannot be acceptable to the consumers.

5.3.1.20   GINGER

The AOA of ginger relies on 6-gingerol and its derivatives. Ginger extract at 2.5% has been found effective as tenderizing, antioxidant, and antimicrobial in smoked spent hen meat (Naveena & Mendiratta, 2001) and 2, 4, and 6% ginger paste as an antioxidant in spent hen meat balls (Rongsensusang et al., 2005).

5.3.1.21   GRAPES

Grapes are used in various forms as seed extract, dietary fiber, pomace, grape wine, etc. Grape seed extract (GSE) is already marketed as an ingredient to the dietary supplementation industry, the quality and price of which are based on its phenolic content. Phenolics in GSE exist as proanthocyanidins in the form of oligomers and polymers of polyhydroxy flavan-3-ols such as catechin and epicatechin (Weber et al., 2007). Antioxidant effect of GSE was evaluated on cooked turkey patties and cooled stored turkey meat (Lau & King, 2003). Grape pomace, a concentrate of grape seeds, stems, and peel, is a rich source of flavonoids including monomeric phenolic compounds such as catechins, epicatechin, and epicatechin-3-O-gallate and dimeric, trimeric, and tetrameric procyanidins. It is used as dietary supplement to increase antioxidant capacity in breast and thigh meat of broiler chickens (Ayerdi et al., 2009).

Grape antioxidant dietary fiber (GADF) was effectively used at 0.5, 1, 1.5, and 2% concentration in raw and cooked chicken breast hamburger. GADF addition resulted in reduction in lightness and yellowness and an increase in redness in raw and chicken hamburgers without affecting the acceptability of the products (Ayerdi et al., 2009).

Grape wine has resveratrol, a strong antioxidant and a free-radical scavenger. It has superior AOA over quercetin, rutin, and carnosine. Antioxidative effectiveness has been reported as BHA > resveratrol > PG > tripolyphosphate > vanillin > phenol > BHT > α-tocopherol (Bekhit et al., 2003).

5.3.1.22   GREEN TEA

Antioxidant properties are attributed to the presence of tea catechins (TC), epicatechins, epigallocatechin, epigallocatechin gallate, and epicatechin gallate. These compounds have high affinity for lipid bilayers of muscle and the radical scavenging activity which prevent lipid oxidation and also have antibacterial action. TC can also reduce the formation of peroxides more effectively than α-tocopherol or BHA in canola oil and chicken fat and fish muscle model system. It has been observed that TC added at a level of 300 mg/kg inhibited the pro-oxidative effect of 1% NaCl and controlled the lipid oxidation for all the cooked patties during refrigerated storage. It (200 or 400 mg/kg) caused discoloration in cooked patties possibly by binding with the iron component of myoglobin.

5.3.1.23   HONEY

Honey alters the water activity, thereby indirectly affecting oxidation rate. Moreover, it facilitates the Maillard reaction during the cooking process and thus the development of an antioxidative effect. Honey (15% wt/wt) was reported to retard lipid oxidation in turkey and chicken meat. However, it is also reported that incorporation of honey has imparted a slightly darker color with lower lightness values but had no effect of redness and yellowness values (McKibben & Engeseth, 2002; Hashim et al., 1999).

5.3.1.24   MARJORAM

Marjoram (*Origanum majorana* L.) essential oil inhibits formation of initial compounds during the oxidation of unsaturated fatty acids (conjugated di-enes) by 50% and the generation of secondary oxidation products of linoleic acid by 80% in a model system (Schmidt et al., 2008). Wild marjoram has also been shown effective in refrigerated and chilled meat patties, however, it has been reported that addition of salt or freezing of samples results in loss of effect. A purified component isolated from marjoram, T3b, a phenolic substance, is a better superoxide anion radical scavenger than BHT, BHA, α-tocopherol, AA and a variety of polyphenolic flavonoids epigallocatechin gallate, quercetin, epicatechin. The inhibitory mechanism of T3b appears to depend on the action of an endogenous enzyme (superoxide dismutase) which destroys the superoxide anion by converting it to H_2O_2.

5.3.1.25   MINT

Spearmint or garden mint (*Mentha spicata L*), family Lamiaceae (Labiatae), is often added to several meat preparations as a color enhancer. It is a rich source of polyphenolic compounds with strong antioxidant properties (Dorman et al., 2003) but its application in meat is yet to be explored. Kanatt et al. (2007) investigated that mint extract (ME) had good total phenolic and flavonoid contents with high superoxide and hydroxyl-scavenging activity but low iron-chelating ability.

5.3.1.26   OLIVE LEAF

Olive leaves extract is known to have antioxidative, antimicrobial, and anti-inflammatory properties and to protect low-density lipoprotein from oxidation, the capacity to lower blood pressure in animals and to inhibit lipid oxidation. Carpenter et al. (2006) found that olive leaf extract (50 µg/ml) strongly protects cells against oxidative stress.

5.3.1.27   ONION

Onion (*Allium cepa* L.) is much valued for its flavoring components and has high quercetin content (284–486 mg/kg). Quercetin, a flavonoid belongs to a group of plant phenolics, known to control rancidity in cooked ground turkey, cooked ground lamb, and oven-cooked turkey breast (Younathan et al., 1980; Karastogiannidou, 1999; Tang & Cronin, 2007). It may be possible to further enhance the antioxidant role of quercetin by using a juice from a high quercetin-yielding onion variety.

5.3.1.28   OREGANO

Oregano (*Origanum vulgare* L.) extracts contain high concentrations of phenols, primarily rosmarinic acid, which can prevent color deterioration (Hernandez et al., 2009). Phenolic carboxylic acids and glycosides are also particularly antioxidative and effectively scavenge superoxide anion radicals. It delays surface discoloration, that is, metmyoglobin formation in ground meat. Camo et al. (2008) compared the effects of direct addition of oregano and rosemary to the use of active films containing oregano and rosemary on the display life of lamb. Active films containing oregano were significantly more efficient than those with rosemary, extending fresh odor and color from 8 to 13 days. Similarly, Grobbel et al. (2006) used oregano in combination with AA in polyvinyl chloride (PVC) film in a high-oxygen modified atmosphere environment.

5.3.1.29   POMEGRANATE

Pomegranate fruit parts have bioactive compounds which are known to possess enormous AOA. Pomegranate juice is effective in prevention of atherosclerosis, low-density lipoprotein oxidation, prostate cancer, platelet aggregation, and various cardiovascular diseases. Ozkal and Dinc (1994) reported the presence of tannins, anthocyanins, and flavonoids in pomegranate rind. Pomegranate peel is a rich source of tannins and other phenolic compounds. The utilization of pomegranate fruits for meat processing and its potential health benefits are not well understood. The meat industry can use these fruits or fruit byproducts as a potential source of phenolics as they have immense nutraceutical value and can be used to produce functional meat products of commercial interest (Naveena et al., 2008; Vaithiyanathan et al., 2011).

5.3.1.30   POTATO PEEL EXTRACT (PPE)

The effective utilization of potato peel, a waste generated in large quantities by the food industry, as an antioxidant in radiation processed lamb meat was investigated by Kanatt et al. (2005). PPE has a high phenolic content (70.82 mg, catechin equivalent/100 g), chlorogenic acid (27.56 mg/100 g of sample) is the major component. The yield of total phenolics and chlorogenic acid increased by 26 and 60%, respectively, when the extract was prepared from γ irradiated (150 Gy) potatoes. The AOA of PPE was found to be comparable to BHT.

5.3.1.31   RAPESEED MEAL

Rapeseed meal contains α-tocopherol (52 µg/g) and phenolic hydroxycinnamic acid derivatives including sinapine (2400–2900 µg/g) and sinapic acid (280 µg/g) which inhibit hexanal formation (≥85%). It is also used in combination with commercial CO_2 extract of rosemary (0.04 g/100 g meat) and was excellent in prevention of oxidation of meat lipids (Salminen et al., 2006).

5.3.1.32   RICE HULL EXTRACT

Rice hull can be an attractive protective source because it contains many easily extractable antioxidant compounds. Furthermore, radiation of rice hull with far infrared (FIR) for 2 h increased the content of phenolic compounds in extract. FIR radiation onto rice hull is reported to liberate and activate covalently bound phenolic compounds that have antioxidant activities. Therefore, rice hull extract treated by FIR can be a good candidate to be used in irradiated meat systems as a natural antioxidant. The antioxidant effect of FIR treated rice hull (Lee et al., 2003) extracts (FRH) was compared with that of sesamol and rosemary oleoresin in irradiated turkey breast meat. FRH significantly decreased TBARS values and volatile aldehydes (hexanal, pentanal, and propanal) and was effective in reducing the production of dimethyl disulfide responsible for irradiation off-odor in irradiated raw and cooked turkey meat during aerobic storage. The AOA of FRH (0.1%, w/w) was as effective as that of rosemary oleoresin (0.1%). However, the addition of FRH increased red and yellow color intensities and produced an off-odor characteristic to rice hull in raw and cooked meat, and cannot be used in meat without further refining process to remove off-color and off-odor compounds to increase its applicability as an antioxidant.

5.3.1.33 ROSEMARY

The AOA of rosemary (*Rosmarinus officinalis* L.) extracts has been known for the last 30 years and is due to presence of phenolic compounds, carnosol, carnosic acid, rosmanol, isorosmanol, rosmariquinone, rosmaridiphenol, and rosmary-diphenol. The phenolic substances react with lipid or hydroxyl radicals and convert them into stable products. Rosemary extracts can also chelate metal ions, such as Fe^{2+}, resulting in a reduced rate of formation of activated oxygen (Formanek et al., 2003). It is four times more effective than BHA and equal to BHT as antioxidants but less than TBHQ. It also improved the color stability of cooked turkey rolls. Rosemary extract has also been used in the combination of various other antioxidants (McBride et al., 2007) to have synergistic effect. However, some of the compounds in rosemary (verbenone, borneol, and camphor) can impart an undesirable rosemary odor to foods, even at low concentrations.

A great deal of research on the antioxidant properties of rosemary extract in different food systems has been carried out, which clearly demonstrated the effectiveness of its bioactives compounds with greater acceptability by the consumers. Moreover, several authors reported that some compounds such as phenolic di-terpenoids present in rosemary extracts have antibacterial activity (Cuvelier et al., 1994).

5.3.1.34 SAGE

Sage (*Salvia officinalis* L.) contains a variety of antioxidative substances including carnosol, rosmanol, rosemadiol, epirosmanol, isorosmanol, galdosol, and carnosic acid. The antioxidative activity of sage oil correlates with the oxygenated diterpene and sesquiterpene concentration. They are more potent in cooked meat than in raw meat (Fasseas et al., 2008). The ethanolic extract of sage can reduce both peroxide oxygen and TBARS values. The polar extracts of the *Salvia* species exhibit better antioxidant activities than its corresponding non-polar sub-fractions and that was comparable to the antioxidant activities of BHT.

5.3.1.35 SEABUCKTHORN

Seabuckthorn (*Hippophae rhamnoides*) berry extracts are rich in antioxidant polyphenols mainly flavonols, which are stable during short-term cooking. Pussa et al. (2008) reported AOA of seabuckthorn in mechanically deboned broiler meat.

5.3.1.36 SESAME OIL

Sesamol (500–2000 µg/ml) of sesame oil educed lipid oxidation in bovine muscle model systems by an average of 97% (Joshi et al., 2005). They further added that sesamol at a concentration of 90 µM inhibited Fe (III)-induced oxidation by 99%. Nam and Ahn (2003) confirmed sesamol at 0.01% as superior antioxidant than gallate, tocopherol, and carnosine in pork homogenate.

5.3.1.37 SOY PROTEIN HYDROLYSATES

These are very frequently utilized in meat products as functional ingredients. Soy protein isolates could inhibit TBARS formation in an iron-catalyzed liposomal system by as much as 65% depending on the proteases and hydrolysis conditions used. Yet, these protein hydrolysates as potential antioxidants in meat products (in situ) have not been validated.

5.3.1.38   THYME

A number of species of thyme (*Thymus vulgaris, T caespititius, T. camphorates, T. mastichina*) have antioxidative activity; all contain 1, 8-cineole, α-terpineol, linalool, carvacrol, and thymol (Lee et al., 2005). The antioxidative activity of *T. caespititius* (250 and 500 mg/l) has been reported to be comparable to that of vit. E and BHT. Among the compounds isolated from thyme, the order of antioxidative activity is thyme oil > thymol > carvacrol > gamma-terpinene > myrcene > linalool > p-cymene > limonene > 1, 8-cineole > α-pinene (Youdim et al., 2002). Thyme essential oil exhibits very strong free radical scavenging ability and inhibits Fe^{2+}/ascorbate and Fe^{2+}/H_2O_2 induced lipid oxidation (Bozin et al., 2006).

5.3.1.39   TOMATO

The active compound of tomatoes is lycopene. It is highly effective antioxidant owing to its ability to act as free radical scavenger and has the highest singlet oxygen quenching rate than all carotenoids tested in biological system. It can be used as tomato paste, sun dried tomato, tomato peel, tomato puree. Osterlie and Lerfall (2005) confirmed the antioxidant effect of tomato paste in meat products and dry tomato peel in dry fermented sausage. Calvo et al. (2008) reported that tomato paste can be successfully used up to a level of 12% without any negative effect on the processing and quality characteristics of the product throughout its storage. It is also effective in limiting the use of nitrite in cured meat products. Moreover, lycopene provides protection against many chronic diseases such as cancer and cardiovascular disease.

5.3.2   ANIMAL SOURCES

5.3.2.1   BONE PROTEIN HYDROLYSATES

These can be prepared by limited alcalase hydrolysis. It possessed significant AOA, and AOA increased with the increasing hydrolysates concentration. Bone protein hydrolysates at 2% application level are able to not only stabilize meat color but also suppress lipid oxidation in pork patties during storage improving sensory quality. Therefore, bone protein hydrolysates can be used as potent natural and cheap antioxidant for meat and meat products.

Further studies are needed to identify the specific compounds in hydrolysate that are responsible for the overall antioxidative capability.

5.3.2.2 CARNOSINE

Carnosine is a naturally occurring skeletal muscle dipeptide, which consists of alanine and histidine. Its function in muscle is not completely understood, but it is thought to act as both a buffering agent and as an antioxidant. The antioxidant mechanism has been postulated to be a combination of its ability to act as a chelator, free radical scavenger, and hydrogen donor. Carnosine is water soluble, thus permitting the inactivation of lipid oxidation catalysts such as heme pigments, iron, lipoxygenase and singlet oxygen, and free radicals in the aqueous phase of muscle. The color protecting effects of carnosine were greater than BHT, α-tocopherol, or sodium tripolyphosphate. Some researchers also investigated the interactions between carnosine and the different redox states of myoglobin and concluded that it does not stabilize oxymyoglobin or significantly catalyze the reduction of met-myoglobin formation. Djenane et al. (2004) concluded that the combination of carnosine (50 mM) with AA (500 ppm) provided the best antioxidant protection for meat during refrigerated storage. Surface application of carnosine or AA combination or alone resulted in an effective delay of oxidative deterioration of meat.

5.3.2.3 CHITOSAN

Chitosan, a linear polymer of 2-amino-2-deoxy-β-D-glucan, is a deacetylated form of chitin, a naturally occurring cationic biopolymer. It occurs as a shell component of crustaceans (crab and shrimp), as the skeletal substance of invertebrates, and as the cell wall constituent of fungi and insects. Chitosan possesses a positive ionic charge which gives its ability to bind with negatively charge fat, lipid, protein, metal ions, and macromolecules. Chitosan is GRAS by the US FDA. Chitosan retards lipid oxidation by eliminating the pro-oxidant activity of ferrous ions. The AOA of chitosan could also be due to its chelating ability with free iron released from myoglobin degradation. Soultos et al. (2008) indicated that chitosan concentration of 1% in pork sausages show 80% decrease in MDA level after 14 days while that retard lipid oxidation upto 70% in meat products after three day storage at 4 °C.

5.3.2.4   MILK PROTEINS

Milk and milk components have been frequently used in the enhancement of nutritional and technological properties of a wide variety of foods. A feasible application of peptides or hydrolysates as antioxidants is being explored especially in muscle foods. The phosphorylated caseinophospho-peptides (CPP) have the ability to scavenge free peroxyl radicals as well as to chelate transition metals such as iron, copper, and zinc (Kim et al., 2007). This is positively correlated with the amounts of histidine, lysine, proline, and tyrosine. Incorporation of casein calcium peptides (2%) inhibited about 70% of lipid oxidation and prevents formation of an off-color in meat products. Rossini et al. (2009) suggested that casein peptides (20 mg/ml) effectively inhibited lipid per-oxidation in ground meat homogenates and mechanically deboned meat. As cooking increases the catalytic activity of iron, the stronger chelating activity of enriched CPP may make them more effective antioxidants in cooked muscle foods (Diaz & Decker, 2004). Whey hydrolysates may also act as potential antioxidants in meat products. Pena-Ramos and Xiong (2003) evaluated the AOA of selected whey hydrolysates in cooked pork patties. The results indicated that at an application level of 2%, the whey protein isolates and their hydrolytic products not only reduced the cooking loss but also suppressed lipid oxidation in cooked pork patties during refrigerated storage. Notably, hydrolysis with protamex improved the capability of whey protein to inhibit early-stage lipid oxidation (formation of hydroperoxides or conjugated dienes) as well as to retard propagation of the oxidation process. Therefore, milk proteins can be superlatively used as nutrient, color enhancer as well as antioxidant in processed muscle foods.

5.4   MODE OF APPLICATION OF ANTIOXIDANTS

5.4.1   DIETARY SUPPLEMENTATION

Dietary supplementation of vit. E and TC is very commonly used to improve the meat quality. The antioxidant activities of many plants, vegetables, herbs and their essential oils (eugenol, thymol, and carvacrol), flavonoids, cyanidine glycosides, etc. are also elucidated. It is considered that the AOA of these compounds is due to their high redox properties and chemical structure, which can be responsible for neutralizing free radicals, chelating transitional metals, and quenching singlet and triplet oxygen by delocalization or decomposing peroxides. Dietary supplementation of rosemary and sage extracts or oregano essential oil in broilers could improve the oxidative stability of raw and precooked meat products during storage. However, combination of oregano and rosemary essential oils had greater effect than those fed individually or α-tocopheryl acetate on inhibiting lipid oxidation. These activities could have resulted from the action of antioxidant compounds such as the phenolic isomers, thymol, and carvacrol found in the extracts and essential oils from the aromatic plants.

Addition of drumstick leaf extracts in feed could reflect concentrations of AA (vit. C), α-tocopherol (vit. E), beta carotene (vit. A precursor), various flavonoids, and other phenolic compounds of broiler meat. So, broiler produced with this leaf revealed potent AOA, and also improve color and flavor of meat products.

The production of lipid peroxides in the carcasses of broiler chickens during storage could be delayed by astaxanthin. Therefore, astaxanthin either synthetic [45 mg/kg feed) or naturally occurring (from the red yeast *Xanthophyllomyces dendrorhous*, 22.5 mg/kg feed)] could be used as an antioxidant as well as a colorant for broiler chickens (Ahn & Nam, 2004). Xanthophyll (natural or synthetic) sources to chicks could improve color intensity of and inhibit lipid oxidation in leg meat. The antibacterial agent, lasalacid sodium salt (LS) added to the diet affected not only their growth performance but also their meat yield and stress response. Withdrawal of LS increased the plasma 2-thiobarbituric acid reactant value, an indicator of lipid per-oxidation.

5.4.2   DIRECTION ADDITION TO THE PRODUCT

Most of the antioxidants whether in the form of extracts, powders, or any other form are added directly. It is the most commonly used method.

5.4.3   SPRAYING

The antioxidants directly or their active principle are sprayed over the surface of the meat. Direct addition of a natural rosemary extract on to the meat surface by spraying 2 ml pure extract diluted in 150 ml n-pentane, according to a ratio of 2 ml solution to 50 g meat, gave rise to a significant decrease of color loss and lipid oxidation spraying of vit. E directly on turkey meat resulted in a lower myoglobin oxidation.

5.4.4   ACTIVE PACKAGE

It is another method that does not involve direct addition of the active agents to the product. Antimicrobial agents in active packaging have been reported (Appendini & Hotchkiss, 2002), but studies on antioxidant active packaging are rarer. Nerín et al. (2006) described the promising results of a new antioxidant active packaging system; a plastic film with an embodied rosemary extract was able to inhibit both myoglobin and lipid oxidation in red meat, leading to enhanced display life of meat. The mechanism of antioxidant active pack is currently under investigation. Generally postulated hypothesis states that mechanism involves inactivation of free radicals either by migration of antioxidant molecules from the active film to the meat or by scavenging of those oxidant molecules from the meat onto the active film. Pezo et al. (2008) has demonstrated that active films react with headspace free radicals. This allows to envisaging antioxidant active packaging with oregano as a promising tool for increasing the display life of lamb and other meats in retail sale. However, the legal regulatory status of active packs is so far not clear and needs to be specifically addressed.

5.4.5   MARINATION

Various marinades according to consumers' acceptance are being incorporated with antioxidants such as turmeric rhizomes, tamarind, lemon grass, etc.

5.4.6   ENROBING

Enrobing is the process of making "further processed products" by applying edible coating on the products in two distinct steps, that is, battering and breading (Ahamed et al., 2007). Enrobing improves display attributes and imparts crispy texture. It provides variety to the meat products with improved juiciness, tenderness, sensory quality, nutritive value, shelf life, and reduced total product cost, moisture, and weight loss. In addition, enrobing material is used as a carrier for various antimicrobial and antioxidant substances (Shelef & Liang, 1982; Giridhar & Reddy, 2001; Biswas et al., 2004). This approach can be used to impart a strong localized functional effect without elevating excessively the overall concentration of an additive in the food (Guilbert et al., 1985).

5.5   SYNERGISTIC EFFECT OF NATURAL ANTIOXIDANTS

Evidence based on pulse radiolysis technique and electron spin resonance studies supports a redox mechanism, involving reduction of the tocopheroxyl radical intermediate by AA to regenerate tocopherol. By this synergistic mechanism, α-tocopherol and AA can mutually reinforce each other by regenerating the oxidized form of the other. Another mechanism of synergism involves metal inactivating effect of AA. A mixture of α-tocopherol, AA, or ascorbyl palmitate and phospholipids are also known for their good synergistic effects leading to inhibition of pigment as well as lipid oxidation. Ascorbate alone enhanced the lipid oxygen radical propagation but its pro-oxidant effect was reversed when TA was added. Yin et al. (1993) reported that 10 ppm of vit. E and 500 ppm of vit. C showed the strongest synergism to inhibit pigment and lipid oxidation in meat. Adding rosemary to vit. E supplemented meat resulted in lower TBARS than either of the compound alone suggesting a synergistic effect (McBride et al., 2007). Mansour et al. (2006) reported synergism of rosemary and marjoram in minced meat. Combination of oil seed by-products such as rapeseed meal (0.5 g/100 g meat) or camelina meal (0.7 g/100 g) along with CO_2 extract of rosemary (0.04 g/100 g meat) showed a higher inhibition of lipid oxidation than that of either alone. A very pronounced synergistic effect exists between rosemary and citric acid. Combinations of oregano and rosemary essential oils (150 mg/kg each) had a greater effect than those fed individually or α-tocopheryl acetate (200 mg/kg) on inhibiting lipid oxidation, and protecting α-tocopheryl acetate concentration in refrigerated meat enriched with n-PUFAs stored for 15 days. This combination of essential oils also proved as effective as α-TA in retaining the sensory qualities of breast meat after 15 days of storage, and was more effective than when these essential oils were fed individually or at 300 mg/kg. There is a possible synergistic effect between oregano and rosemary essential oils in preventing lipid oxidation in stored meat enriched with n-3 PUFAs. However, synergistic effects between various antioxidants need to be exploited in the meat industry.

5.6   MEASUREMENT OF EFFICACY OF AOA

Efficacy of natural antioxidants can be evaluated by various physical and chemical methods. Physical evaluation of meat color and odor can be measured by sensory panelists using score card. However, it is always ensured that the sensory panelists can able to properly recognize the color and odor and able to distinguish them for different meat species. Amongst the chemical methods, total AOA is measured by ferric reducing antioxidant power (FRAP) assay. FRAP assay uses antioxidants as reductants in a redox-linked colorimetric method, employing an easily reduced oxidant system present in stoichiometric excess. At low pH, reduction of ferric tripyridyl-triazine (Fe III–TPTZ) complex to ferrous form (which has an intense blue color) can be monitored by measuring the change in absorption at 593 nm. The reaction is non-specific, in that any half reaction that has lower redox potential, under reaction conditions, than that of ferric ferrous half reaction, will drive the ferrous (Fe III to Fe II) ion formation. The change in absorbance is, therefore, directly related to the combined or "total" reducing power of the electron donating antioxidants present in the reaction mixture. The ability to scavenge DPPH radical by added antioxidant can be measure at 517 nm wavelength to know efficacy of antioxidant compound. DPPH can make stable free radicals in aqueous or ethanol solution. However, fresh DPPH solution should be prepared before every measurement. Super-oxide anion radical scavenging assay is based on the reduction of nitro blue tetrazolium (NBT) in the presence of nicotinamide adenine dinucleotide (NADH) and phenazonium methosulfate (PMS) under aerobic condition at room temperature under the dark. Oxygen radical absorption capacity (ORAC) assay is the measure of the oxidative degradation of the fluorescent molecule (either beta-phycoerythrin or fluorescein) after being mixed with free radical generators such as azo-initiator compounds. Azo-initiators are considered to produce the per-oxiradical by heating, which damages the fluorescent molecule, resulting in the loss of fluorescence. Antioxidants are considered to protect the fluorescent molecule from the oxidative degenera-tion. The degree of protection is quantified using a fluorometer. Fluorescein is currently used most as a fluorescent probe. Equipment that can automati-cally measure and calculate the capacity is commercially available. ABTS method is based on the ability of antioxidants to quench the long-lived ABTS radical cation, a blue/green chromophore with characteristic absorp-tion at 734 nm, in comparison to that of standard antioxidants. ABTS was dissolved in water to a 7 mM concentration. ABTS radical cation (ABTS⁺) was produced by reacting ABTS stock solution with 2.45 mM potassium persulfate (final concentration) and allowed the mixture to stand in the dark at room temperature for 16 h before use. As ABTS and potassium persul-fate react sterio-chiometrically at a ratio of 1:0.5 (mol mol⁻¹), this results in complete oxidation of ABTS. Oxidation of ABTS commenced immedi-ately, but the absorbance was not maximal and stable until 6 h had elapsed.

The radical was stable in this form for more than two days, when stored in the dark at room temperature. Prior to use, the stock solution was diluted with ethanol to an absorbance of 0.70 at t_0 (t = 0 min) and equilibrated at 30 °C exactly 6 min after initial mixing. The TBARS is used to determine primary oxidation products in terms of mg malonaldehyde per kg of meat sample while linoleic acid accelerated by azo-initiators (LAOX) is used to determine oxidation of an aqueous dispersion of linoleic acid accelerated by azo-initiators.

5.7 CONCLUSIONS

Natural compounds enjoy positive consumer image and have applications in development of novel functional meat products. Antioxidants are nature's defense against the damaging effects of free radicals for health and to extend shelf life of food products. Still there is various natures' gift for the mankind need to be explored. The appropriate combinations of natural antioxidants can help in designing of functional foods and also tackle lipid oxidation problems. There are various microbial peptides and plants which need to be explored as antioxidant and scientists are developing niche based technologies, combinations, and delivery vehicles for the incorporation of effective and economic usage of natural antioxidants.

KEYWORDS

- **natural antioxidant**
- **lipid oxidation**
- **chicken meat**
- **turkey**
- **broiler**
- **poultry**
- **phenolics**
- **flavonoids**

196 Natural Antioxidants: Applications in Foods of Animal Origin

REFERENCES

Ahamed, M. E.; Anjaneyulu, A. S. R.; Sathu, T.; Thomas, R.; Kondaiah, N. Effect of Different Binders on the Quality of Enrobed Buffalo Meat Cutlets and Their Shelf Life at Refrigeration Storage (4 ± 1°C). *Meat Sci.* **2007,** *75,* 451–459.

Ahn, D. U.; Nam, K. C. Effects of Ascorbic Acid and Antioxidants on Color, Lipid Oxidation and Volatiles of Irradiated Ground Beef. *Radiat. Phys. Chem.* **2004,** *71,* 151–156.

Antony, S.; Rieck, J. R.; Acton, J. C.; Han, I. Y.; Halpin, E. L.; Dawson, P. L. Effect of Dry Honey on the Shelf Life of Packaged Turkey Slices. *Poult. Sci.* **2006,** *85,* 1811–1820.

Appendini, P.; Hotchkiss. J. H. Review of Antimicrobial Food Packaging. *Innov. Food Sci. Emerg. Technol.* **2002,** *3,* 113–126.

Ayerdi, S. G.; Brenes, A.; Goñi, I. Effect of Grape Antioxidant Dietary Fiber on the Lipid Oxidation of Raw and Cooked Chicken Hamburgers. *LWT - Food Sci. Technol.* **2009,** *42,* 5971–5976.

Aziza A. E.; Quezada, N.; Cherian, G. Antioxidative Effect of Dietary Camelina Meal in Fresh, Stored, or Cooked Broiler Chicken Meat. *Poult. Sci.* **2010,** *89,* 2711–2718

Bekhit, A. E. D.; Geesink, G. H.; Ilian, M. A.; Morton, J. D.; Bickerstaffe, R. The Effects of Natural Antioxidants on Oxidative Processes and Metmyoglobin Reducing Activity in Beef Patties. *Food Chem.* **2003,** *81,* 175–187.

Biswas, A. K.; Keshri, R. C.; Bisht, G. S. Effect of Enrobing and Antioxidants on Quality Characteristics of Precooked Pork Patties under Chilled and Frozen Storage Conditions. *Meat Sci.* **2004,** *66,* 733–744.

Biswas, A. K.; Kondaiah, N.; Anjaneyulu A. S. R. Effect of Spice Mix and Curry (*Murraya koenigii*) Leaf Powder on the Quality of Raw Meat and Precooked Chicken Patties During Refrigeration Storage. *J. Food Sci. Technol.* **2006,** *43,* 438–441.

Bozin, B.; Mimica-Dukic, N.; Simin, N.; Anackov, G. Characterization of the Volatile Composition of Essential Oils of Some Lamiaceae Spices and the Antimicrobial and Antioxidant Activities of the Entire Oils. *J. Agric. Food Chem.* **2006,** *54,* 1822–1828.

Brannan, R. G.; Mah, E. Grape Seed Extract Inhibits Lipid Oxidation in Muscle from Different Species During Refrigerated and Frozen Storage and Oxidation Catalyzed by Peroxynitrite and Iron/Ascorbate in a Pyrogallol Red Model System. *Meat Sci.* **2007,** *77,* 540–546

Calvo, M. M.; Garcia, M. L.; Selgas, M. D. Dry Fermented Sausages Enriched with Lycopene from Tomato Peel. *Meat Sci.* **2008,** *80,* 167–72.

Camo, J.; Beltran, J. A.; Roncales, P. Extension of the Display Life of Lamb with an Antioxidant Active Packaging. *Meat Sci.* **2008,** *80,* 1086–1091.

Carlos, A. M. A.; Harrison, M. A. Inhibition of Selected Microorganisms in Marinated Chicken by Pimento Leaf Oil and Clove Oleoresin. *J. Appl. Poult. Res.* **1999,** *8,* 100–109.

Carpenter, R.; O'Callaghan, Y.; O'Grady, M.; Kerry, J.; O'Brien, N. Modulatory Effects of Resveratrol, Citroflavan-3-ol, and Plant-Derived Extracts on Oxidative Stress in U937 Cells. *J. Med. Foods.* **2006,** *9,* 187–195.

Cuvelier, M. E.; Berset, C.; Richard, H. Separation of Major Antioxidants in Sage by High-Performance Liquid Chromatography. *Sci. Aliments.* **1994,** *1,* 811–815.

Diaz, M.; Decker, E. A. Antioxidant Mechanisms of Caseino phosphopeptides and Casein Hydrolysates and Their Application in Ground Beef. *J. Agric. Food Chem.* **2004,** *52,* 8208–8213.

Djenane, D.; Martínez, L.; Sánchez-Escalante, A.; Beltrán, J. A.; Roncalés, P. Antioxidant Effect of Carnosine and Carnitine in Fresh Beef Steaks Stored under Modified Atmosphere. *Food Chem.* 2004, *85,* 453–459.

Dorman, H. J. D.; Kosar, M.; Kahlos, K.; Holm, Y.; Hiltunen, R. Antioxidant Properties and Composition of Aqueous Extracts from *Mentha* Species, Hybrids, Varieties, and Cultivars. *J. Agric. Food Chem.* **2003,** *51,* 4563–4569.

El-Alim, S. S. L. A.; Lugasi, A.; Hovari, J.; Dworschak, E. Culinary Herbs Inhibit Lipid Oxidation in Raw and Cooked Minced Meat Patties During Storage. *J. Sci. Food Agr.* **1999,** *79,* 277–285.

Evans, P.; Halliwell, B. Micronutrients: Oxidant/Antioxidant Status. *Brit. J. Nutr.* **2001,** *85,* Suppl, S 627.

Fasseas, M. K.; Mountzouris, K. C.; Tarantilis, P. A.; Polissiou, M.; Zervas, G. Antioxidant Activity in Meat Treated with Oregano and Sage Essential Oils. *Food Chem.* **2008,** *106,* 1188–1194.

Formanek, Z.; Lynch, A.; Galvin, K.; Farkas, J.; Kerry, J. P. Combined Effects of Irradiation and the Use of Natural Antioxidants on the Shelf-Life Stability of Overwrapped Minced Beef. *Meat Sci.* **2003,** *63,* 433–440.

Galobart, G.; Barroeta, A. C.; Baucells, M. D.; Guardiola, F. Lipid Oxidation in Fresh and Spray-Dried Eggs Enriched with Omega3 and Omega6 Polyunsaturated Fatty Acids During Storage as Affected by Dietary Vitamin E and Canthaxanthin Supplementation. *Poult. Sci.* **2001,** *80,* 327–337.

Galobart, J.; Barroeta, A. C.; Baucells, M. D.; Guardiola F. Vitamin E Levels and Lipid Oxidation in n3 Fatty Acids Enriched Eggs. In *Materials of the VIII European Symposium on the Quality of Eggs and Egg Products;* Bologna, Italy, September, 1999; pp 19–23.

Giridhar, P.; Reddy, S. M. S. Phenolic Antioxidants for the Control of Fungi. *J. Food Sci.* **2001,** *38,* 397–399.

Goni, I.; Brenes, A.; Centeno, C.; Viveros, A.; Saura-Calixto, F.; Rebole, A.; Arija, I.; Estevez, R. Effect of Dietary Grape Pomace and Vitamin E on Growth Performance, Nutrient Digestibility, and Susceptibility to Meat Lipid Oxidation in Chickens. *Poult. Sci.* **2007,** *86,* 508–516.

Govaris, A.; Botsoglou, N.; Papageorgiou, G.; Botsoglou, E.; Ambrosiadis, I. Dietary Versus Post-Mortem Use of Oregano Oil and/or α-Tocopherol in Turkeys to Inhibit Development of Lipid Oxidation in Meat During Refrigerated Storage. *Int. J. Food Sci. Nutr.* **2004,** *55*(2), 115–123.

Gray, J. I.; Gomaa, E. A.; Buckky, D. J. Oxidative Quality and Shelf Life of Meats. *Meat Sci.* **1996,** *43,* S111–S123.

Grobbel, J. P.; Dikeman, M. E.; Yancey, E. J.; Smith, J. S.; Kropf, D. H.; Milliken, G. A. Effects of Ascorbic Acid, Rosemary, and Origanoxin Preventing Bone Marrow Discoloration in Beef Lumbar Vertebrae in Aerobic and Anaerobic Packaging Systems. *Meat Sci.* **2006,** *72,* 47–56.

Guilbert, S.; Giannakopoulos, A.; Cheftel, J. C. Diffusivity of Sorbic Acid in Food Gels at High and Intermediate-Water Activities. In *Properties of Water in Foods in Relation to Quality and Stability;* Simatos, D., Multon, J. L., Eds.; Martinus Nijhoff Publishing: Dordrecht, 1985; p 343.

Hall, G. Interactions Between Products of Lipid Oxidation and Proteins. *Food Sci. Technol.* **1987,** *1,* 155–158.

Hao, Y. Y.; Brackett, R. E.; Doyle, M. P. Efficacy of Plant Extracts in Inhibiting *Aeromonas hydrophila* and *Listeria monoctyogenes* in Refrigerated, Cooked Poultry. *Food Microbiol.* **1998,** *15,* 367–378.

Hashim, I. B.; McWatters, K. H.; Hung, Y. C. Marination Method and Honey Level Affect Physical and Sensory Characteristics of Roasted Chicken. *J. Food Sci.* **1999,** *64,*163–166.

Hernandez, H. E.; Alquicira, P. E.; Flores, J. M. E.; Legarreta, G. I. Antioxidant Effect Rosemary (*Rosmarinus officinalis* L.) and Oregano (*Origanum vulgare* L.) Extracts on TBARS and Color of Model Raw Pork Batters. *Meat Sci.* **2009,** *81,* 410–417.

Ismail, H. A.; Lee, E. J.; Ko, K. Y.; Ahn, D. U. Effects of Aging Time and Natural Antioxidants on the Color, Lipid Oxidation and Volatiles of Irradiated Ground Beef. *Meat Sci.* **2008,** *80,* 582–591.

Joshi, R.; Kumar, M. S.; Satyamoorthy, K.; Unnikrisnan, M. K.; Mukherjee, T. Free Radical Reactions and Antioxidant Activities of Sesamol: Pulse Radiolytic and Biochemical Studies. *J. Agric. Food Chem.* **2005,** *53,* 2696–2703.

Kanatt, S. R.; Chander, R.; Sharma, A. Antioxidant Potential of Mint (*Mentha spicata* L.) in Radiation-Processed Lamb Meat. *Food Chem.* **2007,** *100,* 451–458.

Kanatt, S. R.; Chander. R.; Radhakrishna, P.; Sharma, A. Potato Peel Extract – a Natural Antioxidant for Retarding Lipid Peroxidation in Radiation Processed Lamb Meat. *J. Agric. Food Chem.* **2005,** *53,* 1499–1504.

Karastogiannidou C. Effects of Onion Quercetin on Oxidative Stability of Cook-Chill Chicken in Vacuum-Sealed Containers. *J. Food Sci.* **1999,** *64*(6), 978–981.

Kim, G. N.; Jang, H. D. Kim, C. I. Antioxidant Capacity of Caseino phosphopeptides Prepared from Sodium Caseinate Using Alcalase. *Food Chem.* **2007,** *104,* 1359–1365.

Konjufca, V. K.; Bottje, W. G.; Bersi, T. K.; Erf, G. F. Influence of Dietary Vitamin E on Phagocytic Functions of Macrophages in Broilers. *Poult. Sci.* **2004,** *83,*1530–1534.

Kumar, D.; Tanwar, V. K. Utilization of Clove Powder as Phytopreservative for Chicken Nuggets Preparation. *J. Stored Prod. Postharvest Res.* **2011,** *2,* 11–14.

Lau, D. W.; King, A. J. Pre- and Post-Mortem Use of Grape Seed Extract in Dark Poultry Meat to Inhibit Development of Thiobarbituric Acid Reactive Substances. *J. Agric. Food Chem.* **2003,** *51,* 1602–1607.

Lee, A. J.; Umano, K.; Shibamoto, T. Identification of Volatile Components in Basil (*Ocimum basilicum L.*) and Thyme Leaves (*Thymus vulgaris L.*) and Their Antioxidant Properties. *Food Chem.* **2005,** *91,* 131–137.

Lee, E. J.; Ahn, D. U. Quality Characteristics of Irradiated Turkey Breast Rolls Formulated with Plum Extract. *Meat Sci.* **2005,** *71,* 300–305.

Lee, S. C.; Kim, J. H.; Nam, K. C.; Ahn, D. U. Antioxidant Properties of Far Infrared-Treated Rice Hull Extract in Irradiated Raw and Cooked Turkey Breast. *J. Food Sci.* **2003,** *68,* 1904–1909.

Leuschner, R. G. K.; Ielsch, V. Antimicrobial Effects of Garlic, Clove and Red Hot Chilli on *Listeria monocytogenes* in Broth Model Systems and Soft Cheese. *Int. J. Food Sci. Nutri.* **2003,** *54,* 127–133.

Lohakare, J. D.; Choi, J. Y.; Kim, J. K.; Yong, J. S.; Shim, Y. H.; Hahn T. W.; Chae, B. J. Effects of Dietary Combinations of Vitamin A, E and Methionine on Growth Performance, Meat Quality and Immunity in Commercial Broilers. *Asian-Australas. J. Anim. Sci.* **2005,** *18,* 516–523.

Mahoney, J. R.; Graf, E. Role of Tocopherol, Ascorbic Acid, Citric Acid and EDTA as Oxidants in Meat Model Systems. *J. Food Sci.* **1986,** *5,* 1293–1296.

Mancini, R. A.; Hunt, M. C.; Seyfert, M.; Kropf, D. H.; Hachmeister, K. A.; Herald, T. J. Johnson, D. E. Effect of Antioxidant Solubility and Concentration on Discoloration of Beef Vertebrae Marrow During Display. *J. Food Sci.* **2006,** *71,* C489–C494.

Mansour, H. A.; Hussein, M. M.; Hania, F. G. E. Improving the Lipid Stability and Sensory Characteristics of Irradiated Minced Beef by Using Natural Herbal Extracts. *Vet. Med. J. Giza.* **2006,** *54,* 737–749.

Mariutti, L. R. B.; Orlien, V.; Bragagnolo, N.; Skibsted, L. H. Effect of Sage and Garlic on Lipid Oxidation in High-Pressure Processed Chicken Meat. *Eur. Food Res. Technol.* **2008,** *227,* 337–344.

McBride, N. T. M.; Hogan, S. A.; Kerry, J. P. Comparative Addition of Rosemary Extract and Additives on Sensory and Antioxidant Properties of Retail Packaged Beef. *Int. J. Food Sci. Technol.* **2007,** *42,* 1201–1207.

McKibben. J.; Engeseth, N. J. Honey as a Protective Agent Against Lipid Oxidation in Ground Turkey. *J. Agric. Food Chem.* **2002,** *50*(3), 592–595.

Mohammad, A. Textbook of Pharmacognosy. 111–115, cited by Kaithwas G, Kumar A, Pandey H, Acharya A J, Singh M, BhatiaD, Mukerjee A. Investigation of Comparative Antimicrobial Activity of Aloe Vera Gel and Juice. *Pharmacologyonline.* **2003,** *1,* 239–243.

Morel, D.; Lin, C. Y. Cellular Biochemistry of Oxysterols Derived from the Diet or Oxidation in Vivo. *J. Nutr. Biochem.* **1996,** *7,* 495–506.

Moyo, B.; Oyedemi, S.; Masika, P. J.; Muchenje, V. Polyphenolic Content and Antioxidant Properties of *Moringa oleifera* Leaf Extracts and Enzymatic Activity of Liver from Goats Supplemented with *Moringa oleifera* Leaves/Sunflower Seed Cake. *Meat Sci.* **2012,** *91,* 441–447.

Mytle, N.; Anderson, G. L.; Doyle, M. P.; Smith, M. A. Antimicrobial Activity of Clove (*Syzgium aromaticum*) Oil in Inhibiting *Listeria monocytogenes* on Chicken Frankfurters. *Food Control.* **2006,** *17,* 102–107.

Nam, K. C.; Ahn, D. U. Use of Antioxidants to Reduce Lipid Oxidation and Off-Odor Volatiles of Irradiated Pork Homogenates and Patties. *Meat Sci.* **2003,** *63,* 1–8.

Naveena, B. M.; Mendiratta, S. K. Tenderisation of Spent Hen Meat Using Ginger Extract. *Br. Poult. Sci.* **2001,** *42,* 344–349.

Naveena, B. M.; Sen, A. R.; Vaithiyanathan, S.; Babji, Y.; Kondaiah, N. Comparative Efficacy of Pomegranate Juice, Pomegranate Rind Powder Extract and BHT as Antioxidants in Cooked Chicken Patties. *Meat Sci.* 2008, 80 (4): 1304–1308.

Nerín, C.; Tovar, L.; Djenane, D.; Camo, J.; Salafranca, J.; Beltrán, J. A. Stabilization of Beef Meat by a New Active Packaging Containing Natural Antioxidants. *J. Agric. Food Chem.* **2006,** *54,* 7840–7846.

Ningappa, M. B.; Dinesha, R.; Srinivas, L. Antioxidant and Free Radical Scavenging Activities of Polyphenol-Enriched Curry Leaf (*Murraya koenigii* L.) Extracts. *Food Chem.* **2008,** *106,* 720–728.

Niu, Z.; Liu, F.; Yan Q.; Li, L. Effects of Different Levels of Selenium on Growth Performance and Immunocompetence of Broilers under Heat Stress. *Arch Anim Nutr.* **2009,** *6,* 56–65.

Nunez de Gonzalles, M. T.; Hafley, B. S.; Boleman, R. M.; Miller, R. K.; Rhee, K. S.; Keeton, J. T. Antioxidant Properties of Plum Concentrates and Powder in Precooked Roast Beef to Reduce Lipid Oxidation. *Meat Sci.* **2008,** *80,* 997–1004.

O'Connell, E. D.; Nolan, J. M.; Stack, J.; Greenberg, D.; Kyle, J.; Maddock, L. Diet and Risk Factors for Age-Related Maculopathy. *Am. J. Clin. Nutr.* **2008,** *87,* 712–722.

Osman, H.; Nazaruddin, R.; Lee, S. L. Extracts of Cocoa (*Theobroma cacao L.*) Leaves and Their Antioxidation Potential. *Food Chem.* **2004,** *86,* 41–46

Osterlie, M.; Lerfall, J. Lycopene from Tomato Products Added Minced Meat: Effect on Storage Quality and Color. *Food Res. Int.* **2005,** *38,* 925–29.

Ozkal, N.; Dinc, S. Evaluation of the Pomegranate (*Punica granatum* L.) Peels from the Standpoint of Pharmacy. *Ankara Univ. Eczacilik Fak. Derg.* **1994,** *22,* 21–29.

Packer, V. G.; Priscilla S.; Melo, P. S.; Bergamaschi, K. B.; Selani, M. M.; Villanueva, N. D. M.; de Alencar, S. M.; Contreras-Castillo, C. J. Chemical Characterization, Antioxidant

Activity and Application of Beetroot and Guava Residue Extracts on the Preservation of Cooked Chicken Meat. *J. Food Sci. Technol.* **2015,** *52,* 7409–7416. DIO: 10.1007/s13197-015-1854-8

Pena-Ramos, E. A.; Xiong, Y. L. Whey and Soya Protein Hydrolysates Inhibit Lipid Oxidation in Cooked Pork Patties. *Meat Sci.* **2003,** *64,* 259–263.

Pezo, D.; Salafranca, J.; Nerín, C. Determination of the Antioxidant Capacity of Active Food Packagings by in Situ Gas-Phase Hydroxyl Radical Generation and High-Performance Liquid Chromatography–Fluorescence Detection. *J. Chromatogr. A* **2008,** *1178,* 126–133.

Püssa, T.; Pällin, R.; Raudsepp, P.; Soidla, R.; Rei. M. Inhibition of Lipid Oxidation and Dynamics of Polyphenol Content in Mechanically Deboned Meat Supplemented with Sea Buckthorn (*Hippophae rhamnoides*) Berry Residues. Food Chem. 2008, 107, 714–721

Reya, A. I.; Hopiab, A.; Kivikarib, R.; Kahkonen, M. Use of Natural Food/Plant Extracts: Cloudberry (*Rubus chamaemorus*), Beetroot (*Beta vulgaris* "Vulgaris") or Willow Herb (*Epilobium angustifolium*) to Reduce Lipid Oxidation of Cooked Pork Patties. *LWT- Food Sci. Technol.* **2005,** *38,* 363–370.

Romero, A. M.; Doval, M. M.; Romero, M. C.; Sturla, M. A.; Judis, M. A. Antioxidant Properties of Dry Soya Sprout Hydrophilic Extracts. Application on Cooked Chicken Patties. *Electron. J. Environ. Agric. Food Chem.* **2008,** *7,* 3196–3206.

Rongsensusang, S. R. K.; Kondal, R. K.; Dhana, L. K. Effect of Ginger on Quality of Frozen Spent Hen Meat Balls. *J. Food Sci. Technol.* **2005,** *42,* 534–539.

Rossini, K.; Noreña, C. P. Z.; Cladera-Olivera, F.; Brandelli, A. Casein Peptides with Inhibitory Activity on Lipid Oxidation in Beef Homogenates and Mechanically Deboned Poultry Meat. *LWT—Food Sci. Technol.* **2009,** *42,* 862–867.

Sallam, K. I.; Ishioroshi, M.; Samejima K. Antioxidant and Antimicrobial Effects of Garlic in Chicken Sausage. *Lebenson. Wiss. Technol.* **2004,** *37,* 849–855.

Salminen, H.; Estevez, M.; Kivikari, R.; Heinonen, M. Inhibition of Protein and Lipid Oxidation by Rapeseed, Camelina and Soy Meal in Cooked Pork Meat Patties. *Eur. Food Res. Technol.* **2006,** *223,* 461–68.

Sakanaka, S.; Tachibana, Y.; Ishihara, N.; Juneja, L. R. Antioxidant Properties of Casein Calcium Peptides and Their Effects on Lipid Oxidation in Beef Homogenates. *J. Agric. Food Chem.* **2005,** *53,* 464–468.

Schmidt, E.; Bail, S.; Buchbauer, G.; Stoilova, I.; Krastanov, A.; Stoyanova, A.; Jirovetz, L. Chemical Composition, Olfactory Evaluation and Antioxidant Effects of the Essential Oil of *Origanum majorana* L. from Albania. *Nat. Prod. Commun.* **2008,** *3,* 1051–1056.

Sepe, H. A.; Faustman, C.; Lee, S.; Tang, J.; Suman, S. P.; Venkitanarayanan, K. S. Effects of Reducing Agents on Premature Browning in Ground Beef. *Food Chem.* **2005,** *93,* 571–576.

Shafi, N.; Khan, L; Khan, G. A. Commercial Extraction of Gel from Aloe Vera Leaves. *J. Chem. Soc. Pak.* **2000,** *22,* 47–48.

Shahzad, K.; Ahmad, R.; Nawaz, S.; Saeed, S.; Iqbal, Z. Comparative Antimicrobial Activity of Aloe Vera Gel on Microorganisms of Public Health Significance. *Pharmacologyonline* **2009,** *1,* 416–423.

Shelef, L. A.; Liang, P. Antibacterial Effect of BHA Against Bacillus Species. *J. Food Sci.* **1982,** *47,* 796–799.

Singh, A.; Sharma, P. K.; Garg, G. Natural Products as Preservatives. *IJPBS.* **2010,** *1,* 601–610.

Smith, L. L.; Johnson, B. T. Biological Activities of Oxysterols. *Free Radic. Biol. Med.* **1989,** *7,* 285–302.

Soultos, N.; Tzikas, Z.; Abrahim, A.; Georgantelis, D.; Ambrosiadis, I. Chitosan Effects on Quality Properties of Greek Style Fresh Pork Sausages. *Meat Sci.* **2008,** *80,* 1150–1156.

Tang, S. Z.; Kerry, J. P.; Sheehan, D.; Buckley, D. J.; Morrissey, P. A. Dietary Tea Catechins and Iron-Induced Lipid Oxidation in Chicken Meat, Liver and Heart. Meat Sci. **2000,** *56,* 285–290.

Tang, S. Z.; Kerry, J. P.; Sheehan, D.; Buckley, D. J.; Morrissey, P. A. Antioxidative Effect of Added Tea Catechins on Susceptibility of Cooked Red Meat, Poultry and Fish to Lipid Oxidation. *Food Res. Int.* **2001,** *34,* 651–57.

Tang, X.; Cronin, D. A. The Effects of Brined Onion Extracts on Lipid Oxidation and Sensory Quality in Refrigerated Cooked Turkey Breast Rolls During Storage. *Food Chem.* **2007,** *100,* 712–718.

Taylor, A.; Vega, L.; Wood, J. D.; Angold, M. In *Extending Color Shelf Life of MA Packed Beef by Supplementing Feed with Vitamin E.* Proceedings of the 40th International Congress of Meat Science and Technology, Vol. IVA, paper 44. 1994.

Tserveni-Gousi, A.; Yannakopoulos, A.; Botsoglou, N.; Chrsitaki, E.; Florou-Paneri, P.; Yannakakis, E. Sensory Evaluation and Oxidative Stability of n-3 Fatty Acid Enriched Eggs in Greece. Materials of the XXII World's Poultry Congress, 8–13 June, 2004. Istanbul, Turkey.2004.

Vaithiyanathan, S.; Naveena, B. M.; Muthukumar, M.; Girish, P. S.; Kondaiah, N. Effect of Dipping in Pomegranate (*Punica granatum*) Fruit Juice Phenolic Solution on the Shelf Life of Chicken Meat under Refrigerated Storage (4°C). *Meat Sci.* **2011,** *88*(3), 409–414.

Wanasundara, U. N.; Shahidi F. Canola Extract as an Alternative Natural Antioxidant for Canola Oil. *J. Am. Oil Chem. Soc.* **1994,** *71,* 817–822.

Weber, H. A.; Hodges, A. E.; Guthrie, J. R.; O'Brien, B. M.; Robaugh, D.; Clark, A. P. Comparison of Proanthocyanidins in Commercial Antioxidants: Grape Seed and Pine Bark Extracts. *J. Agric. Food Chem.* **2007,** *55,* 148–156.

Yadav, A. S.; Pandey, N. K.; Singh, R. P.; Sharma, R. D. Effect of Garlic Extract and Cinnamon Powder on Microbial Profile and Shelf-Life of Minced Chicken Meat. *Indian J. Poult. Sci.* **2002,** *37,* 72–77.

Yaroshenko, F. O.; Surai, P. F.; Yaroshenko, Y. F.; Karadas, F.; Sparks, N. H. C. Theoretical Background and Commercial Application of Production of Se-enriched Chicken. Books Abstract of the XXII World's Poultry Congress, 8–13 June, 2004. Istanbul, Turkey. 2004.

Yin, M. C.; Faustman, C.; Risen. J. W.; Williams, S. N. Tocopherol and Ascorbate Delay Oxymyoglobin and Phospholipid Oxidation in-Vitro. *J. Food Sci.* **1993,** *58,* 1273–1276, 1281.

Youdim, K. A.; Deans, S. G.; Finlayson, H. J. The Antioxidant Properties of Thyme (*Thymus zygis L.*) Essential Oil: an Inhibitor of Lipid Peroxidation and a Free Radical Scavenger. *J. Essent. Oil Res.* **2002,** *14,* 210–215.

Younathan, M. T.; Marjan, Z. M.; Arshad, F. B.Oxidative Rancidity in Stored Ground Turkey and Beef. *J. Food Sci.* **1980,** *45,* 274–275, 278.

Zhang, Z. T.; Du, Y. J.; Liu, Q. G.; Liu, Y. Determination of the Antioxidative Effect of Aloe Vera. *Nat. Prod. Res. Develop.* **2001,** *13,* 45–46.

Zhang, Y.; Xu, W.; Wu, X.; Zhang, X.; Zhang, Y. Addition of Antioxidant from Bamboo Leaves as an Effective Way to Reduce the Formation of Acrylamide in Fried Chicken Wings. *Food Addit. Contam.* **2007,** *24*(3), 242–251.

CHAPTER 6

METHODS AND THEIR APPLICATIONS FOR MEASURING AND MANAGING LIPID OXIDATION: MEAT, POULTRY, AND SEAFOOD PRODUCTS

TOM JONES*

Meat Products, Global Applications and Product Development, Kalsec®, Inc., Kalamazoo, MI 49006, USA

*E-mail: tjones@kalsec.com

CONTENTS

ABSTRACT

The most appropriate analytical method for measuring and managing lipid oxidation in meat, poultry, and seafood products is based on the type of information sought, the precision needed, and the intended use of the data. In most cases, by including a sensory evaluation the significance of the analytical data is enhanced. However, in all cases the early incorporation of an antioxidant(s) is a strict prerequisite for managing oxidation of value-added meat products.

6.1 INTRODUCTION

In every year millions of dollars are lost because of discounting or discarding meat products due to oxidation. For example, the appealing color of pepperoni sausages (Photograph 6.1) can deteriorate due to lipid oxidation. Understanding when and how to use available analytical methods for measuring lipid oxidation is the first step for managing oxidation in meat, poultry, and seafood products.

PHOTOGRAPH 6.1   Oxidative color degradation of sliced pepperoni.

This chapter reviews the principles and applications of chemical and sensory methods for analysis of lipid oxidation and how they can be used to:

1. Estimate the oxidative stability (shelf life) of raw meat materials and finished products.
2. Determine the most effective, and type and usage rate for, oxidation inhibitors.
3. Determine critical "stress" points effecting oxidative stability in the distribution chain.
4. Measure the effect of formulation changes on oxidative stability of value-added meat products.
5. Evaluate and improve upon good oxidation management practices (GOMPs).

6.1.1 SEQUENCE OF CHEMICAL EVENTS, LIPID AUTOXIDATION

Lipid oxidation has also been referred to as lipid peroxidation because peroxides are the initial products formed. Figure 6.1 shows the oxidation of a poly-unsaturated fatty acid (PUFA). The products from lipid oxidation are responsible for color defects and rancid flavors that diminish the value of processed meat products. Fish, poultry, and pork are most susceptible to oxidation (compared to lamb and beef) because higher levels of PUFAs drive lipid oxidation.

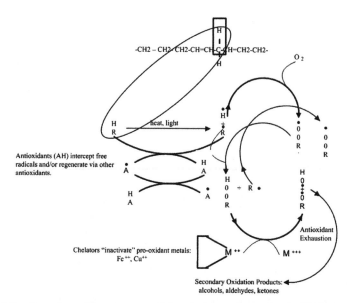

FIGURE 6.1 Overview of the sequence of chemical events associated with the autoxidation of lipids.

6.1.2 RELATIVE OXIDATIVE STABILITY OF MEAT AND FISH LIPIDS

Given that all other processing variables are equal, the presence (not necessarily the concentration) of unsaturated fatty acids drives lipid oxidation. Figure 6.2 shows a comparison of fatty acid profiles for beef, chicken, and salmon, as the fatty acid profile favors PUFAs the susceptibility to oxidation increases and shelf-life decreases. Throughout this chapter data will be presented showing the rate of oxidation in value-added meat products also depends on product form, processing methods, packaging techniques, and storage conditions.

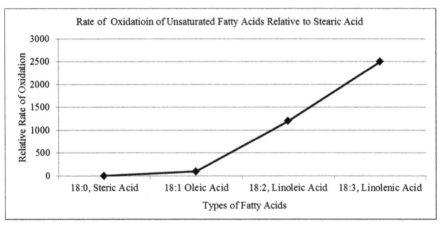

FIGURE 6.2   The susceptibility of a fatty acid to oxidation increases as unsaturation increases. When all other factors are equal, unsaturation dictates the differences in expected shelf life between value added products formulated from lamb, beef, pork, poultry or fish. Modified from http://www.public.iastate.edu/~duahn/teaching/Lipid%20oxidation/Lipid%20 Oxidation%20An%20Overview.pdf

6.1.3 SELECTION CRITERIA AND SAMPLE PREPARATION

Various analytical methods are available for measuring the extent of lipid oxidation. Regardless of the method selected, the more criteria, listed below, that are met, the more valuable the information will be.

1. The method is sensitive, accurate, and precise. The method's limit of quantification (LOQ) distinguishes fresh from oxidized products and less oxidized products from more oxidized products.
2. Data from the analysis are reproducible.

3. Data from the analysis are minimally influenced by interfering substances.
4. Analytical data vary predictably over time.
5. Data are relatable and relevant to the industry segment.
6. Results from the analysis are correlated to sensory (flavor, aroma, and/or color) analysis.
7. Method and technique are easy to use and rapid.

For all methods, sample preparation (collecting samples, grinding, mixing, etc.) is critically important for minimizing errors and uncertainty. Solvent extraction of lipids from meat tissues is of particular importance because solvent selection is critical for complete recovery of the fat for analysis (triacylgylcerides (TAG) and the unsaturated phospholipids (PL)). Lipid oxidation is initiated in the PL fraction and then proceeds to involve lipids in the TAG fraction. Therefore, complete extraction of TAGs and PL fractions will provide the most accurate assessment for the extent of oxidation.

The solvent systems recommended in the procedures described in this chapter for extracting TAG and PL lipid fractions from meat tissues have been determined experimentally by applying the following principles. .

1. Extraction efficiency of the solvent system.
    Solvent system (pairing a polar component, with a nonpolar component) is based on the distribution coefficient, K for a given meat matrix.

$$K = \frac{\text{solubility in lipid phase}}{\text{solubility in aqueous phase}} \quad (6.1)$$

The larger the coefficient K (greater than 1), the greater the yield of fat containing products from lipid oxidation in the sample.

2. Amount of the solvent(s) needed.
    The less solvent that is used the better, in the case of solvent extracted lipids containing secondary oxidation products, less is more. For example,

$$\text{Fraction of lipid extracted} = \frac{1}{(1 + V_s)/((V_m \times nK))^n} \quad (6.2)$$

where K is the distribution coefficient,
    n = the number of extractions,
    V_s = unit of measurement for solvent system (sp. gr. × volume),
    V_m = unit of measurement for meat homogenate (g).

The equation shows that the lipid fraction (containing products of lipid oxidation) extracted from the raw meat homogenate increases with serial extractions, the amount of solvent(s) is minimized when the K is large.

For meat applications, serial extractions are not always practical. Therefore, the meat sample is finely ground under controlled conditions (to minimize further oxidation during preparation) to have the same efficiency as serial extractions:

1. The secondary oxidation products extracted should not react with the solvent system.
2. The secondary oxidation products extracted should be easily recovered; low temperature roto-vaporization under vacuum is often used.
3. The solvent should have low toxicity.

6.2   ANALYTICAL METHODS FOR QUANTIFYING PRODUCTS OF LIPID OXIDATION

6.2.1   PEROXIDES

Peroxides are the initial products formed during lipid oxidation (Fig. 6.1). The peroxides formed are odorless and tasteless. Nonetheless they are widely used as an indicator of the current status of lipid oxidation. As will be discussed later in this chapter, peroxide values (PVs) can be mathematically combined with para-anisidine values (p-AVs) for a more comprehensive approach for determining the oxidative status of meat and meat by-products. Three different methods may be used to determine the PVs on solvent extracted lipids from meat.

1. Iodometric titration (color indicators, stoichiometry).
2. Photo-spectroscopic (Beer's law).
3. Electrochemical (Nernst equation).

The iodometric titration method is a common method for determining PVs on solvent extracted fat from a meat sample. A sample of extracted fat containing a starch indicator is titrated with standardized sodium thiosulfate.

$$RH + O_2 \longrightarrow ROOH$$

$$\text{(lipid)} \qquad\qquad \text{(lipid hydroperoxide)}$$

$$ROOH + 2KI + CH_3COOH \longrightarrow I_2 + K^+ CH3\,COO^- + 2H_2O$$

$$\text{(purple solution)}$$

$$I_2 + 2Na_2S_2O_3 \longrightarrow Na_2S_4O_6 + 2\,NaI$$

$$\text{(yellow translucent solution)}$$

Because volume and normality of the standardized sodium thiosulfate solution are known, the PV is derived stoichiometrically (normality$_1$ × volume$_1$ = normality$_2$ × volume$_2$). This relationship between the reactants and products is shown the Equation 6.3.

$$\frac{(mL\ Na_2S_2O_3) \times (N\ Na_2S_2O_4) \times 0.127\ meq. \times 1000}{wt.(g)\ of\ sample(solvent\ extracted\ fat\ or\ oil)} = \frac{Meq\ O_2\ \ (from\ hydro\ peroxides)}{Kg\ oil} \quad (6.3)$$

PV may also be expressed based on the weight of the sample as long as the exact fat content is known.

Calculation for PVs is based on "equivalents," not equivalent weight. Higher numerical values are indicative of a more oxidized sample. PVs that distinguish rancid foods from non-rancid foods are based on experience and science based historical data. As a guide, specifications for acceptable PVs should be based on the following points:

1. Stage of processing (finished or raw material).
2. Product type (pepperoni, cooked beef roast, fresh sausage, etc.).
3. Required shelf life.
4. Sensory evaluation (aroma, flavor).

The data in Figure 6.3 show that PVs can be used to assess the oxidative quality of freshly rendered lard. PV data provide information about the storage stability of the rendered lard and the effectiveness of natural oxidation inhibitors.

Interpretation of PV data in Figure 6.3 can be used to improve or implement GOMPs by incorporating natural oxidation in combination with a nitrogen (N_2) sparge and/or blanket as well as decreasing the turn over time of the stored lard.

From our experience, the level of quantification (LOQ) has been ± 2 meq kg^{-1} fat or oil. The LOQ for the titrimetric measurement of PVs is based on the precision inherent in the method, the amount of the sample collected for fat extraction, and the efficiency of the fat extraction procedure. Advances

in electrochemistry, that is the development of the understanding of the relationship between the amount of an electrical charge and the quantity of a
chemical compound, led to improvements in faster, more precise analysis
of hydroperoxides in meat samples. Electrochemical methods quantify
compounds produced from lipid oxidation potentiometrically.

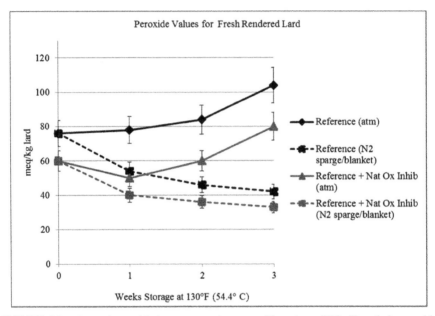

FIGURE 6.3 Assessing oxidative status using peroxide values (PV). Data help provide
information about the storage conditions required to achieve maximum shelf life. Data
modified from internal study.

Measuring the resistance to oxidation (e.g., oxidative stability index of
extracted fats) is an application of coulometric electrochemistry. For our
purposes, more meaningful data depend on quantifying the extent of lipid
oxidation, that is using the potentiometric methods (e.g., oxidation reduction
potentials and determination of PVs).

Ions from the meat matrix form a potential difference at the surface of
the electrode. Because the distance between the inner surfaces of the electrode is small, minute potential differences can be detected. For example, the
potential difference detected by the potentiometer is given in Equation 6.4.

$$\text{Potential difference} = \frac{k \times q}{d^2} = \frac{1\text{mv}}{10^{-6}\,\text{cm}} = 1{,}000{,}000 \text{ mv cm}^{-1} \qquad (6.4)$$

where k is a constant,

mv = millivolt,

q is magnitude of the interfacial electrical potential developed between the meat sample and electrode,

d^2 is square of the interfacial distance between the meat and electrode.

The electrochemical amplification of the potential offers the sensitivity needed when the detection of small differences is needed. This method would be considered when the background color of the sample or the concentration of the analyte is low (limit of detection (LOD) = 0.01 M and LOQ = 0.16 meq. Kg^{-1}). Potentiometric determination of hydroperoxides is based on Equation 6.5, an application of the Nernst principles for electrochemistry.

$$E = Eo - \frac{0.059}{n} \log_{10} \frac{[\text{reduced}]}{[\text{oxidized}]} \qquad (6.5)$$

Application of the Nernst equation relates the electrical potential to aqueous concentrations of a solution under controlled conditions (Karddash-Strochkova & Tur' yan, 2001). The analytical instrumentation is similar to the method for measuring PVs with potassium iodide–starch reagents. Although both are based on titration, the potentiometric method uses a change in an electrical signal whereas the iodometric method depends on a visual color change to determine the concentration of the hydroperoxides in the extracted fat (Table 6.1).

TABLE 6.1 Potentiometric Titration of Hydroperoxide with Sodium Thiosulfate in a Model System.

Potentio metric determination of peroxide values							
Titrant (Liters)	[Analyte] Ox.	[Analyte] Red.	[Titrant] Red.	[Titrant] Ox.	Eo	Constant	n	E system V
0.0005	0.00099	0.098020			1.77	0.0592	2	0.281
0.0050	0.00909	0.081818			1.77	0.0592	2	0.312
0.0100	0.01667	0.066667			1.77	0.0592	2	0.322
0.0150	0.02308	0.053846			1.77	0.0592	2	0.329
0.0200	0.02857	0.042857			1.77	0.0592	2	0.335
0.0240	0.02703	0.035135			1.77	0.0592	2	0.337
0.0249	0.02670	0.033511			1.77	0.0592	2	0.337
0.0250					1.77	0.0592	2	0.200

TABLE 6.1   *(Continued)*

		Potentio metric determination of peroxide values						
Titrant (Liters)	[Analyte] Ox.	[Analyte] Red.	[Titrant] Red.	[Titrant] Ox.	Eo	Constant	n	E system V
0.0253			0.0336	0.003	0.08	0.0592	2	0.050
0.0260			0.0342	0.003	0.08	0.0592	2	0.090
0.0300			0.0375	0.003	0.08	0.0592	2	0.050
0.0350			0.0412	0.004	0.08	0.0592	2	0.050
0.0400			0.0444	0.004	0.08	0.0592	2	0.050

Components for potentiometric titration are:

1. Buret containing the titrant (sodium thiosulfate).
2. Known volume of the sample containing the analyte (extracted fat or oil).
3. Reference electrode (saturated calomel or hydrogen electrode).
4. Electrode for measuring the potential, working electrode.
5. Meter to record the change during titration.

To demonstrate the application of the Nernst equation, titration of hydrogen peroxide (H_2O_2) with sodium thiosulfate ($Na_2S_2O_4$) in a model system was calculated with Microsoft® Excel™ 2010 spreadsheet, E° value sourced from CRC Handbook of Chemistry and Physics 71st edition.

Rxn. in aqueous system $2S_4O_6^{-2} + H_2O_2 - 2e^- <-> 2H_2O + 2e^- + 2S_2O_3^{-2}$

$$K_{eq} = E^o_{cell} = \frac{RT}{nF} \times \ln(K_{eq}) \quad (6.6a); \text{ solving for } K_{\text{(rxn driven strongly to the right)}}$$

$$E_{(Redox)} = (E^o \text{ Red.}) + (E^o \text{ Ox.}) \quad (6.6b)$$

$$E^o_{\text{(analyte at T=0)}} = 0.281 \text{ V}$$

$$\text{Eq. point} = \frac{ne\, E^o + ne\, E^o}{ne + ne} = 0.200; \text{vol. to eq. pt.} = M_1 \times V_1 = M_2 \times V_2 \quad (6.6c)$$

$K_{eq} = 1.43 \times 10^{57}$

$E_{(Redox)}$ 1.69 V

E eq. pt.   0.200 V (estimate)

Potentiometric titration data have been summarized in Figure 6.4.

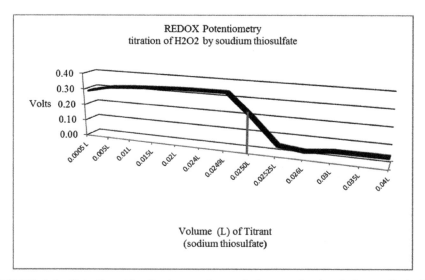

FIGURE 6.4 Potentiometric titration, at the equivalence point (vertical grey line) the amount of hydroperoxide is equivalent to the amount of sodium thiosulfate.

The points of underlying chemistry that make the potentiometric method valuable for quantifying PVs are:

1. The $K_{eq} = 1.43 \times 10^{57}$ for the sodium thiosulfate–hydrogen peroxide system; therefore, the titration reaction is complete and essentially irreversible.
2. The magnitude of the difference between E° for sodium thiosulfate (0.08 V in reduced format) and the E° for hydrogen peroxide (1.77 V in reduced format) provides a large and steep gradient for determining the equivalence point.

Meat scientists requiring a rapid, sensitive, and precise (limit of quantification = 0.16 meq kg^{-1} oil or fat) analytical method will prefer the electrochemical method useful for determining the current status of a meat product.

The relationship between the physical properties of light and the concentration of chemical compound is codified in the Beer Lambert law in Equation 6.3.

$$A_{\lambda} = l \times \epsilon \times C \qquad (6.7)$$

where A_λ is absorbance at a specified wavelength,
      l is path length of light (transverse axis of cuvette) in cm,
      ϵ is extinction coefficient,
      C is concentration.

The Beers Lambert law, illustrated in Figure 6.5, states the absorbance (A) of light at a given wavelength (λ) is related to the concentration of a chemical compound.

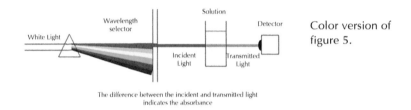

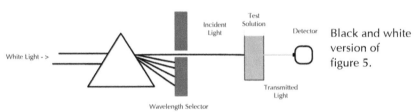

FIGURE 6.5 Beer's law relates the concentration of the analyte of interest to the degree of absorbance at a specified wavelength. The underlying principles for the components in Equation 6.7 are presented in Table 6.2 and Figure 6.6.

TABLE 6.2 The Principle for Quantitative Spectroscopy is Based on the Relationships of Absorbance (A), Path Length of Plane Polarized Light (l), and a Compound's Extinction Coefficient (ϵ). Modified from Data in Food Analysis: Theory and Practice, pp. 63–71.

Principle	Dependent variable y	Independent variable x	R^2	Slope m	Y Intercept
Strong absorbing analyte A (ϵ)	Absorbance	Analyte A	100%	0.25	0
Weak absorbing analyte B (ϵ)	Absorbance	Analyte B	100%	0.06	0
Path length v. absorbance (l)	Absorbance	Path length (dimension of cuvette)	100%	1.24	0
Absorbance v. concentration (C)	Absorbance	[analyte]	100%	0.25	0

Figure 6.6 shows absorbance at a given wavelength and the difference in absorbance is based on the molecular characteristics of a compound.

Practical application of Beer's law starts with developing a standard calibration plot of absorbance (at the specified λ) on the "y" axis and concentration (moles L^{-1} or mg mL^{-1}) on the "x" axis. The linear equation $y = mx + b$ can be used to determine the concentration of hydroperoxides (or secondary oxidation products) by interpolating values from the standard curve.

The spectrophotometry is rapid, sensitive, and can be automated with smaller systemic errors than the titrimetric technique.

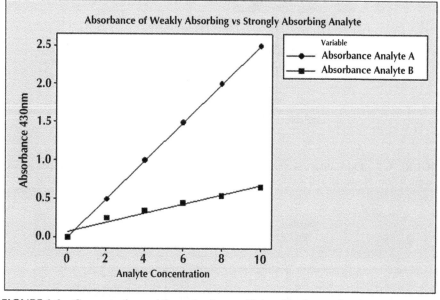

FIGURE 6.6   Concentration and the extinction coefficient (ϵ) of an analyte is a key principle in quantifying the various compounds using spectrophotometric techniques. The linear relationship between the components in Equation 6.7 is the basis of spectroscopic techniques.

Data from hydroperoxide analysis when combined with knowledge about the meat product and process can be used to evaluate oxidation management practices (Fig. 6.7). For example, the Figure 6.8 shows PVs collected over several days for a meat by-product that will be used in a value-added meat product.

The data in Figure 6.8 show a significant amount of the product produced exceeds the specification for PVs (abbreviated USL). Product that does not

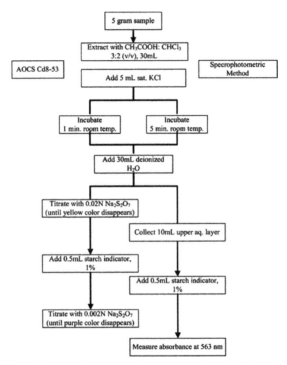

FIGURE 6.7 Flow diagram for spectrophotometric determination of hydroperoxides. Adapted from Research and Reviews, Special Circular 183-02, Ohio State University.

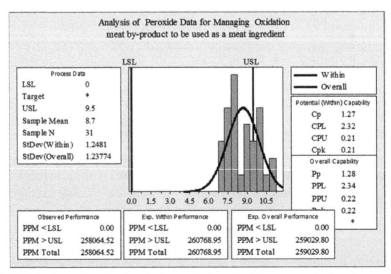

FIGURE 6.8 Peroxide value data in the graph show significant amount of product produced exceeds the specified limit (abbreviated USL). Data from internal spray dry data.

meet specification limits must be "re-worked," discounted in value, or worse case thrown out. However, the data can be used to frame relevant questions for managing oxidation. For example, is the specification realistic or could the process be improved in the selection of raw material quality and/ or production methods.

6.2.2   ANISIDINE VALUES (p-AV)

Secondary oxidation products of lipid oxidation may also be measured using the spectrophotometric technique. AVs are determined spectroscopically at absorbance of 350 nm. p-AVs are dimensionless and based on the absorbance of the chromagen formed through Shiff's base mechanism during the reaction of p-anisidine with cis-hexadecenal as shown in Figure 6.9.

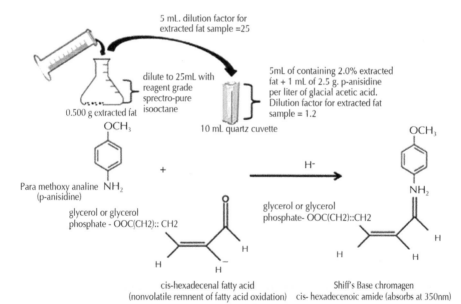

FIGURE 6.9 The flow diagram shows sample preparation of extracted fat for spectrophotometric analysis for secondary oxidation products with para-anisidine. Equation 6.8 is used to calculate p-AV.

$$p\text{-}AV = \frac{25 \times 1.2 \times (A_1 - A_2 - A_3)}{m} \tag{6.8}$$

where the coefficients 25 and 1.2 are dilution factors,

A_1 is the absorbance of the reacted sample,

A_2 is absorbance of the blank,

A_0 is absorbance of unreacted sample,

m is the mass of sample.

Although the coefficient of variation (C of V) for p-AV is relatively precise measurement, it can vary depending on the species source and processing methods in rendering the fat. p-anisidine analysis is a valuable measurement technique for measuring oxidation; however, correlation to sensory evaluation has not been established.

6.2.3 TOTAL OXIDATION (TOTOX) METHOD

A single molecule of a hydroperoxide can produce a variety of different compounds with varying chain lengths and levels of unsaturation during oxidation. To analyze for a single product of oxidation may not enough information. p-AVs and PVs can be combined (Equation 6.9) to provide more information about the oxidative status of meat products.

$$TOTOX = 2PV + p\text{-}AV \qquad (6.9)$$

PVs and p-AVs (a dimensionless value that indicates the level of non-volatile secondary oxidation products) are expressed in different units, the formulation and the significance must be derived empirically. Some food scientists have suggested the using thiobarbituric acid reactive substance (TBARS) values to replace the p-AV values for calculating TOTOX values (Wanasundara & Shahidi, 1995). The justification for the formula 2PV + p-AV is based on an increase of one PV unit corresponding to an increase of two p-AV units (Shahidi et al., 2002). Integrating PVs and p-AVs into one equation for the purpose can result in an accumulation of errors (i.e., uncertainty) by combining values derived from two measurement techniques into a single numerical value. The accumulation or propagation of errors can be estimated for TOTOX, assuming the values for PV and p-AV are normally distributed, independent, and errors are small. The propagation of errors (ϵ) for TOTOX can be calculated using Equation 6.10.

$$TOTOX = 2PV + p\text{-}AV$$

$$\epsilon^2 = (\text{mean of } 2 \times PV)^2 \times (\text{std.dev. of } PV)^2$$

$$+ (\text{mean for } p\text{-}AV)^2 \times (\text{std.dev. of } p\text{-}AV)^2 \qquad (6.10)$$

Therefore,

$$\epsilon = \sqrt{\epsilon^2}$$

For example, the error of uncertainty of TOTOX values can be calculated by using the data given in the Table 6.3.

TABLE 6.3   TOTOX Data Were Collected to Determine the Error of Uncertainty (the Error Propagated when Two Measurements are Combined). Calculating the Error of Uncertainty Assumes the Errors will be Small and the Data Are Normally Distributed. Data from Internal Study.

TOTOX data			
Variable	N	Mean	Std. dev.
$2 \times PV$	7	3.4	1.7
pAV	7	6.5	0.5

Solving for the error of propagation (ϵ) when two factors (PV and p-AV) are incorporated into Equation 6.10.

$$\epsilon^2 = ((3.4)^2 \times (1.7)^2) + ((6.5)^2 \times (0.5)^2))$$

$$\epsilon = \sqrt{(33.6 + 10.6)}$$

$$\epsilon = 6.6$$

The TOTOX value with the calculated error = 14 ± 6.6. If lower values for the error or uncertainty are needed, measurement methods, processing procedure, and/or product specifications need to be reviewed.

The application of p-AV, PV, and TOTOX data is shown in Figure 6.10. Because lard rendering is a lengthy process involving sustained high processing temperatures, analytical techniques that measure the more volatile secondary oxidation products (2-alkenals, 2, 4 alkadienals for TBARS) would be lost during processing. In this situation, p-AV data express the concentration of less volatile secondary oxidation products (less saturated, higher molecular weight fragments).

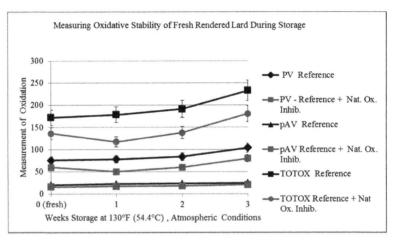

FIGURE 6.10   Data show an improvement in the storage stability of lard with natural oxidation inhibitors (abbreviated as Nat Ox Inhib). PV or p-AV values alone would minimize the importance of the added natural oxidation inhibitors.

TOTOX values (2PV + p-AV) provide more information because it measured the current oxidative status (PV) and oxidation that had occurred during processing (p-AV).

6.2.4 TBARS AND SENSORY EVALUATION

Volatile secondary oxidation products are most likely to be associated with off aromas and off flavors in meat products. Malondialdehyde (MDA) is a secondary oxidation product (from trienes) that at concentrations as low as 2–3 ppm produces off flavors and off aromas in value-added beef, pork, and poultry products (Pearson & Dutson, 1985). The diagram of the reaction shows that a prepared sample containing the secondary oxidation product MDA (compounds in abundance in oxidized tri-ene and tetra-ene fatty acids) reacts with thiobarbituric acid (TBA) to from a chromagen that absorbs strongly at 532 nm (distillation method). However, some TBA reactive aldehydes absorb at 450 nm.

One mole of MDA reacts with two moles of TBA to form a chromagen that absorbs strongly at 532 nm. TBA may react with other compounds present; therefore data form this method are reported as TBARSs. The concentration of thiobarbituric reactive substances or TBARSs (expressed as TBARS or MDA kg^{-1} of sample) is determined by developing a calibration curve using appropriate concentration range (mg mL^{-1} or μmoles mL^{-1})

of a basic solution of reagent grade MDA tetrabutylammonium added to an acidic solution of TBA. The concentration can be determined from absorbance at 532 nm. Beer's law provides a linear relationship between absorbance and concentration of TBARS via the linear equation $y = mx + b$.

From the standard curve $y = mx + b$

y is the absorbance of the test sample,

m is the slope from calibration with reagent grade MDA,

x is the concentration (mg mL^{-1}) of test sample (unknown),

b is the "y" intercept from calibration with reagent grade MDA.

Based on mg mL^{-1} multiplied by the volume of the extract (mL), multiplied again by any dilution factors, divided by the weight of the meat sample, and multiplied 1000 to express data as mg TBARS (as MDA) kg^{-1} of sample. To express in μmoles per unit weight convert it by using 72 μg per μmole.

The LOQ using in this method is 1.0 μmole or 72 μgrams. This level of sensitivity adequately discriminates between fresh and degrees of oxidization between test samples during shelf-life studies.

Acceptable values for TBARS meat type (beef, pork, poultry, and fish), formulation, and process specific. Table 6.4 shows the variation in acceptable levels of TBARS for a few types of value-added meat products.

TABLE 6.4 Acceptable Values for TBARS Varies from Product to Product. Adapted from Handbook for Meat Chemists, Edward Koniecko.

Average acceptable TBARS values for selected meat products	
TBARS expressed as mg MDA 1000 g^{-1} of meat	
Meat product	**Avg. TBARS**
Beef bologna	0.222
Fresh Italian sausage	0.195
Smoked sausage	0.273
Beef weiner	0.197
Hard salami	0.429
Chicken roll	1.100
Boiled ham	0.105
Pepperoni	0.343
Fresh sausage patties	0.280
Bacon, sliced vacuum packaged	0.052

Variation in the values for TBARS can be unduly influenced by the method of preparation (Figs. 6.11 and 6.12).

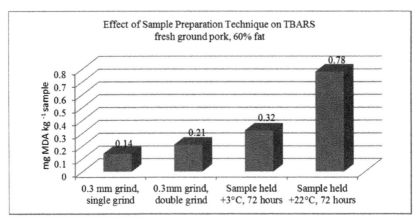

FIGURE 6.11   Sample preparation can cause TBARS to vary within samples from the same treatment of between treatments. Adapted from Handbook for Meat Chemists, E.S. Koniecko, (1979). Non-meat ingredients commonly used in formulating value-added meat products can be a source of variation in the values for TBARS.

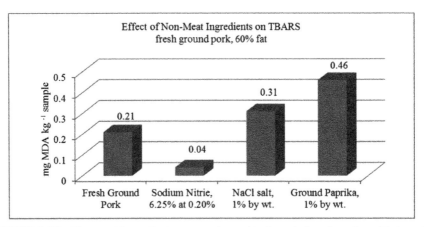

FIGURE 6.12   Non-meat ingredients commonly used in formulations for value added meats can increase or decrease TBARS values. Adapted from Handbook for Meat Chemists, E.S. Koniecko (1979).

Non-meat ingredients can increase or decrease the values for TBARS. Ingredients that react with TBA and absorb plane polarized light at 530 nm increase TBARS. Ingredients that react with MDA and block the reaction with TBA (e.g., oximes) depress TBARS. In other cases co-pigmentation occurs (red–orange colors of carotenoid pigments) or Maillard reaction products absorb light at or in proximity to 530 nm adding to the absorbance of MDA–TBA adduct in the sample (Decker, 1998).

In addition to interferences from the chemical reactions, values for TBARS are decreased by "dilution" by non-meat ingredients in the formulation (water, starch, soy, etc.). Figures 6.13 and 6.14 show formula for a chicken sausage (nitrites not added) containing 24% non-meat ingredients and the comparison of TBARS, respectively.

The data in the Figure 6.14 show values for the TBARS for the two treatments, one based on sample weight the other on meat weight. TBARSs expressed on meat weight were higher than TBARSs expressed on sample (formulation) weight.

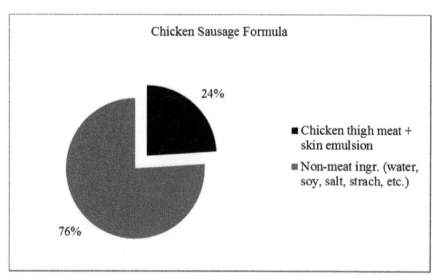

FIGURE 6.13   Formulation for a chicken sausage 76% meat ingredients and 24% non-meat ingredients.

A study to determine the oxidative stability of frozen, minced ocean-run pink salmon utilized the TBARS method (University of Alaska, Alaska Fishery Development Foundation, Kalsec®, Inc.) and sensory evaluation. The data in Figure 6.14 show the rate of change in oxidation of salmon mince during storage. The TBARSs are expressed as μmoles MDA per 100 g of minced salmon (Fig. 6.15). The minced salmon will be frozen then thawed as needed for processing value-added products (nuggets, patties, sausages). Processing accelerates lipid oxidation; therefore, meat to be further processed into value-added products will need have lower values for TBARS than that of finished products.

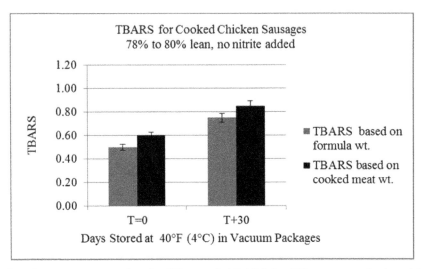

FIGURE 6.14 Understanding the difference in TBARS for different types of value added meat products requires knowledge about the formulation (high levels of non-meat ingredients or lower level of nonmeat ingredients) to understand the significance of the data.

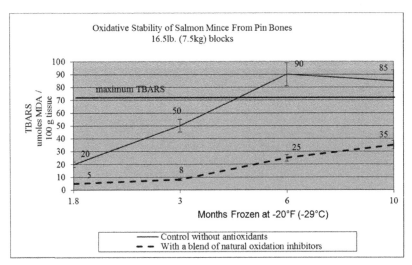

FIGURE 6.15 TBA values for frozen salmon mince show that the maximum storage time prior to thawing and further processing needs to be at some point in time prior to reaching the maximum TBA value of 70. Data modified from the pink salmon study conducted by Alaska Fishery Development Foundation, University of Alaska and Kalsec®, Inc. ca 1995.

Two methods are used for analyzing 2-TBARSs in meat products that are widely used in the meat industry (Fig. 6.16). The distillation method involves partitioning secondary oxidation products in the meat sample into the appropriate solvent then separating these compounds by distillation. The direct method determines the concentration of secondary oxidation products directly without partitioning the secondary oxidation products or

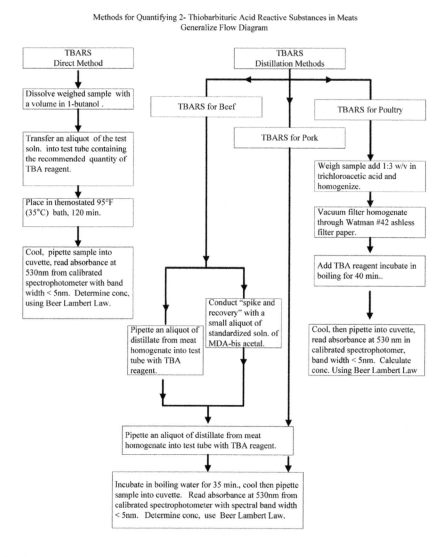

FIGURE 6.16 Methods and techniques for the analysis of TBARS (thiobarbituric acid reactive substances).

distillation. The distillation method is most widely used because the assay is relatively inexpensive, values are related to off-flavor, off-aromas, and color degradation and the data are familiar to the meat industry. Since the sample preparation and analysis differ for both methods, it is important to confirm which method is used.. TBARS data in this chapter are from the distillation method, this method is commonly used because the assay is relatively easy and inexpensive. The direct method for TBARS is problematic for samples containing carbohydrates and proteins, the distillation method uses the distillate which avoids this problem.

Data in the Figures 6.17(a) and 6.17(b) below show the application of the distillation method to evaluate the comparative shelf life between a natural and synthetic antioxidant for raw frozen pork sausages. TBARS values increase at different rates between treatments during frozen storage.

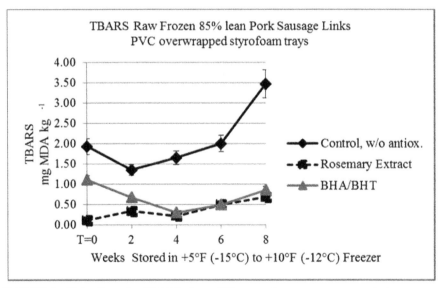

FIGURE 6.17(a) Data for TBARS show differences between treatments during storage time. Data from internal storage stability study.

TBARS values in the study correspond inversely with changes in the color of the raw pork sausages.

Some have suggested combining PVs and TBARS as a derivative of the previously mentioned TOTOX equation to be TOTOX $_{(modified)}$ = 2PV + TBA. This derivative of the standard TOTOX equation would need further testing to assure the data are valid.

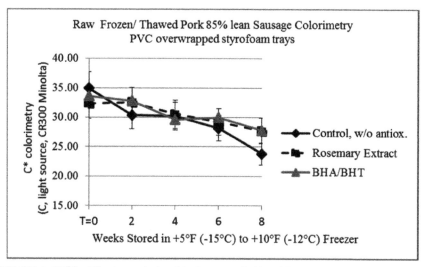

FIGURE 6.17(b)   The interrelationship between lipid oxidation and raw meat pigment color is inverse. As meat lipids oxidize, meat color also oxidizes. Data from internal storage stability study.

Values for TBARS of raw frozen pork sausage links can be transformed into a linear model to compare slope (rate of oxidation) and "y intercept." The Figure 6.17(c) shows pork sausages without an oxidation inhibitor ($R^2 = 90.9\%$) are initially more oxidized ("y" intercept) and oxidize most rapidly during storage, pork sausages with a natural oxidation inhibitor ($R^2 = 75.7\%$) are initially less oxidized and the rate of oxidation is inhibited with effectiveness comparative to BHA/BHT ($R^2 = 93.2\%$). Predictive application of the linear model is reserved for interpolation, extrapolation of the values for TBARs beyond eight weeks, in the data below, is risky. Note linear transformation does not alter the overall results.

The TBARS method provides valuable information about the shelf-life meat products and can be used to improve formulations, alter packaging materials, and review raw material specifications. However, good information can be compromised if used incorrectly. For example, using TBARS data to measure oxidation during a storage stability study under accelerated conditions can provide information in a relatively short period of time. However, using TBARS data from an accelerated shelf-life test to predict shelf life under different conditions carries a great deal of risk. The schematic representation in Figure 6.18 shows one possible error associated with predicting shelf life.

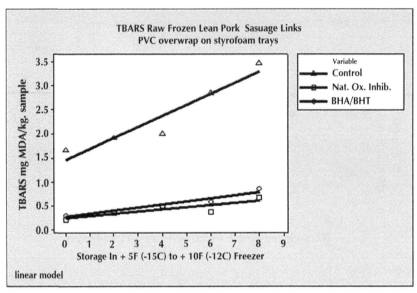

FIGURE 6.17(c) Transforming TBARS data into a linear model can provide information about the trend between and within the treatments. Creating a linear model has limitations and should be used only when appropriate.

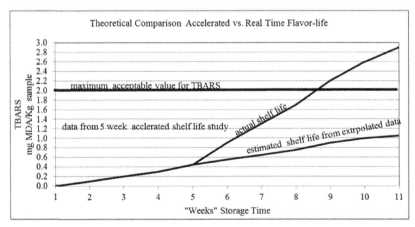

FIGURE 6.18 The schematic generalized representation showing one possible type of error in using valid data to predict shelf life by extrapolating data from accelerated shelf life studies.

As important as color is for consumer appeal of raw frozen sausages, it is critically important for fresh, minimally processed packaged meat products. Consumer demand for meat products with nutritional claims demanded analytical methods that would help to establish the oxidative stability (raw meat color and flavor) of these products. High oxygen (80%) modified atmosphere packaging (HOX MAPackaging) was developed to maintain the freshness by maintaining levels of oxygenated myoglobin pigments during lengthy distribution and storage at retailers.

6.2.5 ANALYSIS OF OXIDATION OF RAW MEAT PIGMENTS

HOX MAPackaing provides an oxygen rich headspace to maintain high levels of oxymyoglobin (the pigment responsible for the bright pink or red color of ground meat and sausages). The data in the Figure 6.19 show how headspace gases in HOX MAPackaged ground beef change during storage. The biochemically active meat absorbs oxygen from the headspace to convert myoblobin (a purple meat pigment) to oxymyoglobin (oxygenation) and also promotes lipid oxidation. As oxygen is depleted in the headspace, meat color decreases as oxymyoglobin levels decrease and lipid oxidation increases.

Photographs 6.2(a), 2(b), and 2(c) of HOX MAPackaged ground turkey (70% ground skinless thigh meat, 30% breast trimmings) show the change in color during storage in a cooler (Pettersen et al., 2004). From left to right the HOX MAPackaged ground turkey becomes progressively more oxidized, raw meat color becomes progressively less appealing (i.e., lower colorimetric a* values). At some point, the deterioration of color would result in discounting the price of the meat product to the consumer or discarding the product.

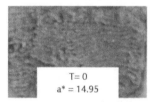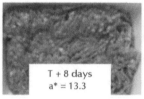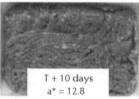

T= 0
a* = 14.95

T + 8 days
a* = 13.3

T + 10 days
a* = 12.8

PHOTOGRAPHS 6.2(a), (b), and (c), left to right. For minimally processed HOX MAPackaged products, raw color and lipid oxidation are inversely related. As meat color degrades to brown, lipid oxidation (as measured by TBARS) increases. Photographs and colorimetry data from internal study.

Photographs 6.3(a) and 3(b) of 80% lean HOX MAPackaged ground beef show color degradation over time. Colorimetry data (colorimetric a*) are higher in beef than that of turkey due to the increased inherent concentration of oxymyoglobin in 80% lean beef compared to that of ground turkey with 30% breast muscle trimmings.

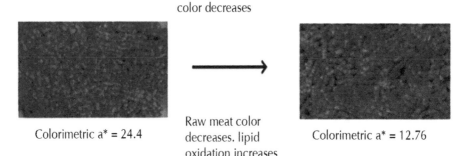

Headspace oxygen decreases, raw red meat color decreases

Colorimetric a* = 24.4

Raw meat color decreases. lipid oxidation increases

Colorimetric a* = 12.76

PHOTOGRAPHS 6.3(a) and (b), left to right   Raw meat color changes during storage and is an indication of the development of oxidized flavors. By regulation (standard of identity as stated in CFR Chpt. III, Part 319, §319.15), the types of oxidation inhibitors in HOX MAPackaged meats is restricted. Photographs and supporting colorimetry data from internal study. Photographs from internal study.

During the color life of HOX MAPackaged 80% lean ground beef, head-space gases as well color were monitored. The data in Figure 6.18 show the change in the concentration of oxygen (O_2) is associated with color degradation.

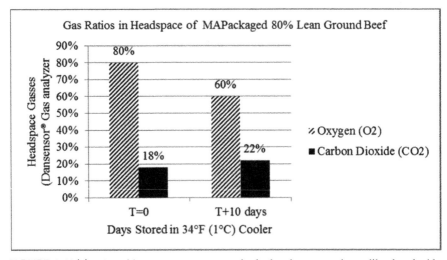

FIGURE 6.19(a)   At cold temperatures oxygen in the headspace gas is readily absorbed by the raw meat. The oxygen absorbed participates in several biochemical reactions including oxygenation of myoglobin and oxidation of lipids.

The interrelationship between meat color and lipid oxidation (Greene, 1969) places importance on CIE (International Commission on Illumination), L*a*b* reflectance colorimetry as well as diffuse reflectance spectroscopy is an important consideration when evaluating the oxidative stability in value-added raw meat products.

Reflectance spectroscopy requires a light source, an object (meat surface), and a sensor. In this case, the colorimeter describes raw meat color in terms of L* (lightness and darkness), a* (redness), and b* (yellowness). Fresh meat has initial high a* values (redness) (Table 6.5).

TABLE 6.5 Reflectance Spectroscopy is Used to Monitor the Raw Meat Color Stability During Storage. Raw Meat Color and Lipid Oxidation in Raw Meats are Interrelated, as Meat Color Changes from Red (Relatively High a*/b* in Combination with Higher L* Values) to Brown (Relatively Low a*/b* in Combination with Low L* Values) Lipid Oxidation Increases. Data from Internal Fresh Beef Storage Stability Study.

CIE L*a*b* reflectance colorimetry							
Sample description	L*	a*	b*	a*/b*	C*	Hue angle	a* × b*
Fresh meat color (red)	57.42	24.40	13.48	1.81	27.88	28.93	328.91
Oxidized meat color (redish brown)	52.94	12.76	7.51	1.70	14.81	30.49	95.83

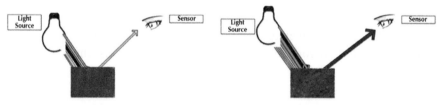

FIGURE 6.19(b) Reflectance colorimetry has three main components, a light source, an object (i.e. meat surface) and a sensor. The sensor detects changes in redness (a*, a* ÷ b* or a* ÷ b*), color saturation $C* = \text{chroma} = \sqrt[2]{(a* + b*}$, hue angle or color (in the CIE color sphere. For diffuse reflectance spectroscopy the intensity of light at a particular wavelength is detected (Modified from Hunt, M.; King, A. *Meat Color Guideline Measurements*; American Meat Science Association: Savoy, Illinois, 2012.)

The color of meat is based on the relative concentrations of oxymyoglobin (red), myoglobin (purple), and metmyoglobin (brown) (Fig. 6.20). For raw packaged meats (e.g., ground beef, pork, turkey) higher levels of oxymyoglobin are associated with freshness and quality. Meat containing higher levels of metmyoglobin is associated with oxidized or rancid flavors.

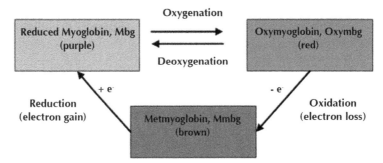

FIGURE 6.20   The three main types of raw meat pigments are in equilibrium during storage. The equilibrium shifts away from red meat pigments to brown meat pigments as the meat and lipid oxidize. (Modified from American Meat Science Association, Meat Color Measurement Guidelines. www.meatscience.org.)

Scientists depend on reflectance colorimetry methods for assessing meat color because it relates well to consumer preferences. Two methods for evaluating surface meat color frequently used are CIE L*a*b* colorimetry described earlier and diffuse reflectance spectroscopy. Both methods relay on light reflectance; however, diffuse reflectance spectroscopy is more discriminating because the method measures the intensity of light at specific wavelengths for oxymyoglobin, myoglobin, and metmyoglobin (610, 473, and 572nm wavelengths, respectively). Diffuse reflectance spectroscopy (Mancini et al., 2003) describes the oxidative status of meat pigments and indirectly the extent of lipid oxidation.

The colorimetric reflectance data in the Figures 6.21(a) and 6.21(b) show how the distribution for the principle raw meat pigments changes during storage.

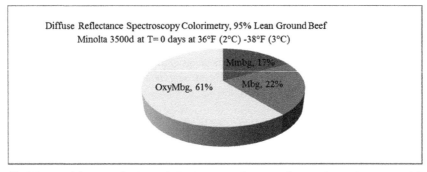

FIGURE 6.21(a)   Distribution of fresh meat pigments favors the red oxymyoglobin (OxyMbg) pigments.

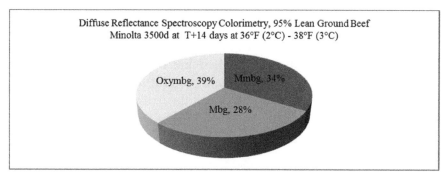

Diffuse Reflectance Spectroscopy Colorimetry, 95% Lean Ground Beef
Minolta 3500d at T+14 days at 36°F (2°C) - 38°F (3°C)

Oxymbg, 39%   Mmbg, 34%
Mbg, 28%

FIGURE 6.21(b)   Colorimetry data show a shift in the concentration of brown metmyoglobin (Mbg) during storage.

Generally, for raw ground meat products, development of oxidized or rancid flavors lags behind degradation of raw meat pigments. Pairing colorimetry with methods for measuring the level of secondary oxidation products (e.g., TBARS or gas chromatography mass spectroscopy (GCMS) headspace) defines whether the meat needs to be discounted or discarded.

Although visual appearance and flavor are the main drivers for consumer acceptance of value-added meat products, nutritional value and wholesomeness are becoming influential considerations. The inherent nutritional value of meat products (low caloric content and high biological value proteins and source of essential minerals and B vitamins) is known; however, some conventional meat products are being reformulated to respond to consumer's interests. For example, nutritionists and physicians are advising that calories derived from fat be reduced to 25–35%, with less than 10% of the calories from saturated fats (SFA), the ω-6 to ω-3 fatty acids be reduced to improve the thrombogenicity and atherogenicity indicies (abbreviated IT and AI, respectively). Based on these guidelines (American Dietetic Association, 2007), the demand for conventional value-added meats could be marginalized unless addressed through product reformulation and placement (Fig. 6.22).

The objectives for reformulating value-added meats to improve the following.

1.   The ω-3 to ω-6 fatty acid ratios
2.   IT and AI
3.   Shelf-life stability

Sodium reduction in value-added meats is of considerable interest to consumers and health professionals. However, the role of salt (NaCl) as an essential ingredient in processed meat products and is beyond the scope of this chapter.

As has been discussed, changes in meat formulations can significantly alter the oxidative stability, that is shelf life of the product. Analytical methods for measuring the oxidative status in meats have taken on renewed interest in product development (evaluating the oxidative stability of reformulated meat products).

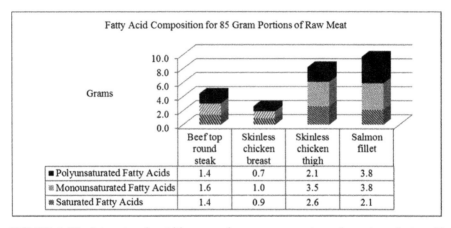

Fatty Acid Composition for 85 Gram Portions of Raw Meat				
	Beef top round steak	Skinless chicken breast	Skinless chicken thigh	Salmon fillet
■ Polyunsaturated Fatty Acids	1.4	0.7	2.1	3.8
▨ Monounsaturated Fatty Acids	1.6	1.0	3.5	3.8
▨ Saturated Fatty Acids	1.4	0.9	2.6	2.1

FIGURE 6.22 Interests of nutrition conscious consumers toward meat products with lower levels of ω-6 fatty acids and less saturated fatty acids has shifted the demand for less conventional products. Data sourced from USDA food composition table.

As the data in the chapter show, meat with improved "health indices" is also more susceptible to lipid oxidation and, consequently, shorter shelf lives. Therefore, it is critical to determine and confirm the oxidative stability of more healthful meat products. For example, HOX MAPackaged grass fed ground beef has been in particular demand by consumers because the meat is higher in ω-3 fatty acids, value, and convenience.

Before proceeding to the methods used the comparative shelf-life study for HOX MAPackaged ground beef formulated from grass fed beef (high ω-3 fatty acid content) versus grain fed beef (low ω-3 fatty acid content), the anatomical features of α-linolenic is reviewed in Figure 6.23. Fats and oils can be described in terms of their chemical, physical, and sensory properties. The diagram of the anatomy of a α-linolenic acid describes several physical characteristics that determine nutritional value and oxidative stability.

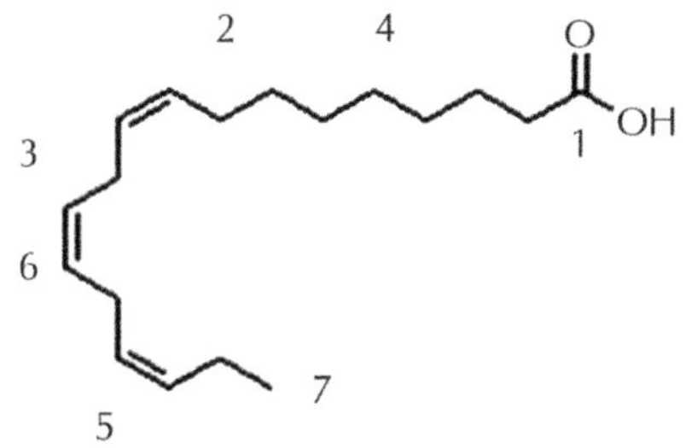

FIGURE 6.23 Alpha linolenic (ω-3 fatty acid) fatty acid. This fatty acid is 4–5 times higher in grass fed beef than grain fed beef, and is highly susceptible to oxidation. Modified from ChemDraw® software CambridgeSoft®/Perkin-Elmer®.

The ω-3 fatty acids have seven important structural features.

1. The alpha carbon is associated with the carboxyl end of the molecule.
2. The approximate reactive site (allylic carbon) where peroxidation and scission begin with the abstraction of the carbon atom.
3. The approximate reactive site (bis-allylic carbon) where peroxidation and scission are most likely to occur during initial phases of oxidation.
4. Hydrocarbon backbone (cis-$C_n H_{2n}$) of the fatty acid
5. Unsaturated portion of the fatty acid (cis-$C_n H_{n-2}$)
6. ω-6 carbon
7. ω-3 carbon

To preserve the status of meat products in the consumer's daily diet, some meat processors offer conventional products reformulated with grass fed. The data in Table 6.6 compare the fatty acid profiles between grass fed/grass finished (95% lean) and grain fed (90% lean) beef forequarter and show grass fed beef has a higher ω-3 fatty acid content. Small changes in levels of unsaturated fatty acids significantly impact the oxidative stability (Wood et al., 2003). Using analytical methods, oxidative stability can be monitored when there are changes in the fatty acid composition in the meat formula. Indices based on the fatty acid composition in raw beef, pork, poultry, and finfish are shown. Indices that are positively correlated to nutritional value also present shelf-life challenges.

Indices of nutritional value (bottom of Table 6.6) are IT, AI, and ratio of desirable fatty acids (DFA) to the undesirable fatty acids (OFA) that elevates low density lipoproteins (LDL).

$$IT = \frac{(C14:0 + C16:0 + C18:0)}{(0.5 \times MUFA) + (0.5 \times \omega - 6\ PUFA) + (3 \times \omega - 3\ PUFA) + \dfrac{\omega - 3\ PUFA}{\omega - 6\ PUFA}} \quad (6.11)$$

$$A = \frac{(4 \times C14:0) + (C16:0 + C18:0)}{\sum MUFA + \sum PUFA\ \omega - 6 + \sum PUFA\ \omega - 3} \quad (6.12)$$

$$\frac{DFA}{OFA} = \frac{desirable\ fatty\ acids}{undesirable\ fatty\ acids} = \frac{\sum(PUFA + C18:0)}{\sum(SFA - C18:0)} \quad (6.13)$$

$$\frac{PUFA\ \omega - 3}{\omega - 6\ PUFA} \quad (6.14)$$

where C:14, C:16, C:18 are carbon notations for the fatty acids,
PUFA is polyunsaturated fatty acids,
MUFA is monounsaturated fatty acids,
SAT is saturated fatty acids.

The nutritional indices (Garaffo et al., 2011) were calculated and are presented in Table 6.6. The data show grass fed beef DFA/OFA (Moawad et al., 2013) is more favorable than grain fed beef because of the higher levels of unsaturated fatty acids. The table also shows PUFA ω-3/PUFA ω-6

TABLE 6.6 Data Show the Fatty Acid (FA) Profile that is Nutritionally Valuable (Percent C:183, IT, AI, DFA/OFA) are also Fatty Acids that Reduce Oxidative Stability (APE, BAPE, IV). Data from Internal Study.

Fatty acid analysis (FAME)							
Calculated derivative values for oxidative stability and nutritional value							
Fatty acid carbon notation							
% FA (normalized)	C14:0	C16:0	C18:0	C18:1	C18:2	C18:3	Total
Grass fed	6.3%	39.5%	23.4%	27.5%	2.00%	1.4%	100%
Grain fed	11.1%	43.2%	9.60%	33.9%	2.00%	0.30%	100%
FA distribution (millimoles)							
Grass fed	0.27	1.54	0.82	0.98	0.07	0.05	3.73
Grain fed	0.49	1.68	0.34	1.20	0.07	0.05	3.83

TABLE 6.6 *(Continued)*

Fatty acid analysis (FAME)							
Calculated derivative values for oxidative stability and nutritional value							
Fatty acid carbon notation							
% FA (normalized)	C14:0	C16:0	C18:0	C18:1	C18:2	C18:3	Total
FA distribution (mole fraction)							
Grass fed	7.3%	41.3%	22.0%	26.1%	1.90%	1.40%	100%
Grain fed	12.9%	44.4%	8.90%	31.7%	1.90%	0.20%	100%
Stability and Nutritional value	Calc. IV	% IV (from C 18:3)	APE	BAPE	IT	AI	DFA/ OFA
Grass fed	31.0	0.12%	62	5	0.47	1.06	1.53
Grain fed	33.0	0.02%	72	2	1.84	2.38	0.62

ratios are more favorable for grass fed beef. Similarly, IT and AI data (not shown) are more desirable for grass fed beef. Beef in the form of HOX MAPacked ground fresh beef offers consumers nutritional value, quality, and convenience they seek. However, HOX MAPackages have a finite shelf life because, during distribution and storage, meat becomes less red and lipid oxidation products increase. The relationship between TBARS and sensory (raw meat color) between raw HOX $_{(80\% \text{ oxygen, } 20\% \text{ carbon dioxide})}$ MAPackaged ground boneless forequarter meat from grain fed versus grass fed/grass finished beef.

Color (CIE L*a*b*) and TBARS were used for both grass fed and grain fed MAPackaged raw, lean ground beef shelf-life study. Using both methods provided information about what consumers value most—color or flavor. The colorimetric data in the Figure 6.24 show grass fed beef; particularly grass fed beef with the incorporation of rosemary natural oxidation inhibitor is more color stable than grain fed beef.

However, the TBARS data (Fig. 6.25) show the necessity of early addition of an oxidation inhibitor (TBARS value at the "y" intercept for grass fed beef without an oxidation inhibitor at T = 0). The data also show it is the reactivity of the fatty acids not the fat content that is responsible for driving oxidation. Information from both methods provides the information needed to identify and address pre-emptively shelf-life challenges during the development of products from grass fed beef, for example, raw HOX MAPackaged ground beef.

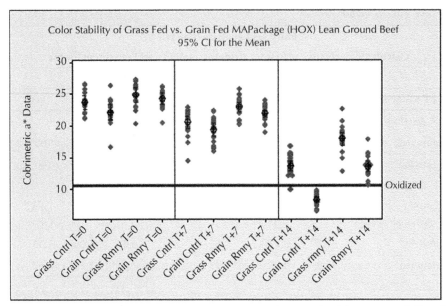

FIGURE 6.24 Color stability of HOX MAPackages of ground beef from "grass fed" and "grain fed" was studied. MAPackages of ground beef from grass fed cattle were more color stable than MAPackages of ground beef from grain fed cattle.

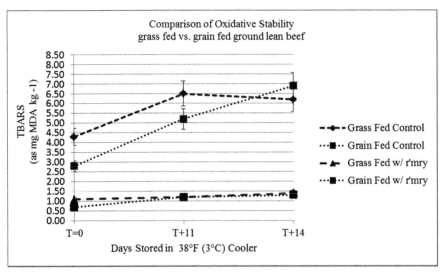

FIGURE 6.25 When TBARS are paired with colorimetry (Fig. 24), the susceptibility to oxidation reflects the changes in the composition of grass fed beef, in particular α linolenic.

6.2.6 *OXIDATION REDUCTION POTENTIAL (ORP) METHOD*

ORP can be an important method for measuring the general oxidative status of meat products. The ORP method does not measure specific products of oxidation. It can be used for evaluating oxidation inhibitors (Fig. 6.26).

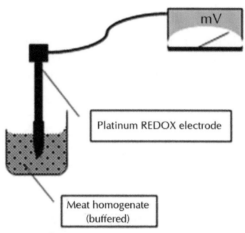

FIGURE 6.26   Oxidation reduction potentials (ORPs) can be used to broadly measure the oxidative status of meat products. Negative values for ORPs indicate reducing (favorable) environment. Diagram by Tom Jones.

The ORP method is a direct measurement of the all the oxidized and non-oxidized compounds in the meat sample (Vahabzadeh, 1986). The electrical potential measured is related to the concentration of oxidized and non-oxidized species through the Nernst equation.

$$E = E_o + \frac{0.059}{n} \log_{10} \left[\text{oxidized and nonoxidized species} \right] \qquad (6.15)$$

where E is the reading taken at the potentiometer in mV,

E_o is reference electrode in mV,

0.059 is a constant calculated by mathematically combining the universal gas constant, temperature (K), Faraday's constant,

n is the number of electrons (unknown) involved in the oxidation reduction process, since meat systems have dilute ionic concentrations (or activities) n can be ignored.

The read-out is based on the following equation.

$$\Delta E_\mathrm{h} \text{ (ref. Pt. electrode)} + E_\mathrm{h} \text{ (meat homogenate)} \qquad (6.16)$$

Reducing conditions will register more negative values for E_h while oxidizing conditions will register more positive or less negative values for E_h.

The ORP method is used for measuring the effectiveness of two oxidation inhibitors for stabilizing raw meat color in HOX MAPackaged ground beef. The two oxidation inhibitors differ in their respective water/fat partition coefficients, molecular weight, and size. The data show significant differences between the ORP values for the oxidation inhibitors (Fig. 6.27).

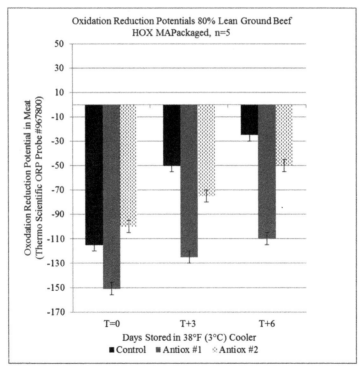

FIGURE 6.27 Oxidation reduction potentials are more negative (reducing conditions) when an antioxidant is added.

Colorimetry was also conducted, the data in the Figure 6.28 show the oxidation inhibitor with the most negative ORP value over time, also has a positive effect on raw meat color stability.

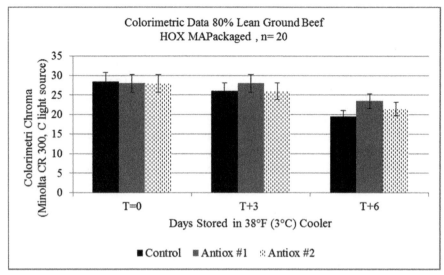

FIGURE 6.28   Colorimetry data (C*, Chroma) shows the addition of an antioxidant improves the stability of raw meat pigments during storage. Negative oxidation reduction potentials relates to higher color stability.

The ORP method for determining the oxidative status of meat products is precise to within 5%; however, the following factors can reduce precision and repeatability:

1.   Temperature and ionic strength
2.   Concentration of myoglobin (i.e., heme-iron) in the meat sample
3.   Presence of oxygen in the homogenate
4.   Time for potentiometer to reach a steady state

6.2.7   HEADSPACE ANALYSIS OF VOLATILE LIPID OXIDATION PRODUCTS (GCMS)

TBARS method for quantifying secondary oxidation products in value-added meats has stood the test of time, the need for faster turn-around, sensitivity, and precision has become a bigger. GCMS headspace advantages include reduced sample preparation time, increased sensitivity (LOQ), and reduced analytical time. The GCMS method is gaining wider acceptance in research and industry. Coupling the components of GC and MS allows for the identification and quantification of a wide range of secondary oxidation products in a meat sample.

Meat samples are prepared by first comminuting meat in a 38 °F (3 °C) room, the appropriate solvent is added to the meat to convert mixture into a form compatible for the analysis. Controlled heating volatilizes the extracted compounds prior to injection via an inert gas onto the column. The column can be thought of as consisting of a structural support (e.g., stainless steel) lined with a stable film (stationary/liquid phase) that separates the complex mixture of volatiles and secondary oxidation products in the meat sample. Separation is best achieved when the polarity of the stationary phase is similar in polarity to the compounds being analyzed.

Retention time, archived data, and experience provide the information needed to identify "marker compounds" (aldehyde A and aldehyde B in the Fig. 6.29).

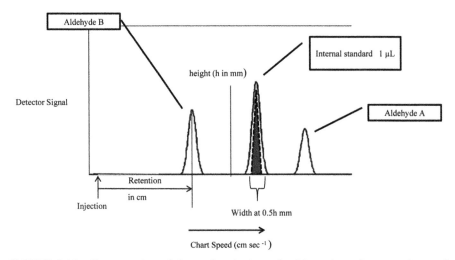

FIGURE 6.29 Concentration of the analyte is determined by using a known volume of an internal standard. Retention time of the analytes separates from the sample for ease of quantification and identification. Diagram by Tom Jones.

The (calibrated) detectors used for quantifying marker compounds are selected based on their sensitivity, signal stability, and linear responsiveness over a wide concentration range; these characteristics are critically important for precise quantification of secondary oxidation products using GCMS (Fig. 6.30).

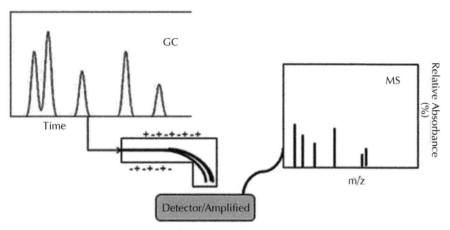

FIGURE 6.30 GCMS is the combination of gas chromatography and mass spectroscopy (GCMS). When used in combination, one or several types of compounds from lipid oxidation can be measured.

Using the read-out from the recorder, the concentration of the marker compounds can be calculated using the following steps.

1. A known amount (usually 1 µL) of an appropriate standard is injected into the column.
2. Although peak measurement for chromatographic quantitative analysis has been computerized, the areas of peaks in the diagram have been calculated.

Area of eluted peak (mm²) = (ht. of peak) × (peak w at 0.5 × h)   (6.17)

(0.5 x h to reduce error from "tailing" at the base of the peak)

3. The peak from 1 µL of the standard is 101.5 mm; therefore, the concentration of the eluted marker compound(s) can be calculated (Table 6.7).
4. The separated compounds exiting the GC are bombarded by a high energy electron to produce charged particles necessary of mass analysis.

$$CH_3 (CH_2)_4 \ CHO + 1 \ e^- \longrightarrow CH_3 (CH_2)_4 \ CHO^+ + 2 \ e^-$$
(hexanal from GC)   (electron beam)   (+charged particle for mass analysis)

TABLE 6.7 Concentration (Usually ppm) can be Calculated Based on the Known Quantity of an Internal Standard. The Amount of Individual Compounds from Oxidation (e.g., Hexanal, Propanal, Nonenal, etc.) may be Calculated or Grouped Together to Calculate Total Aldehydes).

	Internal std.	Aldehyde A	Aldehyde B	Total aldehydes
Peak h, mm	29.0	15.0	20.0	
Peak w at 0.5h, mm	3.50	2.50	3.00	
Peak area, mm^2	101.5	37.5	60.0	97.5
Peak area ratio	1.00	0.37	0.59	0.96
Concentration (μL)	1.00	0.37	0.59	0.96

Charged particles are accelerated and directed through the mass spectrometer by controlling and varying (referred to as scanning potentiometry) the applied voltage. The electron bombardment produces singly charged particles; therefore, the path of a charged particle will depend only on mass or mass charge ratio (m/z). The mass analyzer segregates the charge particles based on m/z and the detector records the abundance (frequency) of each m/z.

There is not just one single pathway for fatty acid oxidation, the types and concentration of secondary oxidation products produced are differed and concentrations of each can vary. In addition, the type of meat (beef, pork, poultry, and fish) and processing conditions (cooking method, drying, aging, fermenting, etc.) will also affect the profile of secondary oxidation products produced from oxidation.

When samples of oxidized beef were analyzed by GCMS, hexanal, 2-hexanal, 2-nonenal, 2-heptenal, 2-octenal, nonanal, and decanal were identified (Flores et al., 2013). These "marker compounds" increase over time to provide information about the rates and extent of lipid oxidation during shelf-life studies.

The mass spectrograph data in Figure 6.31 show several compounds that were detected in oxidized ground chicken and the internal standard shown is a part of the calibration procedure.

The GCMS method was used to monitor the oxidative stability of raw frozen restructured beef steaks through the distribution chain. The data from the study would provide information about the stages in distribution (processing, storage, cooking) would have the most the greatest impact on lipid oxidation. Restructured steaks are made by comminuting underutilized beef cuts (chuck, round, sirloin, etc.), then formed into steak or cutlet shapes.

Restructured steaks have the appearance and texture "whole-muscle" steaks and cutlets (see Photograph 6.4 below) and can be engineered to meet any portion weight, shape, lean content, or seasoning profile. Archived data were available for a variety of value-added meat products (Fig. 6.32).

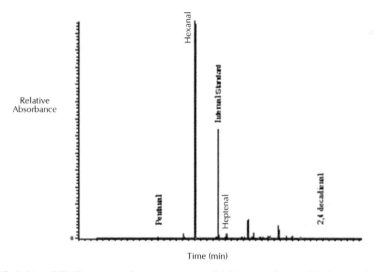

FIGURE 6.31 GCMS was used to measure multiple secondary oxidation products or "marker" compounds. These compounds were monitored during storage stability tests to determine maximum acceptable shelf life of cooked chicken patties.

PHOTOGRAPH 6.4 Raw restructured steak.

What effect on would the method for the unique method for comminuting meat and the use of non-barrier packaging during storage and distribution have on shelf life little restructured beef steaks and is the amount of the natural oxidation inhibitor adequately protect the product?

The GCMS method was used to determine the oxidative stability of restructured beef steaks during storage for the following reasons.

1. Raw restructured steaks will have low levels of secondary oxidation products, the sensitivity of GCMS will differentiate between fresh and low levels of lipid oxidation.
2. The low variability in the analysis (as measured by the C of V) translates to fewer samples that are needed for analysis without sacrificing the quality of the information.
3. GCMS data correlate well with raw meat color and sensory.

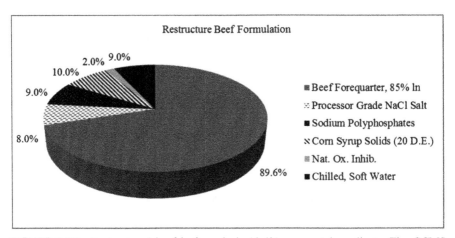

FIGURE 6.32 Restructured beef is formula is 10.4% nonmeat ingredients. The GCMS method provided the sensitivity for measuring oxidative stability in value added raw meat. In addition, the polyphosphates and corn syrup solids could interfere if TBARS method was used. Formulation developed internally for a study.

A study was conducted to determine at what stage during distribution in commerce affected shelf life the most. Identifying the points of greatest oxidative stress will provide information needed to manage oxidation.

1. Stored raw, non-barrier film in a + 20 °F (–7 °C) freezer for 15 days (processor abuse, meat is held at –15 °F (–26 °C) to –25 °F (–32 °C).

2. Cooked to 165 °F (74 °C) at 325 °F (163 °C) and held three hours in a135 °F (57 °C)warming oven according to hazardous analysis critical control point (HACCP) procedures (product abuse #1).
3. Cooked to 165 °F (74 °C), stored at 35—38 °F (2 °C–3 °C) for 48 hours then reheated (product abuse #2).

The GCMS data in Figure 6.33 identify that cooking and reheating stage in the distribution process greatly accelerates oxidation in the steaks.

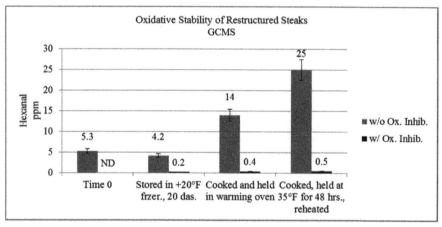

FIGURE 6.33 Data show reheating precooked restructured steaks accelerates oxidation more than freezer storage or cooking and holding in a warming oven. The data also indicate the type and levels of the oxidation inhibitor met product requirements.

Sensory evaluation was conducted using a nine point hedonic scale (higher scores represent more intense flavor) and was conducted by 14 trained panelists on five aroma characteristics for restructured beef steaks. Sensory data indicate the significant differences ($\alpha = 0.05$) between steaks with a natural oxidation inhibitor and steaks without an oxidation inhibitor occurs when the steaks were reheated. The sensory data are presented in Figures 6.34(a), 6.34(b), and 6.34(c).

Note the flavor threshold for hexanal is 14–25 ppm and for TBARS (MDA) is 1.5–2.5 ppm. Pairing chemical analysis (GCMS) and sensory confirms a natural oxidation inhibitor is needed; and incidence for consumer "abuse" would be of greatest concern.

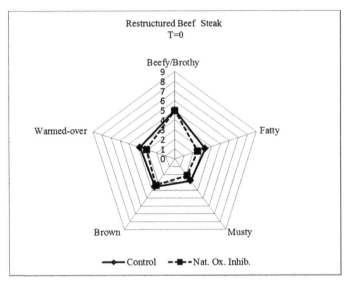

FIGURE 6.34(a)   Sensory evaluation at time zero (t = 0) of cooked restructured steaks immediately after cooking. Data shows little difference between steaks with or without (control) an oxidation inhibitor. Sensory evaluation conducted internally with trained panel.

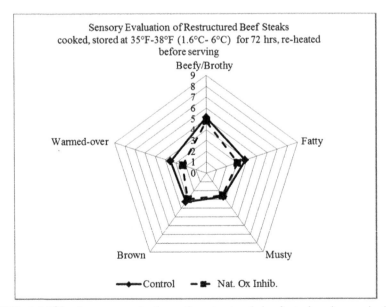

FIGURE 6.34(b)   Sensory evaluation of restructured beef steaks shows significant development of warm over aroma (descriptor for oxidized meat aroma) in cooked then reheated steaks. Sensory evaluation conducted internally with trained panel.

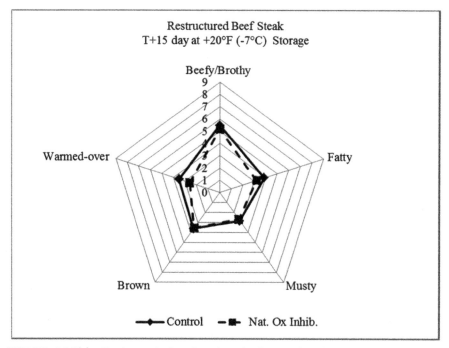

FIGURE 6.34(c)   Steaks cooked and evaluated after 15 days in +20°F (−7°C) freezer (storage abuse during distribution). Sensory evaluation conducted internally with trained panel.

The change in color of the steaks between time zero and after 16 days storage in the freezer is likely due to the presence of NaCl salt in the formulation. Sodium chloride in the meat has two functions—solubilizes myofibril protein for binding the chunks of raw meat together and enhances meaty flavors.

Based on analytical data, the critical stage in the distribution chain is reheating of the precooked steak (e.g., reheating left overs), the incorporation of a natural oxidation inhibitor effectively inhibits oxidation at all stages of commerce (Fig. 6.35).

Raw meat color of the restructure steaks was also evaluated during frozen storage.

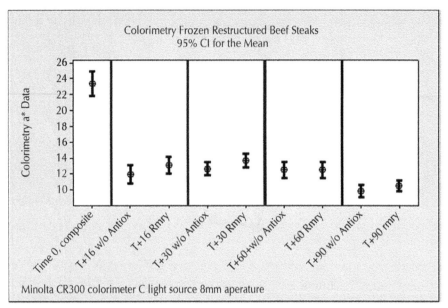

FIGURE 6.35 Colorimetry data (a*) complement GCMS and sensory data. Raw meat color of frozen restructured steaks in non-barrier packaging decreases over time. The data suggest when raw meat color is a critical consumer attribute, barrier packaging may be required.

6.3 MEASURING THE CONSEQUENCES OF LIPID OXIDATION

6.3.1 WATER HOLDING CAPACITY (WHC)

Consequences of lipid oxidation is not limited to flavor, aroma, and color, WHC is also adversely affected (Figure 6.36). As WHC decreases during frozen storage, meat products will lose moisture during cooking resulting in dry, "tough," and chewy steaks.

Studies (Lund et al., 2007) conducted by meat scientists have determined that myofibrillar proteins are susceptible to chemical modification due to oxidation. It is myofibrillar protein oxidation that has been a factor in observed loss of WHC in raw and cooked value-added meats. Our study was designed to determine the effect of natural antioxidants on meat protein oxidation in raw frozen restructured beef steaks during 90 days of frozen (–10 °F, –23 °C) storage. After 30, 60, and 90 days of frozen storage the restructured steaks were thawed at 38 °F (6 °C) overnight and evaluated for raw meat color (oxidation of sarcoplasmic protein) and WHC (a measure of myofibrillar protein function). The trend in the data shows WHC decreases during frozen storage for the control ($R^2 = 100.0\%$), natural oxidation inhibitor ($R^2 = 100.0\%$) with the rate of loss (slope) for WHC was similar for all treatments. The data suggest that inhibiting oxidation of meat proteins will require further study, perhaps a blend of oxidation inhibitors or modified atmosphere packaging techniques or the quality of raw meat can be considered.

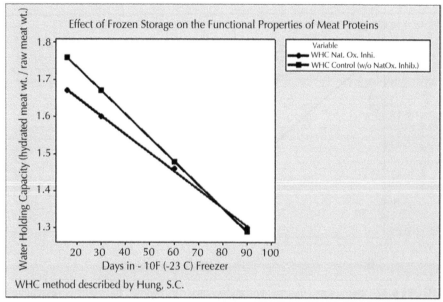

FIGURE 6.36 Meat protein functionality (WHC) decreases during frozen storage. As WHC decreases in the raw meat, the amount of moisture retained in the meat after cooking also decreases. The lack of moisture in the cooked steak is associated with undesirable changes in texture and palatability. Data from study conducted internally.

6.3.2 OXIDATIVE STABILITY OF SEASONINGS DURING STORAGE (OLEORESIN PAPRIKA)

Inhibiting oxidation in value-added meat products often begins with an assessment of the non-meat ingredients added during processing. For example, dry soluble pepperoni seasoning contains some or all of the following ingredients oleoresin paprika, dextrose, processors grade NaCl salt, and an antioxidant (s). During storage and prior to use, the seasoning can lose significant amount of color (oleoresin paprika). The data in the Figure 6.37 show the rate of oxidative degradation of oleoresin paprika (critical component in pepperoni seasonings), during storage in a high density polyethylene lines multi-ply bag stored at three temperatures. Information can be used to determine the optimum storage time and temperature to assure the color-life requirements for a pepperoni seasonings is met at the time of use.

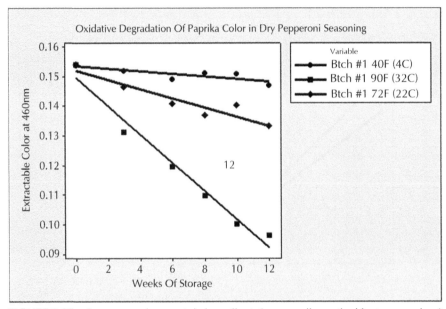

FIGURE 6.37   Some seasonings contain ingredients (e.g., paprika, antioxidants, seasonings) that are susceptible to oxidation during storage and prior to use.

6.3.3   MEASUREMENT OF THE RATE OF ANTIOXIDANT DEPLETION IN MEAT

The rate of an antioxidant which is consumed in ground beef stored under different storage conditions is also a measure of the rate of oxidation. In this case the rate in the disappearance of the antioxidant is measured instead of the production of oxidation products. Data in Figure 6.38 show there are issues in meeting the targeted usage level. The use of barrier packaging and cold storage temperatures reduces the rate of oxidation and maximizes the effectiveness of the antioxidant.

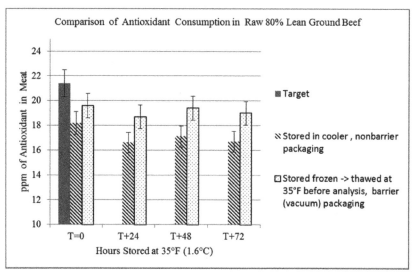

FIGURE 6.38 The analysis confirms the addition of the oxidation inhibitor missed the targeted level and frozen storage inhibits oxidation.

6.4   CONCLUSION

The methods and their application for processed meat products are described in the chapter to emphasize there is no such thing as a routine analysis. The information presented reviews the importance of understanding which method provides the information appropriate for a given situation, the importance of following good technique, the relationship between sensory and analytical data; and the effect of ingredients have on analytical data.

As the reliance on analytical data for making good decisions increases, so will the need for more analytical precision and reduced analytical time. For example, advantages by modifying an existing analytical method (described in Section 6.2.1) using electrochemical technology are illustrated in Figures 6.39(a) and 6.39(b).

Figure 6.39(b) shows an improved precision (i.e., a standard deviation of five) by technical modification decreases the number of samples (thereby the time for analysis) collected for analysis without sacrificing the quality of the information.

The scope of the chapter includes description of methods that can also be used to measure shelf life, improve processing efficiency (e.g., WHC of meat); and, improve the quality of formulations for value-added meat products (e.g., color stability of paprika in Italian sausage seasoning).

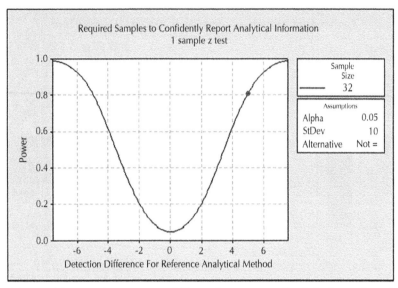

FIGURE 6.39(a) For the purpose of illustration, a generalized "reference" analytical method, with a standard deviation of 10 requires that 32 samples be collected and analyzed.

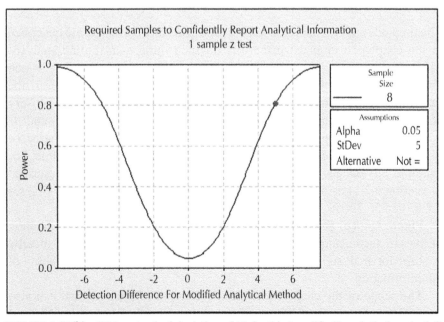

FIGURE 6.39(b) With a standard deviation of 10 for this "modified" analytical method, 8 samples instead of 32 provides the quality of information needed.

6.5   ATTACHMENTS

Attachment 1. Statistical basis for the inferences made for data in Figure 6.40. Making inferences about peroxide values collected to describe the effectiveness in managing oxidation requires samples be randomly selected and the data are normally distributed. Figures 6.40 and 6.41 show the data collected are normally distributed.

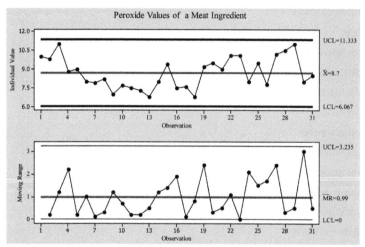

FIGURE 6.40   PV data taken from samples collected at different times during production show acceptable variation. This data can be used to determine if the peroxide values fit the normal distributed curve.

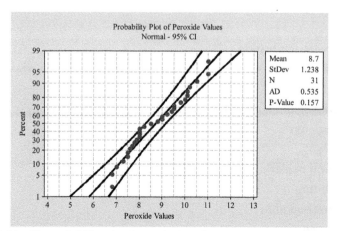

FIGURE 6.41   The probability plot shows the data fits a normal distribution. Therefore, the data can be used to make inferences about how well the analytical method for peroxide values or their specification provide the degree of control for managing oxidation.

KEYWORDS

- absorbance
- aldehyde
- antioxidants
- analyte
- autoxidation
- Beer's law
- carbon dioxide
- carotenoid
- chelation or chelators
- chroma
- colorimeter
- cured meats
- decanal
- electrochemistry
- fatty acid(s) flavor threshold
- formulations
- ground beef
- ground turkey headspace
- hexanal
- heptanal
- hue
- iodometric titration
- Maillard
- meat protein
- metmyoglobin
- myoglobin
- mince
- natural oxidation
- inhibitors
- nernst equation
- nitrogen

- **nonanal**
- **nutritional indices**
- **octenal**
- **oxidative stability**
- **oxygen**
- **oxymyoglobin**
- **pepperoni**
- **propanal**
- **redox**
- **reflectance spectroscopy**
- **restructured steaks**
- **rendered pork fat**
- **sausage**
- **sensory**
- **sparge**
- **titrant titrimetry**
- **vacuum packages**

REFERENCES

American Dietetic Association; *Position of the American dietetic association and Dietitians of Canada: Dietary Fatty Acids;* Journal of American Dietetic Association, September, 2007.

Decker, A. E.; Chan, K. M.; Faustman, C. In *TBA as an Index of Oxidative Rancidity in Muscle Foods,* 51st Annual Reciprocal Meat conference Proceedings, Storrs, CT,1998; American Meant Science Association: Savoy, Illinois, Vol. 51, 1998.

Flores, M.; Olivares, A.; Dryahina, K.; Spanel, P. Real Time Detection of Aroma Compounds in Meat Products by SIFT-MS and Comparison to conventional Techniques (SPME-GC-MS). *Curr. Anal. Chem.* **2013,** *9,* 622–630.

Garaffo, A. M.; Vassallo-Agius, R.; Nengas, Y.; Lembo, E.; Rando, R.; Maisano, R.; Dugo, G.; Giuffrida, D. Fatty Acids Profile, Athergenic (IA) and Thrombogenic (IT) Health Lipids, of Raw Rose of Blue Fin Tuna (Thunnus thynnus L.) and their Salted Product "Bottarga". *Food Nutr. Sci.* **2011,** *2,*736–743.

Greene, B. E. Lipid Oxidation and Pigment Changes in Raw Beef. *J. Food Sci.* **1969,** *34,*110–113.

Hung, S. C.; Zayas, J. F. *J. Food Qual.* **1992,** *15,* 153–157.

Hunt, M.; King, A. *Meat Color Guideline Measurements;* American Meat Science Association: Savoy, Illinois, 2012.

Jones, T. Science Fellow, Kalsec®, Inc., Diagrams drawn using Visio® software.

Karddash-Strochkova, E.; Tur'yan Ya, I. Redox-Potentiometric Determination of Peroxide Values in Edible Oils without Titration. *Talanta.* **2001,** *54,* 411–416. www.elsevier.com/locate/talents.

Koniecko, E. *Handbook for Meat Chemists;* Avery Publishing Group, Inc.: Wayne, NJ, 1979; pp 68–69.

Ladikos, D.; Lougovois, V. Lipid Oxidation in Muscle Foods: A Review. *Food Chem.* **1990,** *35,* 295–314.

Li, C. T.; Marriott, N. G.; McClure, K. E. *Dietary Intake of Vitamin E Affects the Peroxide Value of Subcutaneous Lamb Fat;* Research and Reviews: Meat, Special Circular No. 172–199, Ohio State University: Columbs, OH, 2007.

Lide, D. R. *Handbook of Chemistry and Physics;* 71st ed.; CRC Press Inc.: US, 1990–1991.

Lund, M. N.; Hvid, M. S.; Skibsted, L. H. The Combined Effect of Antioxidants and Modified Atmosphere Packaging on Protein and Lipid Oxidation in Beef Patties during Chill Storage. *Meat Sci.* **2007,** *76,* 226–253.

Mancini, R. A.; Hunt, M. C.; Kropf, D. H. Refectance at 610nm Estimates Oxymyoglobin Content on the Syrface of Ground Beef. *Meat Sci.* **2003,** *64,* 157–162.

Manuela, A. G.; Vassallo-Agius, R.; Nengas, Y.; Lembo, E.; Rando, R.; Maisano, R.; dugo, D.; Giuffrida, D. Fatty Acids Profile, Atherogenic (IA) and Thrombogenic (IT) Health Lipid Indices, of Raw Rose of Blue Fin Tuna (*Thunnus L.,*) and their Salted Product "Bottarga". *FNS.* **2011,** *2,* 736–743.

Moawad, R. K.; Mohamed, G. F.; Ashour, M. M. S.; Enssaf, M. A.; El-Hamzy, A. *J. Appl. Sci. Res.* **2013,** *9* (8), 5048–5059.

Ohio State Extension Research; *Research and Reviews,* Meat 2001, Special Circular No. 183–02, Ohio State Extension Research: OH, 2001.

Pearson, A. M.; Dutson, T. R. *Advances in Meat Research, Restructured Meat and Poultry Products;* Nostramd Reinhold Company: New York, Vol. 3, 1985.

Pettersen, M. B.; Mielnik, T. E.; Skrede, G.; Nilsson, A. Lipid Oxidation in Frozen, Mechanically Deboned Turkey Meat as Affected by Packaging Parameters and Storage Conditions. *Poultry Sci.* **2004,** *83,* 1240–1248.

Pietrzyk, D. J.; Frank, C. W. *Analytical Chemistry, an Introduction Chapters 7 and 15;* Academic Press, Inc.: New York, NY, 1979.

Pomeranz, Y.; Meloan, C. E. *Food Analysis: Theory and Practice;* The Avi Publishing Company, Inc: Baton Rouge, Louisiana, pp 64–71, 1971.

Witte, V. C.; Krause, G. F.; Baily, M.E. A New Extraction Method for Determining 2-Thiobabituric Acid Values of Pork and Beef During Storage. *J. Food Sci.* **1970,** *35,* 582–585.

Ross, C. F.; Smith, D. M. *Use of Volatiles as Indicators of Lipid Oxidation in Muscle Foods, Comprehensive Reviews in Food Science and Food Safety;* Institiute of Food Tecnologists: Chicago, Illinois, 2006; Vol. 5.

Shahidi, F.; Wanasundara, U. N.; Akoh, C. C.; Min. D. B.; Eds.; *Food Lipoids: Chemistry, Nutrition and Biotechnology;* Marcel Dekker, Inc.: New York, 2002; pp183–1987.

Surendranath, P. S.; Hunt, M. C.; Mahesh, N. N.; Rentfrow, G. Improving Beef Color Stability: Practical Strategies and Underlying Mechanisms. *Meat Sci.* **2014,** *98,* 490–504.

Vahabzadeh, F. *A Dissertation Submitted in Partial Fulfillment of the Requirements for the Degree of Doctor of Philosophy in Nutrition and Food Sciences;* Utah State University: Logan, UT, 1986.

Wanasundara, U. N.; Shahidi, F. *J. Food Lipids.* **1995,** *2,* 73–68.

Wood, J. D.; Richardson, R. I.; Nute, G. R.; Fisher, A.V.; Campo, M. M.; Kasapidou, P. R.; Shead, E. M. Effects of Fatty Acids on Meat Quality: A Review. *Meat Sci.* **2003,** *66,* 21–32.

CHAPTER 7

APPLICATION OF NATURAL ANTIOXIDANTS IN DAIRY FOODS

NEELAM UPADHYAY[1,*], VEENA N.[2], SANKET BORAD[1],
ASHISH KUMAR SINGH[1], SUMIT ARORA[1], and MINAXI[3]

[1]*Dairy Technology Division, ICAR-National Dairy Research Institute, Karnal 132001, Haryana, India*

[2]*Dairy Chemistry Division, College of Dairy Science and Technology, Guru Angad Dev Veterinary and Animal Sciences University, Ludhiana 141001, Punjab, India*

[3]*Agricultural Structures and Environmental Control Division, ICAR-Central Institute of Post- Harvest Engineering and Technology, Ludhiana 141004, Punjab, India*

Corresponding author. E-mail: neelam_2912@yahoo.co.in

CONTENTS

ABSTRACT

Milk, a nature's perfect food containing high quality of almost all nutrients (proteins, lipids, carbohydrates, vitamins, and minerals), is a polyphasic secretion of the mammary glands. It is considered to be one of the indispensable ingredients for the preparation of functional foods due to the presence of myriad bioactive components. Although, besides containing some antioxidants (like lactoferrin, α-tocopherol, etc.), milk also contains few oxidants (like polyunsaturated fatty acids (PUFA) of triglycerides, phospholipids, cholesterol ester, riboflavin, ferri-porphyrin, etc.) that may lead to development of off-flavors in milk and milk products. Some of the oxidation products especially of lipid oxidation are toxic and reactive. Thus, several studies have been conducted on the addition of antioxidants, both synthetic and natural, in milk and milk products. However, natural antioxidants are gaining wide acceptance as synthetic antioxidants are associated with potential health hazards. Also, natural antioxidants have positive health implications and enhance functionality. Several studies have been conducted on the applications of the natural antioxidants in milk and several dairy products. These antioxidants can be added either in the form of the essential oil or in the form of extract (of polar and non-polar solvents) from skins, seeds, peels, pomace, bark, or leaf of the natural source. This chapter highlights the applications of these natural antioxidants in dairy foods.

7.1 INTRODUCTION OF MILK AND DAIRY PRODUCTS

Milk and milk products contribute a very significant proportion in our daily diet. Milk, considered to be closest to the nature's perfect food, is an excellent source of calcium, a good source of minerals and high-quality proteins, the only source of lactose, and a source of lipids, the most valuable component, which also forms the basis of milk pricing (Upadhyay et al., 2014). The milk proteins are a good source of essential amino acids and have a high biological value. The majority of nitrogen in milk is distributed in casein (around 80%) which is present as colloidal dispersions called as casein micelles; and whey proteins (around 20%) which are present as true solution. However, in addition to these proteins, milk also contains two other groups of proteins or protein-like materials, the proteose-peptone fraction and the non-protein nitrogen (NPN) fraction (Fox & McSweeney, 1998). As per the protein quality ranking, casein and whey proteins are reported to have protein efficiency ratio, biological value, net protein utilization, and protein digestibility corrected amino acid score (PDCAAS) as 2.5, 3.2; 77, 104; 76, 92; and 1.00, 1.00, respectively, while these values for milk are 2.5, 91, 82, and 1.00, respectively. As can be noticed, the PDCAAS for casein, whey protein and milk is 1.00 which is highest for a protein (Puranik & Rao, 1996; Sarwar, 1997; United States Dairy Export Council, 1999; Hoffman & Falvo, 2004). The biological function of casein is to carry calcium and phosphate. It forms a gel or clot in the stomach which leads to continuous but slow release of amino acids into the blood stream (Boirie et al., 1997; Hoffman & Falvo, 2004). Some bioactive peptides are also released from casein on digestion that have physiological significance like, antithrombotic peptides, antihypertensive peptides, opioid peptides, immune modulatory peptides, antimicrobial peptides, casein phosphopeptides, glycomacropeptides, and so forth. Unlike casein, the plasma appearance of amino acids is fast, high, and transient upon ingestion of whey proteins (Boirie et al., 1997; Hoffman & Falvo, 2004). Whey proteins are rich source of branched chain amino acids (BCAAs) that are metabolized directly into the muscle tissue leading to replenishment of the exhausted levels followed by repairing and rebuilding of lean muscle tissue. These are, thus important for athletes and alike. Whey proteins are also rich and balanced source of sulfur-containing amino acid cysteine that help boost body's antioxidant levels as cysteine, along with glycine and glutamic acid is a precursor of glutathione which is the potent intracellular antioxidant and it gets oxidized to glutathione disulfide (GSSG) (oxidized glutathione) leading to removal of reactive oxygen species (ROS), thus regulating the level of ROS in the cells (Haug et al., 2007; Smithers, 2008).

The milk lipids, similar to the milk proteins, are important dietary components for supplying nutrients as contain certain bioactive components like short chain fatty acids, conjugated linoleic acid (CLA), branched chain fatty acids, and so forth, and are present as emulsified droplets in globules coated with membrane. Butyric acid (4:0) is reported to be a modulator of gene function (German, 1999), besides inhibiting colon and mammary tumors (Parodi, 2003). Caprylic (8:0) and capric (10:0) acids are reported to have antiviral activities; however, caprylic acid is also reported to delay tumor growth (Thormar et al., 1994). Sun et al. (2002) reported that lauric acid (12:0) may have antiviral and antibacterial function, while Schuster et al. (1980) reported that it may act as an anticaries and antiplaque agent. An interesting observation was reported by Henry et al. (2002) that capric and lauric acid inhibit cyclooxygenase, that is, COX-I and COX-II. Milk fat naturally contains CLA which is shown to be anticarcinogenic and anti-atherogenic and it has effects on body composition and fat metabolism

(Ip et al., 1999; Pariza et al., 1999; Truitt et al., 1999; Benjamin & Spener, 2009). Phytanic acid, a branched chain fatty acid presents in milk fat, has been shown to increase insulin-independent glucose uptake by cells and to decrease liver triglyceride accumulation in some mouse models (Hein et al., 2002; Hellgren, 2010; Palmquist, 2010). Besides fat and protein, milk is naturally a good source of lactose and contains ~5% w/w. Lactose, present as true solution, promotes the absorption of calcium like other sugars and it is a ready source of energy, providing 30% of the caloric value of bovine milk. Also, it accounts for 50% of the osmotic pressure of milk, which is isotonic with blood and hence is essentially constant (Fox & McSweeney, 2009). The minor components present in milk include minerals (calcium, selenium, iodine, magnesium, and zinc) and vitamins (vitamin A, vitamin E, riboflavin, folate, and vitamin 12) whose nutritional significance is well established.

Milk and other dairy products have been undoubtedly considered as nature's perfect functional foods owing to the presence of wide myriad of bioactive components. Milk has been advocated by the food formulators for the development of novel dairy foods because of its richness in nutrients and bioactives and molecules that assist in development of excellent sensory characteristics. Since time immemorial milk and milk nutrients have been utilized for the commercial manufacture of numerous products like cream, butter, butter oil, cheese, condensed milk, dried milks, and indigenous dairy products that contain components only from milk. Besides this, certain other food products are prepared from milk where milk components form the major ingredients like kheer, khoa/mawa, rabri, khurchan, kulfi, shrikhand, ice cream, and so forth. Probably, it is the only raw material that has been exploited to such a great extent, not only for the manufacture of value added food products, but also to harness the valuable ingredients for food and pharmaceutical sector.

7.2 NATURALLY OCCURRING OXIDANTS AND ANTIOXIDANTS IN MILK AND DAIRY PRODUCTS

The different dairy food components, namely polyunsaturated fatty acids (PUFA), proteins, vitamins, and pigments undergo oxidative changes, which could be both desirable and undesirable, during processing and storage. The rate and extent of oxidation depends on the concentration and activity of oxidizing agents, which may form the part of food naturally or used in the food product as ingredient or may be formed during processing and subsequent storage. Thus, the knowledge of the principal oxidants and anti-oxidants occurring in the foodstuff is indispensable for understanding the oxidative reactions that may occur in the food components and their implication on food quality and safety. Moreover, it is must to have the information pertaining to enzymatic and non-enzymatic catalysts of oxidation.

The various oxidants present in the dairy products include PUFA attached to triglycerides, phospholipids, cholesterol esters, riboflavin, proteins, and so forth. Phospholipids are considered to be pro-oxidant because of the presence of monounsaturated and PUFA group attached to diglyceride or sphingosine in it (Table 7.1). However, Chen and Nawar (1991) reported that phospholipids can act as either pro-oxidant or antioxidant in dairy products depending on the pH and ratio of water to phospholipid species in it. Similarly, the thiol groups (-SH) in milk are reported to act as pro-oxidant or antioxidant depending upon the conditions. Yee and Shipe (1982) observed in a model system, free thiol groups in the presence of copper promoted oxidation of emulsified methyl linoleate, while it acted as antioxidant component in the presence of haem. Milk and milk products may contain metallic components (such as transition metal ions- cupric, ferric, and haem proteins) that normally act as pro-oxidant by catalyzing the decomposition of preformed hydroperoxides to initiate the new oxidation chain reactions (Labuza, 1971; Korycka-Dahl & Richardson, 1980; Pokorny, 1987). The ferri-porphyrin proteins, together with their juxtaposition with lipids in the milk fat globule membrane (MFGM) have powerful pro-oxidative properties (Kendrick & Watts, 1969; O'Connor & O'Brien, 2006). Juxtaposition of copper-protein complex with the phospholipids of the MFGM is also an important factor in development of oxidized flavor in liquid milk (Samuelsson, 1966). The water-soluble vitamin, that is, riboflavin, is a potent photosensitizer and is associated with photo-oxidation of milk. Besides riboflavin, porphyrins and chlorophylls are also reported to be involved in the photo-oxidation of certain milk products like cheese (Wold et al., 2005). It is reported that concentrations of ascorbic acid above those in normal milk (~20 mg/L) could provide antioxidant protection; however, at the normal concentrations in milk, ascorbic acid acts as a pro-oxidant (O'Connor & O'Brien, 2006). Caseins are more effective as antioxidant than whey proteins. The antioxidant activity of casein may be attributed to the hydrophobic nature of the same (Taylor & Richardson, 1980; Eriksson, 1982) and orientation of potential antioxidant side-chains of constituent amino acids at the lipid interface. Among the whey proteins, lactoferrin has also been reported to inhibit peroxidation induced by Fe^{2+} by chelating it (Gutteridge et al., 1981; Allen & Wrieden, 1982).

TABLE 7.1   Oxidants and Antioxidants in Milk.

Oxidants	Antioxidants	Both (oxidants and antioxidants)
Polyunsaturated fatty acids of triglycerides, phospholipids, cholesterol esters	Casein	Phospholipids
Riboflavin	Lactoferrin	Thiol group
Proteins	α- tocopherol	
Transition metal ions- cupric, ferric	Carotenoids (β-carotene)	
Haem proteins	Maltol (in heated milk)	
Ferri-porphyrin		
Copper-protein complex with phospholipids		
Xanthine oxidoreductase		
Sulfhydryl oxidase		

Milk also contains appreciable amount of antioxidants. Vitamins including α-tocopherol, principal tocopherol of the eight vitamers of the vitamin E, act as a free radical scavenger. The concentration of α-tocopherol in cow's milk ranges from 3.0 to 5.0 mg/L (Hendricks & Guo, 2006). α-tocopherol is also reported to quench singlet oxygen ($^1O_2$) *via* a charge-transfer scavenging mechanism (Yamauchi & Matsushita, 1977; Burton & Ingold, 1981). Vitamin A and carotenoids act as antioxidant due to the hydrophobic chain of polyene units that quench $^1O_2$, neutralize thiyl radicals and combine with peroxyl radicals and stabilize the same (Palace et al., 1999). Beta-carotene also acts as antioxidant in milk. It is reported to offer good protection against light-induced lipid oxidation and it inhibits the chlorophyll-sensitized photo-oxidation of methyl linoleate (Terao et al., 1980). In addition, it provides protection against riboflavin degradation by competing for absorption of light. However, this protective effect disappears outside the absorption band of carotene (i.e., <366 nm) as the protective mechanism is not a quenching effect of radical or $^1O_2$, but absorption (filter) effect of the incident light (Hansen & Skibsted, 2000). Some of the indigenous enzymes of milk also act as antioxidants, while others may play a role in promoting the lipid oxidation like Xanthine oxidoreductase. Sulfhydryl oxidase is proposed to cause oxidation of thiols in ultra-high temperature (UHT) milk, in conjunction with lactoperoxidase (to destroy the resultant

H_2O_2) by doing away with the pro-oxidants resulting from auto-oxidation of thiols (Swaisgood & Abraham, 1980).

The products and by-products obtained from processing of milk can also serve as potent antioxidant. Steps associated with preparation of cream and during pre-heating of milk for milk powder manufacture, the antioxidative activity gets increased due to enhanced activation of free sulfhydryl groups (Farkye, 2006; Keogh, 2006). It is proposed that buttermilk solids are effective in reducing the severity of lipid oxidation during chain propagation stage; however, it is ineffective in delaying the onset of lipid oxidation. Thus, incorporation of buttermilk solids into the food matrices can be an approach to stabilize against lipid peroxidation (Wong & Kitts, 2003). The antioxidant activity of buttermilk may be attributed to the presence of sulfhydryl content. Few of the workers (Dugan, 1980; Eichner, 1980; Eriksson, 1982) reported that carbonyl-amine reactions between lactose and milk proteins produce intermediates having potent antioxidative activity and may play role in stabilization of milk fat (Wyatt & Day, 1965). Maltol, an important flavor component in the heated milk also exhibit powerful antioxidant activity (O'Brien, 2009). However, it is also reported that browning reaction products may exert anti-nutritional and toxicological effects (O'Brien & Morrissey, 1989). Thus, a balance between pro-oxidant and antioxidant factors is critical for the oxidative stability of milk (Stapelfeldt et al., 1999; Morales et al., 2000).

7.3 OXIDATION IN DAIRY PRODUCTS

Oxidation of dairy products involves the addition of oxygen atom to or the abstraction of hydrogen atom from the different chemical moieties present in the milk, which is further associated with conversion of primary hydroxyl groups (alcohols) to aldehyde and finally to carboxylic acid functionality by either chemically or biochemically mediated oxidation. In few instances, the oxidation reaction is desirable and may lead to an improvement in the product quality, such as oxidative cross-linking of proteins to manipulate viscosity and gelation in dairy products. However, in majority of the cases, food oxidation leads to decrease in consumer acceptance, minimizes the shelf life of the product and in few may even be associated with formation of anti-nutrients and toxicity as well. The two principal types of oxidation that contribute to food deterioration are autoxidation of unsaturated fatty acids and enzyme-catalyzed reaction. The residues susceptible for oxidation are also present on proteins and carbohydrates, besides being present on lipids. Radicals formed on oxygen or sulfur (on proteins) and oxygen excited-state species ($^1O_2$) generated by heat, light, or metal catalyzed reactions target double bonds and generate reactive forms of these substrates (homoradicals). Subsequent cross-reactivity of intermediate species produces complex product matrices including auto-oxidative molecular products (Bennett et al., 2013).

In a mixed food system, lipids are considered to be most vulnerable to oxidation and also to initiate attack on non-lipid substances. In milk and dairy products, lipids, besides serving as the precursor of certain flavorful compounds like methyl ketones (Kinsella, 1969a) and lactones (Wadhwa & Jain, 1985), also induce the formation of few undesirable compounds that cause off-flavor defects via hydrolytic and oxidative rancidity. The oxidative flavors are products of autoxidation of unsaturated fatty acids of triglycerides, phospholipids, and cholesterol esters, which are essentially a free-radical chain reaction involving initiation, propagation, and termination stages. Unsaturated fatty acids oxidize to form unstable odorless and tasteless hydroperoxide that degrade to yield flavorful carbonyls and other compounds (O'Connor & O'Brien, 2006). It is assumed that during autoxidation, initially oxygen attaches itself in loose linkage to a double bond of the unsaturated fatty acid (Gunstone & Hilditch, 1945). This results in an activation of the adjacent (α-) methylene group, from which a hydrogen atom or proton gets loosened leading to drawing the oxygen atom towards it to form hydroperoxide and not peroxide, the process creates a new double bond at a different place; which has a *trans* configuration as opposed to the *cis* configuration of natural unsaturated fatty acids. Thus, the reaction proceeds via free radical mechanism. It is reported that cow ghee has lesser tendency in comparison to buffalo ghee for autoxidation because cow ghee absorbs oxygen slowly in comparison to buffalo ghee (Rangappa & Achaya, 1975). The oxidized unsaturated fatty acids, particularly oleic, linoleic, linolenic, and arachidonic acid lead to formation of n-alkanals, alk-2-enals, and alk-2,4-dienals. The type of off-flavor perceived depends upon the quality and quantity of carbonyls formed. Kinsella (1969b) reported that C_7–C_{10} alkanals possess oily and tallowy odor, C_7–C_{11} alk-2-enals exhibit oxidized painty odors, alkanals (C_5, C_6, and C_8) and alk-2-dienals may exhibit nutmeg spicy odors. The potent flavor compound oct-1,3-diene produced by autoxidation of linoleic acid is responsible for metallic flavor. Unsaturated alcohol pent-1-ene-3-ol produced from linoleic acid is responsible for an oily and grassy aroma (Forss, 1972).

Riboflavin has been traditionally considered as an active photosensitizer occurring naturally in dairy products like, milk, cheese, butter and its presence makes these products susceptible to photosensitized oxidation.

Photosensitizers are the substances that absorb light and become excited to one or more higher energy-rich state(s). They promote the photo-oxidation of diverse substrates, when foods are exposed to visible light (Dalsgaard et al., 2007). Photosensitizers are reported to have two excited states: singlet and triplet. The triplet-excited state has a longer lifetime and initiates the oxidation. Photo-oxidation by a photosensitizer can proceeds through two types of reactions, that is, either type I or type II. In type I reaction, the excited sensitizer (Sen*) undergoes internal reactions that ultimately results in the oxidative alteration of a second molecule primarily by free radical mechanism on the exposure of the primary substrate to UV radiation. In type I reactions, transfer of hydrogen atoms or electrons occurs via interaction of the triplet excited state of the sensitizer with the target, while in the type II reactions, the excited triplet sensitizer reacts with ground state oxygen to produce $^1O_2$ by energy transfer. $^1O_2$ is a strong oxidant because of its higher reactivity and has the potential to damage proteins, lipids, and so forth (Airado-Rodríguez et al., 2011). The two reactions can occur simultaneously, in a competitive mode (Spikes, 1988) and the same has been observed in milk (Lee & Min, 2009). However, at low oxygen concentrations, type I reactions are most common. After photodegradation, riboflavin is reported to break down to lumiflavin (under alkaline condition) or lumichrome (under acidic conditions) (Ahmad et al., 2006) and probably formylmethylflavin. Among these, lumichrome is reported to be a strong photosensitizer (Parks & Allen, 1977). Riboflavin has three absorption bands. The band with maxima between 430 and 460 nm is the main band responsible for the photo-oxidation of food, especially milk and dairy products (O'Connor & O'Brien, 2006). Wold et al. (2006) reported the presence of five photosensitizers in butter other than riboflavin: protoporphyrin, hematoporphyrin, a chlorophyll a-like molecule, and two unidentified tetrapyrroles. Chlorophyll and porphyrin molecules absorb light in the UV and violet region with absorption peaks of ~410 nm (the soret band) along with the absorption of light in the red above 600 nm, and therefore, they may be responsible for the formation of off-flavors in dairy products when exposed to light having wavelengths longer than 500 nm (Wold et al., 2006). Chlorophyllic compounds have also been suggested to contribute prominently to the major part of photo-oxidation in cow's milk (Airado-Rodríguez et al., 2011).

Proteins are very complex molecules organized in large structures and oxidation of proteins may have severe consequences on product quality, their functionality and nutritional qualities like loss of essential amino acids. Structurally, protein oxidation may lead to a number of modifications either on its side chains or on the backbone, including amino acid changes, protein cleavage, and promotion of cross-linking. Protein oxidation can be classified as direct or indirect oxidation. Direct oxidation involves attack of the radical species or hydrogen abstraction from protein leading to formation of a protein radical or anion. However, in indirect oxidation, the protein interacts with other components like secondary lipid oxidation products, such as aldehydes or reducing sugars, which may lead to the formation of carbonyl compounds on amino acid side chains or formation of protein carbonyl groups (Baron, 2013). Oxidation of proteins at the side chain often leads to development of protein carbonyls, alcohols, and peroxides. In dairy products, methionine sulfoxide, dityrosine and cysteic acid have been detected as the oxidation products of methionine, tyrosine, and cysteine, respectively (Toran et al., 1996; Østdal et al., 2000).

It is believed that metal-catalyzed oxidation of protein is a predominant mechanism in foods, because of abundance of transition metals, such as iron and copper in the food matrix. Østdal et al. (2000) showed that in milk, the lactoperoxidase can transfer a radical to other milk proteins, such as β- lactoglobulin, casein, and serum albumin and has been suggested to be a key element in oxidation of milk proteins. Milk proteins are also reported to be susceptible to photo-oxidation. Among milk proteins, casein is more susceptible to photo-oxidation than the globular proteins, α-lactalbumin, β-lactoglobulin, and lactoferrin (Baron, 2013).

Various dairy products are reported to show flavor defects due to the occurrence of oxidation of different components in the products like fat phase in ice cream besides the positive effects on the flavor, is also reported to be associated with the flavor defect like cardboard, painty, metallic, or oxidized that may be due to the auto-oxidation or lipolysis of fat (Marshall et al., 2003). Auto-oxidation of fat is also reported in milk powders and the dry milk deposits may self-ignite to cause explosion in milk powder factories (Walker & Jackson, 1978; Knipschildt, 1986). The oxidation of milk fat and the reducing sugar-protein browning (i.e., Maillard browning) are the most significant deteriorative changes that occur in dry milk products that result in the defects in sensory and functional properties of the product (Farkye, 2006). Stapefeldt et al. (1997) reported that low-heat powders are more susceptible to severe oxidative changes than medium-heat and high-heat powders during storage. The unsaturated fatty acids in whole milk powder oxidize to generate saturated aldehydes (Boon et al., 1976) which are responsible for stale, tallow, cardboard, and painty flavors, while polar lipids oxidize to generate unsaturated aldehydes and ketones which are responsible for oxidized flavor even at a very low level (Kinsella, 1969a; Keen et al., 1976; Hall & Anderson, 1985).

As mentioned earlier, light also influences milk flavor due to riboflavin-sensitized effects on milk proteins via oxidation of methionine to methional (3-methylthiopropionaldehyde) (Patton, 1954; Tada et al., 1971; Sattar et al., 1977) and also oxidation of other amino acids in the presence of light and riboflavin. Allen and Parks (1975) reported that exposure of milk serum to fluorescent light chemically modified 10 amino acids in immunoglobulins. Dimick (1976) reported the considerable loss (~80%) in the activity of milk lipase due to riboflavin-photosensitized oxidation when the milk was exposed to sunlight for 30 min. Light- and riboflavin-induced changes in cheese have also been reported (Deger & Ashoor, 1987).

7.4   QUALITY AND SAFETY ISSUES OF OXIDATION

The secondary oxidation products of lipid oxidation especially α, β-unsaturated aldehydes, such as 4-hydroxynonenal (4-NHE) or trans-4-hydroxy-2-hexanal (HHE) are reported to be toxic and very reactive. These can easily react to proteins or peptides. The interactions between sugar/lipid oxidation/degradation products and protein often result in addition of carbonyl groups to protein via covalent binding between the advanced glycation end products (AGEs), carboxymethyllysine was first AGE to be identified in milk by Tauer et al. (1999) or advanced lipid oxidation end products (ALE). There are evidences of the role of these interactions on the pathologies of several diseases (Baron, 2013).

7.5   METHODS OF MEASUREMENT OF OXIDATION

The methods of detection and measurement of the extent of oxidation, pertaining to the oxidation of lipid, in the dairy foods are mainly associated with the measurement of hydroperoxides which is one of the classical methods for quantifying the lipid oxidation. A number of methods are based on iodometric titration, that is, hydroperoxides oxidize and liberate iodine from potassium iodide, the test commonly referred to as peroxide value determination or spectrophotometric methods, that is, hydroperoxides oxidize the ferrous to ferric iron in the presence of ammonium thiocyanate to produce ferric thiocyanate, which can be spectrophotometrically quantified at 505 nm (Loftus Hills & Thiel, 1946), also hydroperoxides of the oxidized fat reacts with 1,5-diphenyl-carbohydrazide to yield red-colored products (Hamm et al., 1965). The progress of auto-oxidation is measured spectrometrically at 532 nm by the reaction based on the condensation of two molecules of thiobarbituric acid (TBA) with one of the end products of auto-oxidation, malondialdehyde which results in the red-colored complex (Dunkley & Jennings, 1951). The other traditional methods include m-phenylenediamine test, IR value, Anisidine value, Kreis test (for aldehydes), methods based on the carbonyl content of oxidized fats, and measurement of oxygen uptake by manometry or polarography (Henick et al., 1954; Tappel, 1955; Mehlenbacher, 1960). Recent methods of measurement include the use of instruments like electron spin resonance (ESR) spectrometry, static and dynamic GC/MS, head space GC/MS (electronic nose), and high performance liquid chromatography (HPLC) (Kim & Morr, 1996; Nielsen et al., 1997; Stapelfeldt et al., 1997; Kristensen & Skibsted, 1999). Certain tests are employed to determine the proneness of the fat to autoxidize that can be measured by studying the degree of resistance that a fat bears to autoxidation and it can be seen by iodine value, active oxygen method, rancimat method, induction period (by measuring manometric reading for macro work and by using Warburg apparatus for micro studies), and so forth.

Photoreactions in milk can be measured by monitoring the formation of $^1O_2$ as a function of light exposure using $^1O_2$ fluorescent probe. One of the probes, available under the trade name singlet oxygen sensor green (SOSG) reagent, is highly selective for $^1O_2$, and it does not show any appreciable response to hydroxyl radical or superoxide. SOSG is reported (Molecular Probes, 2004; Airado-Rodríguez, 2011) to emit weak blue fluorescence peaking at 395 and 416 nm for excitation at 372 and 393 nm. After reaction with $^1O_2$, it emits a green fluorescence similar to that of fluorescein (excitation/emission maxima ~504/525 nm) that is measured.

7.6 CHARACTERISTICS, TYPES, AND MECHANISM OF ACTION OF ANTIOXIDANTS

Antioxidants are the substances that inhibit, retard, or interfere with the formation of free radicals in fat-rich foods, thus terminating the oxidative reaction in its initial stage. From a practical standpoint, it means that when an antioxidant system is properly selected and correctly applied to meet the needs of a particular food item, will help to maintain the original freshness, flavor, and odor of the product for a longer period of time than would otherwise be possible. In the food industry, a substance having the technical function of delaying the oxidation of nutrients, such as lipids, sugars, and proteins, whose oxidation leads to an inevitable deterioration of the organo-leptic qualities of a food due to the formation of undesirable substances like aldehydes, ketones, and organic acids that yield off-flavors, is considered to be an antioxidant (Saad et al. 2007; Andre´ et al. 2010). However, to be used in foods, antioxidants must be nontoxic, inexpensive, effective at low concentrations (0.001–0.02%), capable of surviving processing (carry-through), stable in the finished products, and devoid of undesirable color, flavor, and odor effects (Shahidi & Zhong, 2010). These days, there is an extensive use of natural and synthetic antioxidants in the food and pharma-ceutical industry.

Antioxidants can be divided into primary or chain-breaking antioxidants and synergists or secondary antioxidants based on their mechanisms of action. Primary antioxidants include hindered phenols and secondary aryl amines, while secondary antioxidants include organophosphites and thioes-ters. All the primary antioxidants commonly used in foods, have either two—OH groups or one—OR group in the ortho or para positions (Hudson, 1990). They are effective at extremely low concentrations of 0.01% or less and for some of them the effectiveness decreases as concentration is increased. It is reported that at high concentrations, they may act as pro-oxidant due to their involvement in the initiation reactions (Cillard et al., 1980). Phenolic (primary) antioxidants, whether naturally occurring, for example, tocoph-erols or flavonoids or permitted synthetic compounds, such as hindered phenolic (e.g., BHT, BHA, and TBHQ) and polyhydroxy phenolic (e.g., gallates), inhibit chain reactions by acting as hydrogen donors or free radical acceptors, resulting in the formation of more stable products. They interfere directly with the free radical propagation process and thus block the chain reaction.

Secondary antioxidants or synergist can be accounted for metal chela-tion (Khokhar & Owusu Apenten, 2003). They have little direct effect on the autoxidation of lipids but are able to enhance considerably the action of primary antioxidants. Chelating agents and sequestering agents like citric acid and isopropyl citrate, amino acids, phosphoric acid, tartaric acid, ascorbic acid and ascorbyl palmitate, ethylenediaminetetraacetic acid (EDTA) chelate metallic ions (such as copper and iron) that promote lipid oxidation through a catalytic action. The chelators are referred to as syner-gists since they greatly enhance the action of phenolic antioxidants. Thus, antioxidants slow down the oxidation rates of foods by a combination of mechanisms like, scavenging free radicals; chelating pro-oxidative metals; quenching $^1O_2$ and photo-sensitizers, and inactivating lipoxygenase (Thorat et al., 2013). The effectiveness of antioxidants to scavenge free radicals in foods depends on the bond dissociation energy between oxygen and a phenolic hydrogen, reduction potential, and delocalization of the antioxidant radicals (Choe & Min, 2006; Cao et al., 2007). Phenolic compounds primarily inhibit lipid oxidation through their ability to scavenge free radicals and convert the resulting phenolic radicals into a low-energy form that does not further promote oxidation. Flavonoids are known to exhibit a strong metal chelating activity in addition to their antioxidant properties, with the arrangement of 4-keto and 5-OH, or 3' and 4'-OH substituents resulting in the formation of chelating complexes between flavonoids and divalent cations (Cheng & Breen, 2000). Carotenoids with nine or more conjugated double bonds are good $^1O_2$ quenchers by energy transfer. The $^1O_2$ quenching activity of carotenoids depends on the number of conjugated double bonds in the structure (Min & Boff, 2002; Foss et al., 2004) and the substituent in the β-ionone ring (Di Mascio & Sies, 1989). Beta-carotene and lycopene which have 11 conjugated double bonds are more effective $^1O_2$ quenchers than lutein which has 10 conjugated double bonds (Viljanen et al., 2002). The reaction mechanisms of a primary antioxidant, AH (Antunes et al., 1999) and secondary antioxidant BH, is shown in Figure 7.1.

a) Reaction of primary antioxidant (AH) with lipid radical

AH + ROO• → ROOH + A•

RH + A• → AH + A•

AH + ROO• → [ROO•AH] Complex

b) Termination reaction

[ROO•AH] → non-radical product

A• + A• → AA

A• + R• → RA

A• + ROO• → ROOA

c) Regeneration of primary antioxidant

A• + BH → AH + B•

FIGURE 7.1 Reaction mechanism of primary antioxidant with free radical. AH, antioxidant; ROO•, lipid peroxyl radical; ROOH, hydroperoxide; A•, antioxidant free radical; RH, unsaturated lipid; R•, lipid radical; ROO• AH, stable compound (non-radical product); BH, secondary hydrogen donor; B•, secondary antioxidant free radical (Antunes et al., 1999).

Antioxidants are naturally present in a wide variety of raw food materials; still there is a need to add antioxidants into foods so as to provide additional protection against oxidation. The antioxidants which are added to food products can be natural or synthetic compounds depending on their availability and preparations (Yanishlieva-Maslarova, 2001). Natural antioxidants such as polyphenols are primarily derived from plants, while the synthetic antioxidants are chemically produced. Antioxidants containing a phenol group play a prominent role in biological and food system (Shui & Leong, 2004).

The naturally occurring antioxidant substances are at times associated with the beneficial effects of foods (Vision et al., 1999). They are available in complex forms, which include tocopherols, lycopenes, flavonoids, nordihydroguaiaretic acid (NDGA), sesamol, gossypol, vitamins, provitamins and other phytochemicals, enzymes (catalase, glutathione peroxidase, and super oxide dismutase), minerals (Zinc and Selenium), and lecithin (Cuppett, 2001). α-tocopherol (vitamin E) is well known as one of the most efficient naturally occurring lipid-soluble antioxidants (McCarthy et al., 2001). The most important natural antioxidants commercially exploited are tocopherols, ascorbic acid and more recently plant extracts, such as, from sage (Djarmati et al., 1991), rosemary (Tena et al., 1997), green tea (Chen et al., 2004), spinach (Aehle et al., 2004), grape (Baydar et al., 2004), and marigold (Cetkovic et al., 2004) are also gaining acceptance. These extracts contain mainly phenolic compounds (e.g., flavonoids and phenolic acids), and they are well known for their antioxidant (López et al., 2001; Gülçin et al., 2003), antimicrobial (Gülçin et al., 2003), anti-ulcer, anti-carcinogenic (Chen et al., 2004), anti-mutagenic, and anti-inflammatory (Caillet et al., 2006) properties, as well as for reducing the risk of cardiovascular diseases (Cetkovic et al., 2004; Louli et al., 2004).

The antioxidant effects in milk rely primarily on endogenous compounds (Brien & Connor, 2003). However, synthetic antioxidant compounds are also widely used to inhibit progress of lipid oxidation. Some of the popular synthetic antioxidants used in many countries including India are Butylated Hydroxyanisole (BHA), t-butylhydroquinone (TBHQ), and esters of gallic acid (Yanishlieva-Maslarova, 2001). These are mainly phenolic compounds whose structure allows them to form low-energy radicals through stable resonance hybrids that prevent the further propagation of the oxidation reaction (Karovicova & Simko, 2000). Butylated hydroxytoluene (BHT) is very effective in animal fats, low-fat food, fish products, packaging materials, paraffin, and mineral oils but is less effective in vegetable oils, and is reported to be lost during frying because of its steam volatility (Gordon & Kourimska, 1995).

7.7   APPLICATION OF SYNTHETIC ANTIOXIDANTS AND THEIR EFFECTS

The synthetic antioxidants are lipid soluble and terminate free-radical chain reactions by donating hydrogen atoms or electrons to free radicals and thus, converting them to more stable structures (Frankel, 1998). The use of synthetic antioxidants in the prevention or retardation of autoxidation in lipids and lipid containing food products has been the subject of numerous investigations. The use of synthetic antioxidants in dairy products is prohibited in most countries. However, certain studies on the use of these in dairy products reveal that their effectiveness varies in different products. While NDGA inhibits the development of oxidized flavor in liquid milk, it promotes autoxidation in milk fat (Hammond, 1970). Tocopherols are very effective inhibitors of spontaneous or copper-induced oxidation in liquid milk (Dunkley et al., 1967; King, 1968) but have little effect in whole milk powder (Abbot & Waite, 1965). Other antioxidants that have been shown to exert protective effects are dodecyl gallate in spray-dried whole milk (Abbot & Waite, 1962), ascorbyl palmitate in lactic butter (Koops, 1964) and propyl gallate and quercetin in butter oil (Wyatt & Day, 1965). Anhydrous bovine or buffalo milk fats (ghee) may be stabilized when stored in a hot climate by combinations of phenolic antioxidants (BHA, BHT, and propyl gallate) and ascorbic acid (Helal et al., 1976). Wade et al. (1986) reported that BHA and BHT were effective in retarding oxidation of anhydrous milk fat but DL-α-tocopherol acted as a pro-oxidant.

There are a number of controversies surrounding the use of synthetic antioxidants. Since food additives are subjected to the most stringent toxicological testing procedures, only a few synthetic antioxidants have been used in foods for any length of time. Since the toxicity of some synthetic antioxidants is not easily assessed, as a result of which, a chemical may be considered safe by a country, tolerated in another country and forbidden in a third one (Thorat et al., 2013). For example, TBHQ is authorized as an antioxidant in the United States, while it is forbidden in the European Union countries. The legal limit for the addition of BHA and BHT to most foods in the United States is 200 mg/kg of fat. When added in combination, a total of 200 mg/kg of fat is permitted (CFR, 2001). In India, according to the Food Safety and Standards Regulations (2011), no antioxidant other than lecithin, ascorbic acid, and tocopherol shall be added to any food. However, in ghee and butter, BHA may be added in a concentration not exceeding 0.02%. It is interesting to note that the addition of these artificial chemicals is restricted by the FDA because of food safety concerns, not to mention emerging trends for consumer preferences toward more "green" food processing applications (Yue et al., 2008). BHA has been revealed to be carcinogenic in animal experiments. Similarly, at high doses, BHT is reported to cause internal and external hemorrhaging, leading to death in some strains of mice and guinea pigs (Ito et al., 1986). Natural antioxidants are, thus, generally recognized as safe when used in accordance with food manufacturing practices and therefore not limited in most foods (CFR, 2001). The addition of α-tocopherol, ascorbic acid, and ascorbyl palmitate to milk is permitted and no legal limit exists for the use of the same. However, the presence of these must be noted on the label and the same must not be used in higher concentrations as it may lead to pro-oxidative effects (Frankel, 1998).

7.8   NEED OF NATURAL ANTIOXIDANTS IN DAIRY PRODUCTS

The potential health hazards of synthetic antioxidants have prompted researchers to search for natural antioxidants from plant source (Lee & Shibamoto, 2002). Additionally, recent trends in the marketplace have focused on natural and organic products that do not utilize synthetic additives which have further spurred this research. Foods manufacturers have also been motivated to carry out research on the use of natural antioxidants because studies have shown that such compounds are not only beneficial to the shelf life of food products but also as preventive medicine. Reports revealing toxic and carcinogenic effects of BHA and BHT; and the higher manufacturing costs and lower efficiency of natural antioxidants such as tocopherols, together with the increasing consciousness of consumers with regard to food additive safety, created a need for identifying alternative natural and probably safer sources of food antioxidants (Wanasundara & Shahidi, 1998; De Oliveira et al., 2009; Prasad et al., 2009; Gutteridge & Halliwell, 2010). The replacement of synthetic antioxidants by natural ones may have benefits due to health implications and functionality of the natural antioxidants such as solubility in both oil and water that could be of interest in preparations of emulsions in food systems. Natural antioxidants from plant products are reported to be more effective in reducing ROS levels compared to synthetic single dietary antioxidants due to the synergistic actions of a wide range of biomolecules such as vitamins C and E, phenolic compounds, carotenoids, terpenoids, and phytomicronutrients (Podsędek, 2007; Serrano et al., 2007; Pérez-Jiménez et al., 2008).

The interest for the natural antioxidants can be explained either because of their capacity to ameliorate the quality of food and cosmetic products, and also more recently because of their potential role *in vivo* against free radicals, via feeding (health-food notion) or medication. The great diversity of antioxidant sources observed in the plant world can be explained for the phenolic compounds are largely expanded, and are found in every part of plant: flowers, fruits, grains, leaves, bark, and roots. The efficacy, however, depends not only on the amount of phenolic acids and flavonoids, but also of tocopherol (vitamin E), ascorbic acid (vitamin C), and β-carotene content. Sources of β-carotene include fruits, vegetables, dairy products, eggs, and a few seafoods. Fresh fruits and green vegetables are particularly rich in vitamin C. The crude oils, mainly that of wheat germ and nuts, are very good sources of vitamin E. The active components of natural plant-derived antioxidants are polyphenolic compounds. The most effective antioxidants are those that contain two or more phenolic hydroxyl groups (Dziedzic & Hudson, 1984; Shahidi et al., 1992). Plant phenolics compounds can either act as reducing agents, free radical terminators, metal chelators, or $^1O_2$ quenchers.

The antioxidants obtained from spices and herbs (oregano, thyme, dittany, marjoram, lavender, and rosemary) were reported to have limited applications in spite of their high antioxidant activity, as they impart a characteristic herb flavor to the food, thus deodorization steps are required (Reglero et al., 1999). Since many plant-derived substances often have a strong, distinctive taste of their own, plant-derived antioxidants must not only be tested for their ability to retard oxidation but also for any sensory characteristics they impart to the food product. Naturally occurring antioxidant substances also need safety testing. Caution regarding an assumption of safety of natural antioxidants has been repeatedly advised, since the fact that an antioxidant comes from a natural source does not prove its assumed safety. Hattori et al. (1998) summarized the requirements that antioxidants must satisfy to be used as food additives.

7.9   APPLICATION OF NATURAL ANTIOXIDANTS IN VARIOUS DAIRY PRODUCTS

Dairy products contain lipids rich in PUFA and their esters are easily oxidized by molecular oxygen over time. Deleterious changes in dairy products caused by lipid oxidation include not only loss of flavor or development of off-flavors, but also loss of color, nutrient value, and the accumulation of compounds, which may be detrimental to the health of consumers. One of the most effective ways of retarding lipid oxidation in dairy products is to incorporate antioxidants (Gad & Sayd, 2015). The growing interest in the study of natural antioxidant compounds has been accompanied by an increase in the market presence of functional foods or nutraceuticals or health foods. Fortification of dairy products with bioactive components (natural antioxidants compounds) enhances their antioxidant activity and anti-inflammatory properties, which prevent the damaging effects of free radicals (Berardini et al., 2005) and provide various health benefits.

Herbs, fruits, vegetables, spices, and other plant materials rich in phenolic compounds are of growing interest in the food industry because they retard oxidative degradation of lipids and thereby improve the quality and nutritional value of food (Wojdyło et al., 2007). Numerous herbs have the potential to retard lipid oxidation during storage of foods which is usually mediated through their intrinsic antioxidant activity. The reports on the addition of herb and spice extracts in milk and milk products is evolving rapidly (Pokorny & Korczak, 2001; Pawar et al., 2012) and the same is discussed as in the next section.

7.9.1 MILK AND MILK BASED BEVERAGES

Several approaches for antioxidant incorporation in milk have been used in an attempt to reduce lipid oxidation. Jung et al. (1998) added ascorbic acid (from 200 to 1000 mg/kg) directly to milk and, by using dynamic headspace analyses and gas chromatography, concluded that formation of dimethyl disulfide decreased. The result of sensory evaluation revealed that milk containing ascorbic acid (and therefore less dimethyl disulfide) improved in flavor in the presence of ascorbic acid.

In a subsequent study, the effect of antioxidants, added in a single initial dose or in weekly additions to extend shelf life of milk, was evaluated for over six weeks of lighted storage at 4 °C (van Aardt et al., 2005). Light induced oxidation was measured by determining pentanal, hexanal, heptanal, and 1-octen-3-ol contents. Weekly addition of a combination of BHA and BHT (100 mg/kg of milk fat, each) maintained heptanal content of milk at levels comparable to light-protected milk, whereas an initial single addition of α-tocopherol significantly decreased hexanal content over the first four weeks of storage. Odor-active compounds associated with light-induced oxidation included 2,3-butanedione, pentanal, dimethyl disulfide, hexanal, 1-hexanol, heptanal, 1-heptanol, and nonanal. The results revealed that the addition of BHA and BHT in a single initial addition resulted in decrease in pentanal and hexanal odor, but not in heptanal and 1-heptanol odor, whereas the addition of α-tocopherol and ascorbyl palmitate decreased pentanal and heptanol odor, but not hexanal and heptanal odor.

Antioxidant capacity of blends with 13% (w/w) bilberry and black currant extract in low-fat milk or low-fat fermented milk were assessed by Skrede et al. (2004). Antioxidant capacity in 13% berry/fermented milk blends packed in glass or cardboard cartridges and stored for three weeks in the dark or under fluorescent light was determined. Anti-radical power (ARP) and oxygen radical absorbance capacity (ORAC) values of most fruit preparations greatly exceeded those of plain milk. Milk blends with berry extract increased ARP values 5–13-folds and ORAC values by 40–100%. Packaging material, illumination, and storage period had no consistent effects on antioxidant capacity.

The use of oregano extract (OE) and oregano essential oil (OEO) as anti-oxidants in dairy beverages enriched with 2 g/100 g linseed oil was studied by Boroski et al. (2012). OE and OEO reduced light- and heat-induced oxidation of omega-3 fatty acids and change in color during storage of dairy beverages enriched with linseed oil. OE showed better antioxidant properties than OEO. Physical stability of dairy beverages was not affected by the addition of OE or OEO. It was concluded that natural antioxidants can be added to dairy beverages enriched with omega-3 fatty acid to effectively inhibit oxidation during storage.

Jung et al. (2015) investigated the physicochemical and antioxidant properties of milk supplemented with red ginseng extract (RGE) (at 0.5, 1, 1.5, and 2%) during storage at 4 °C. The antioxidant activity of milk samples was determined using the 1,1-diphenyl-2-picrylhydrazyl (DPPH) method, β-carotene bleaching assay, and ferric thiocyanate assay. It was observed that the antioxidant activity of milk samples supplemented with RGE was higher than that of the control sample.

7.9.2 GHEE/CLARIFIED BUTTER FAT

Ghee, the most famous traditional dairy product in India, Egypt, and many countries in Middle East, undergoes oxidative degradation during storage, resulting in alteration of major quality parameters affecting its suitability for consumption. Development of rancidity reduces the shelf life of the product, which ultimately affects consumer acceptability (Mehta, 2006; Mariod et al., 2010; Pawar et al., 2012). A number of natural antioxidants have been added during processing and reported to have elongated the shelf life and oxidative stability of stored products.

Merai et al. (2003) reported that water insoluble fraction of Tulsi (*Ocimum sanctum* L.) leaves possess good antioxygenic properties and phenolic substances present in Tulsi leaves were the main factors in extending the oxidative stability of ghee (Butterfat). Pankaj et al. (2013) reported that addition of ethanolic extract of Arjuna (*Terminalia arjuna* Wight & Arn.) bark at 7% by the weight was highly effective in retarding the auto-oxidation of both cow and buffalo ghee during storage. However, the ability of ethanolic extract of Arjuna to enhance the antioxidant potential of ghee was observed to be more pronounced in case of cow ghee than in buffalo ghee. Shelf life of the Arjuna ghee samples was eight days at 80 ± 1 °C as compared to two days in the control. A study was conducted by Pawar et al. (2012) for evaluating the effect of Shatavari *(Asparagus racemosus)* on storage stability of ghee. It was observed that the samples incorporated with ethanolic extract of Shatavari showed a strong activity in quenching DPPH radicals than the aqueous extract of the same herb. Gandhi et al. (2013) carried out study on oxidative stability of ghee added with Vidarikand (*Pueraria tuberosa*) (both aqueous and ethanolic) extracts and compared the same with BHA, TBHQ, rosemary, and green tea using β-carotene bleaching assay, DPPH, and the rancimat method. Phenolic content and antioxidative activity of ethanolic extract of Vidarikand was more compared to its aqueous extract. Ethanolic extract of the Vidarikand was more effective in preventing the development of the peroxide value and conjugated diene value in ghee compared to its aqueous extract. Aqueous and ethanolic extracts of Vidarikand were found to be capable of retarding oxidative degradation in ghee but were less effective than natural (rosemary and green tea) and synthetic (BHA and TBHQ) antioxidants. Similar work was carried out by Pawar et al. (2014) on antioxidant activities of Vidarikand (*P. tuberosa*), Shatavari (*A. racemosus*), and Ashwagandha (*Withania somnifera*) extracts (aqueous and ethanolic) which were evaluated and compared with BHA. Antioxidant activity of the herbs decreased in the order Vidarikand > Ashwagandha > Shatavari. Thus, the ethanolic extract of Vidarikand was found to have the maximum antioxidant activity among all the herbs.

A huge amount of plant biomass wastes are produced yearly as by-products from the agro-food industries. These wastes are attractive sources of natural antioxidants. In one of the studies, natural antioxidants found in peanut skins (PS), pomegranate peels (PP), and olive pomace (OP) were extracted using ethanol (80 %), ethyl acetate, and n-hexane and the oxidative stability of ghee during storage under thermal oxidative conditions was reported. Ethanol extract showed slightly better antioxidant characteristics compared with ethyl acetate and hexane extracts. It could be due to the reason that extracts obtained from higher-polarity solvents were more effective radical scavengers than those obtained using lower-polarity solvents. Extracts obtained from PS exhibited strong antioxidant capacity in all assays, followed by PP and OP extracts (El-Shourbagy & El-Zahar, 2014).

In a subsequent study, Asha et al. (2015) evaluated the antioxidant activities of BHA and orange peel powder extract in ghee stored at different storage temperatures during the storage period of 21 days. Ghee incorporated with orange peel extract (OPE) showed stronger activity in quenching DPPH radicals and least development of peroxide value, free fatty acid content and TBA than ghee incorporated with BHA and control. The study revealed that orange peel could be a good natural source of antioxidants which could be used in fat rich food products like ghee to retard oxidative deterioration.

7.9.3 FERMENTED MILK PRODUCTS

Yogurt is among the most common dairy products consumed around the world (Saint-Eve et al., 2006). Yoghurt with added antioxidants from natural sources appears to be a convenient food format to satisfy consumer interest in terms of beneficial effects of starter cultures, and health benefits of added antioxidants over original yoghurt nutrients. For this reason, several attempts to produce yoghurts fortified with natural antioxidant-rich extracts have been studied, including supplementation with polyphenol-rich wine extract (Howard et al., 2000), *Hibiscus sabdariffa* extract (Lwalokun & Shittu, 2007), pycnogenol from French marine bark extract (Ruggeri et al., 2008), green bell pepper juice (Halah & Mehanna, 2011), quince scalding water (Trigueros et al., 2011), apple polyphenols (Sun-Waterhouse et al., 2012), grape and grape callus extracts (Karaaslan et al., 2011), tea infusions (Najgebauer-Lejko et al., 2011), grape seed extracts (Chouchouli et al., 2013), berry polyphenols (Sun-Waterhouse et al., 2013) and pomegranate peel extracts (PPE) (El-Said et al., 2014).

Chouchouli et al. (2013) evaluated the potential of using grape seed extracts from two Greek wine grape varieties for the production of anti-oxidant-rich full-fat and non-fat yoghurts. Fortified yoghurts contained more polyphenols and exhibited higher antiradical and antioxidant activity than controls, even after 3–4 weeks of cold storage. The degradation of polyphenols and the decrement of yoghurts' antiradical and antioxidant activities followed first order kinetics, with full-fat yoghurts exhibiting higher deterioration rates and lower half-lives than the non-fat ones. In one of the studies, stirred yoghurt was prepared from reconstituted skim milk powder fortified with 5, 10, 15, 20, 25, 30, and 35% of the aqueous extract of whole PPE, before and after inoculation with the traditional yoghurt starter (El-Said et al., 2014). Addition of PPE before inoculation with the starter resulted in stirred yoghurt of higher antioxidant activities than that with PPE added after inoculation with the starter. Also, increasing the percentage of the added PPE increased significantly the antioxidant activities of stirred yoghurt up to 25% and further increase in the percentage of added PPE led no significant effect.

Najgebauer-Lejko et al. (2014) evaluated the sensory quality and antioxidant capacity of yoghurts with addition of selected vegetables (carrot, pumpkin, broccoli, and red sweet pepper) (at 10% w/w) during two weeks of refrigerated storage. The highest ability to scavenge DPPH radicals was stated for yoghurts with broccoli and red sweet pepper. All vegetable yoghurts were characterized by higher ferric reducing antioxidant power (FRAP) values (measured directly after production) than the natural yoghurt. However, the level of this parameter significantly decreased after storage. Authors concluded that the red sweet pepper fortified yoghurt was the most beneficial in terms of antioxidant properties and organoleptic acceptance of the studied yoghurts.

Singh et al. (2013) carried out a study on the effect of strawberry polyphenol extract (0.5 mg/mL) addition on physicochemical and antioxidant properties of stirred dahi preparation. This fortification resulted in a 7-fold increase in the antioxidant activity of polyphenol-enriched stirred dahi, while pH, acidity, water-holding capacity, and viscosity remained comparable with the control. Sensory analysis indicated that the product was acceptable up to two weeks when stored at 7–8 °C with no significant difference ($p > 0.05$) in the antioxidant activity and total phenolic content (TPC).

7.9.4 CHEESE

The essential oils obtained from aromatic plants are natural products. Many essential oils have demonstrated antioxidant properties; specifically, the essential oils of oregano, rosemary, laurel, and other plants have been studied as potential natural antioxidants (Kulisic et al., 2004; Olmedo et al., 2008; Olmedo et al., 2009; Olmedo et al., 2012; Asensio et al., 2011).

Gad and Abd El-Salam (2010) monitored the effect of addition of different concentration of rosemary extract (RE) to skim milk during processing for production of soft cheese. The antioxidant activity of the blends of skim milk and REs was improved by heat treatment. The addition of calcium chloride and pasteurization further significantly increased the phenol content and the antioxidant activity of skim milk, whereas addition of sodium chloride and homogenization of the skim milk-REs did not affect antioxidant activity. Skim milk with high rosemary concentration was reported to have high antioxidant activity.

In a subsequent study, ultra-filtered (UF)-soft cheese was prepared from UF milk retentate (1.5% fat) and supplemented with 1–5% RE and cold stored for 30 days (El-Din et al., 2010). The TPC and antioxidant capacity were evaluated using DPPH and FRAP methods in retentate before and after pasteurization and salting along with the resultant cheese. Fortification of retentate with RE increased its phenolic content and consequently, its anti-oxidant activity. Pasteurization increased the TPC and antioxidant activity. It was observed that addition of 3% NaCl reduced slightly the radical scavenging activity (RSA), TPC, and FRAP values. Moreover, it was noticed that UF-soft cheese fortified with 1% RE retained more TPC and antioxidant activity; also, increasing the concentration of RE to 5% had more acceptable flavor, body and texture, and antioxidant activity until 30 days. Furthermore, the rate of decrease in TPC, RSA (%), and FRAP values in cheese samples containing RE after 30 days of storage were less as compared to control cheese (without RE) (El-Din et al., 2010).

A functional cheese product containing polyphenolic compounds was developed, and the polyphenolic retention efficiency and antioxidant property of the product was evaluated by Han et al. (2011). Single phenolic compounds, including catechin, epigallocatechin gallate (EGCG), tannic acid, homovanillic acid, hesperetin and flavones, and natural crude compounds, such as grape extract, green tea extract, and dehydrated cranberry powder, were added as functional ingredients to the prepared cheese. Cheese curds with polyphenolic compounds at a concentration of 0.5 mg/mL showed effective free radical-scavenging activity and showed high retention coefficient ranging between 0.74 and 0.87. The nutritional value of cheese product was reported to be improved by adding bioactive phenolic compounds to the cheese curd.

Olmedo et al. (2013) evaluated the preservative effect of oregano and rosemary essential oils on the oxidative and fermentative stabilities of flavored cheese prepared with cream cheese base. The addition of oregano and rosemary essential oils was reported to improve the oxidative and fermentative stability along with preventing the lipid oxidation and the development of rancid and fermented flavors in the said cheese. As consequence, these essential oils prolonged the shelf life of the product.

7.9.5 OTHER DAIRY PRODUCTS

Herbs fortified dairy products serve as a good source of antioxidant provided the antioxidant capacity of the dairy product and herbs are not depleted through oxidation–reduction reactions upon mixing and storage of the products. Bandyopadhyay et al. (2007) carried out a study on the antioxidant activities of beet (*Beet vulgaris*), mint (*Mentha spicata* L.), and ginger (*Zingiber officinale* L.) alone and in combination after fortification in sandesh (a heat desiccated product of coagulated milk protein mass). The same was compared with the synthetic antioxidants like TBHQ, BHA, and BHT. Among the natural sources, ginger and combination of ginger with mint showed excellent results and value was almost equal to TBHQ (200 mg/kg). It was suggested that the use of BHA and BHT can be substituted by all the natural sources (beet, mint, and ginger) alone or in combined form. Among the four forms of herbs such as paste, tray-dried powder, freeze-dried powder, and solvent extracted form, addition of solvent extracted form in sandesh preparation showed highest antioxidant level than any other form. In their subsequent study, Bandyopadhyay et al. (2008) demonstrated the antioxidant activity, peroxide value, and ultra-violet absorbance to evaluate the effectiveness of natural antioxidants in reducing lipid oxidation in sandesh as compared to synthetic antioxidants. All the natural sources and their combinations significantly improved the oxidative stability of sandesh and their effectiveness was comparable with synthetic antioxidant TBHQ and a combination of BHA and BHT. Among the natural sources, although ginger had the highest antioxidant activity but mint showed better effectiveness in the inhibition of lipid oxidation. Regarding antioxidant activity and lipid oxidation, combination of mint or ginger with beet showed better result as compared to beet alone.

The antioxidant activities of peels of pomegranate, lemon, and orange extracts were studied in paneer samples. Among the three extracts pomegranate exhibited a high percentage of antioxidant activity and phenolic content in comparison to lemon and OPE. The ability to prevent peroxide formation in paneer sample was in the order of pomegranate peel > lemon peel > orange peel (Singh & Immanuel, 2014) (Table 7.2).

TABLE 7.2 Application of Natural Antioxidants in Milk and Dairy Products.

Product	Natural Antioxidant	Form	Reference
Milk and milk based beverages	Bilberry and Black currant	Extract	Skrede et al. (2004)
	Oregano	Extract and essential oil	Boroski et al. (2012)
	Red ginseng	Extract	Jung et al. (2015)
Ghee/clarified butter fat	Tulsi	Water insoluble fraction of leaf	Merai et al. (2003)
	Arjuna (*Terminalia arjuna* Wight & Arn.)	Ethanolic extract of bark	Pankaj et al. (2013)
	Shatavari (*Asparagus racemosus*)	Ethanolic extract	Pawar et al. (2012)
	Vidarikand (*Pueraria tuberosa*)	Aqueous and ethanolic	Gandhi et al. (2013)
	Ashwagandha (*Withania somnifera*)	-Do-	Pawar et al. (2014)
	Peanut skins, pomegranate peels, olive pomace	Extract using ethanol, ethyl acetate, and n-hexane	El-Shourbagy and El-Zahar (2014)
	orange peel	Extract	Asha et al. (2015)
Fermented milk products	*Hibiscus sabdariffa*	-Do-	Lwalokun and Shittu (2007)
	Polyphenol-rich wine extract	-Do-	Howard et al. (2000)
	Pycnogenol from French marine	Extract	Ruggeri et al. (2008)
	Green bell pepper	Juice	Halah and Mehanna (2011)
	Quince scalding water	-Do-	Trigueros et al. (2011)
	Apple polyphenols	-Do-	Sun-Waterhouse et al. (2012)
	Grape and grape callus	-Do-	Karaaslan et al. (2011)
	Tea	Infusions	Najgebauer-Lejko et al. (2011)
	Grape seed	Extracts	Chouchouli et al. (2013)

TABLE 7.2 *(Continued)*

Product	Natural Antioxidant	Form	Reference
	Berry polyphenols	-Do-	Sun-Waterhouse et al. (2013)
	Pomegranate peel	-Do-	El-Said et al. (2014)
	Strawberry polyphenol	-Do-	Singh et al. (2013)
Cheese	Rosemary	-Do-	Gad and Abd El-Salam (2010)
Paneer	Pomegranate, lemon, and orange peels	-Do-	Singh and Immanuel (2014)
Sandesh	Beet (*Beet vulgaris*), mint (*Mentha spicata* L.), and ginger (*Zingiber officinale* L.)	-Do-	Bandyopadhyay et al. (2007)

7.10 CONCLUSION

Milk and milk products are considered to be complete food due to the presence of almost all the macro and micronutrients; and thus are also regarded as an indispensable part of the diet. However, due to the presence of several oxidants, they are prone to oxidation in spite of the presence of naturally occurring antioxidants. Hence, addition of antioxidants into the dairy products is gaining importance these days. The most widely used synthetic antioxidants in food (BHT and BHA) are very effective in their role as antioxidants but their use in food products has been failing off due to their instability, as well as due to the suspected action as promoters of carcinogenesis. Consequently, there has been considerable interest in the use of natural antioxidants on account of safety and acceptability. Most of the naturally occurring antioxidants not only do keep the food stable against oxidation but can also be effective in controlling microbial growth. However, better understanding of the role of natural antioxidants on food stability and human health is required; and toxicological studies are mainly carried out to establish the no-effect level for an acceptable daily intake for humans as the origin of the antioxidant from a natural source does not prove its assumed safety.

KEYWORDS

- milk
- oxidants
- synthetic antioxidants
- natural antioxidants
- extracts
- essential oil

REFERENCES

Abbot, J.; Waite, R. The Effect of Antioxidants on the Keeping Quality of Whole Milk Powder. I. Flavones, Gallates, Butylhydroxyanisole and Nordihydroguaiaretic Acid. *J. Dairy Res.* **1962,** *29,* 55–61.

Abbot, J.; Waite, R. The Effect of Antioxidants on the Keeping Quality of Whole Milk Powder. II. Tocopherols. *J. Dairy Res.* **1965,** *32,* 143–146.

Aehle, E.; Raynaud-Le Grandic, S.; Ralainirina, R.; Baltora-Rosset, S.; Mesnard, F.; Prouillet, C.; Mazière, J. C.; Fliniaux M. A. Development and Evaluation of an Enriched Natural Antioxidant Preparation Obtained from Aqueous Spinach (Spinacia oleracea) Extracts by an Adsorption Procedure. *Food Chem.* **2004,** *86,* 579–585.

Ahmad, I.; Fasihullah, Q.; Vaid, F. H. Effect of Light Intensity and Wavelengths on Photo-degradation Reactions of Riboflavin in Aqueous Solution. *J. Photoch. Photobio. B.* **2006,** *82,* 21–27.

Airado-Rodríguez, D.; Intawiwat, N.; Skaret, J.; Wold, J. P. Effect of Naturally Occurring Tetrapyrroles on Photooxidation in Cow's Milk. *J. Agr. Food Chem.* **2011,** *59* (8), 3905–3914.

Allen, C.; Parks, O. W. Evidence for Methional in Skim Milk Exposed to Sunlight. *J. Dairy Sci.* **1975,** *58* (11), 1609–1611.

Allen, J. C.; Wrieden, W. L. Influence of Milk Proteins on Lipid Oxidation in Aqueous Emulsion. II. Lactoperoxidase, Lactoferrin, Superoxide Dismutase and Xanthine Oxidoreductase. *J. Dairy Res.* **1982,** *42,* 249–263.

Andre´, C.; Castanheira, I.; Cruzb, J. M.; Paseiro, P.; Sanches Silva, A. Analytical Strategies to Evaluate Antioxidants in Food: A Review. *Trends Food Sci. Technol.* **2010,** *21,* 229–246.

Antunes, F.; Barclay, L. R. C.; Ingold, K. U.; King, M.; Norris, J. Q.; Scaiano, J. C.; Xi, F. On the Antioxidant Activity of Melatonin. *Free Radical Biol. Med.* **1999,** *26,* 117–128.

Asensio, C. M.; Nepote, V.; Grosso, N. R. Chemical Stability of Extra-Virgin Olive Oil Added with Oregano Essential Oil. *J. Food Sci.* **2011,** *76,* 445–450.

Asha A.; Manjunatha, M.; Rekha, R. M.; Surendranath, B.; Heartwin, P.; Rao, J.; Magdaline, E.; Sinha, C. Antioxidant Activities of Orange Peel Extract in Ghee (Butter Oil) Stored at Different Storage Temperatures. *J. Food Sci. Technol.* **2015,** *52,* 8220–8227. DOI 10.1007/s13197-015-1911-3.

Bandyopadhyay, M.; Chakraborty, R.; Raychaudhuri, U. A Process for Preparing a Natural Antioxidant Enriched Dairy Product (Sandesh). *LWT-Food Sci. Technol.* **2007,** *40,* 842–851.

Bandyopadhyay, M.; Chakraborty, R.; Raychaudhuri, U. Antioxidant Activity of Natural Plant Sources in Dairy Dessert (Sandesh) under Thermal Treatment. *LWT-Food Sci. Technol.* **2008,** *41,* 816–825.

Baron, C. P. Protein Oxidation in Foods and Its Prevention. In *Food Oxidants and Antioxidants Chemical, Biological, and Functional Properties*; Bartosz, G., Ed.; CRC Press Taylor and Francis Group: Boca Raton, FL, 2013; pp 115–136.

Baydar, N. G.; Özkan, G.; Sağdiç, O. Total Phenolic Contents and Antibacterial Activities of Grapes (Vitis vinifera L.) Extracts. *Food Control.* **2004,** *15,* 335–339.

Benjamin, S.; Spener, F. Conjugated Linoleic Acids as Functional Food: An Insight into Their Health Benefits. *Nutr. Metab.* **2009,** *6,* 36–50.

Bennett, L.; Logan, A.; Shiferaw-terefe, N.; Singh, T.; Warner, R. Measuring the Oxidation potential in Foods. In *Food Oxidants and Antioxidants Chemical, Biological, and Functional Properties*; Bartosz, G.; Ed. CRC Press Taylor and Francis Group: Boca Raton, FL, 2013; pp 47–78.

Berardini, N.; Knodler, M.; Schieber, A.; Carle, R. Utilization of Mango Peels as a Source of Pectin and Polyphenolics. *Innov. Food Sci. Emerg. Technol.* **2005,** *6,* 442–452.

Boirie, Y.; Dangin, M.; Gachon, P.; Vasson, M. P.; Maubois, J. L.; Beaufrere, B. Slow and Fast Dietary Proteins Differently Modulate Postprandial Protein Accretion. *Proc. Nat. Acad. Sci. USA.* **1997,** *94,* 14930–14935.

Boon, P. M.; Keen, A. R.; Walker, N. J. Off-Favour in Stored Whole Milk Powder. II. Separation and Identification of Individual Monocarbonyl Components. *New Zeal. J. Dairy Sci. Technol.* **1976,** *11,* 189–195.

Boroski, M.; Giroux, H. J.; Sabik, H.; Petit, H. V.; Visentainer, J. V.; Matumoto-Pintro, P. T.; Britten, M. Use of Oregano Extract and Oregano Essential Oil as Antioxidants in Functional Dairy Beverage Formulations. *LWT - Food Sci. Technol.* **2012,** *47,* 167–174.

Brien, N.; Connor, T. Lipid Oxidation. In *Encyclopedia of Dairy Science*; Fuquay, J. W., Fox, P. F., Eds.; Academic Press: New York, 2003; pp 1600–1604.

Burton, G. W.; Ingold, K. U. Auto-Oxidation of Biological Molecules. I. The Antioxidant Activity of Vitamin E and Related Chain Breaking Phenolic Antioxidants in Vitro. *J. Am. Chem. Soc.* **1981,** *103,* 6472–6477.

Caillet, S.; Salmiéri, S.; Lacroix, M. Evaluation of Free Radical-Scavenging Properties of Commercial Grape Phenol Extracts by a Fast Colorimetric Method. *Food Chem.* **2006,** *95* (1), 1–8.

Cao, W.; Chen, W.; Sun, S.; Guo, P.; Song, J.; Tian, C. Investigating the Antioxidant Mechanism of Violacein by Density Functional Theory Method. *J. Mol. Str. Theochem.* **2007,** *817,* 1–4.

Cetkovic, G. S.; Djilas, S. M.; Canadanovic-Brunet, J. M.; Tumbas, V. T. Antioxidant Properties of Marigold Extracts. *Food Res. Int.* **2004,** *37,* 643–650.

Chen, D.; Daniel, K. G.; Kuhn, D. J.; Kazi, A.; Bhuiyan, M.; Li, L.; Wang, Z.; Wan, S. B.; Lam, W. H.; Chan, T. H.; Dou, Q. P. Green Tea and Tea Polyphenols in Cancer Prevention. *Front. Biosci.* **2004,** *1,* 2618–2631.

Chen, Z. Y.; Nawar, W. W. Prooxidative and Antioxidative Effects of Phospholipids on Milk Fat. *J. Am. Oil. Chem. Soc.* **1991,** *68* (12), 938–940.

Cheng, I.; Breen, K. On the Ability of Four Flavonoids, Baicilein, Luteolin, Naringenin, and Quercetin, to Suppress the Fenton Reaction of the Iron-ATP Complex. *Biometals.* **2000,** *13,* 77–83.

Choe, E.; Min, D. B. Mechanisms and Factors for Edible Oil Oxidation. *Compr. Rev. Food Sci. Food Saf.* **2006**, *5,* 169–186.

Chouchouli, V.; Kalogeropoulos, N.; Konteles, S. J.; Karvela, E.; Makris, D. P.; Karathanos, V. T. Fortification of Yoghurts with Grape (*Vitis vinifera*) Seed Extracts. *LWT - Food Sci. Technol.* **2013**, *53,* 522–529.

Cillard, J.; Cillard, P.; Cormier, M.; Girre, L. α–Tocopherol Prooxidants Effect in Aqueous Media: Increased Autoxidation Rate of Linoleic Acid. *J. Am. Oil Chem. Soc.* **1980**, *57,* 252–255.

Code of Federal Regulations (CFR). Code of Federal Regulations, US Government Printing Office: United States, Washington, DC, 2001.

Cuppett, L.; *Susan.* The Use of Natural Antioxidants in Food Products of Animal Origin. In *Antioxidants in Food Practical Applications*; Pokorny, J., Yanishlieva, N., Gordon, M., Eds.; CRC press, Woodhead publishing Ltd.: Cambridge, 2001; pp 284–310.

Dalsgaard, T. K.; Otzen, D.; Nielsen, J. H.; Larsen, L. B. Changes in Structures of Milk Proteins upon Photo-Oxidation. *J. Agr. Food Chem.* **2007**, *55* (26), 10968–10976.

De Oliveira, O. C.; Valentim, I. B.; Silva, C. A.; Bechara, E. J. H.; De Barros, M. P.; Mano, C. M.; Goulart, M. O. F. Total Phenolic Content and Free Radical Scavenging Activities of Methanolic Extract Powders of Tropical Fruit Residues. *Food Chem.* **2009**, *115,* 469–475.

Deger, D.; Ashoor, S. H. Light-Induced Changes in Taste, Appearance, Odor, and Riboflavin Content of Cheese. *J. Dairy Sci.* **1987**, *70* (7), 1371–1376.

Di Mascio, P.; Kaiser, S.; Sies, H. Lycopene as the Most Efficient Biological Carotenoid Singlet Oxygen Quencher. *Arch. Biochem. Biophys.* **1989**, *274,* 532–538.

Dimick, P. S. Effect of Fluorescent Light on Amino Acid Composition of Serum Proteins from Homogenized Milk. *J. Dairy Sci.* **1976**, *59* (2), 305–308.

Djarmati, Z.; Jankov, R. M.; Schwirtlich, E.; Djulinac, B.; Djordjevic, A. High Antioxidant Activity of Extracts Obtained from Sage by Supercritical CO_2 Extraction. *J. Am. Oil Chem. Soc.* **1991**, *68* (10), 731–734.

Dugan, L. R. Natural Antioxidants. In *Auto-Oxidation in Food and Biological Systems;* Simic, M. G., Karel, M., Eds.; Plenum Press: New York, 1980; pp 261–282.

Dunkley, W. L.; Jennings, W. G. A Procedure for Application of the Thiobarbituric Acid Test to Milk. *J. Dairy Sci.* **1951**, *34* (11), 1064–1069.

Dunkley, W. L.; Ronning, M.; Franke, A. A.; Robb, J. Supplementing Rations with Tocopherol and Ethoxyquin to Increase Oxidative Stability of Milk. *J. Dairy Sci.* **1967**, *50,* 492–499.

Dziedzic, S. Z.; Hudson, B. J. F. Phenolic Acids and Related Compounds as Antioxidants for Edible Oils. *Food Chem.* **1984**, *14,* 45–51.

Eichner, K. Anti-Oxidative Effect of Maillard Reaction Intermediates. In *Auto-Oxidation in Food and Biological Systems;* Simic, M. G., Karel, M., Eds.; Plenum Press: New York, 1980; pp 267–285.

El-Din, H. M. F.; Ghita, I. E.; Badran, S. M. A.; Gad, A. S.; El-Said, M. W. Manufacture of Low Fat UF-Soft Cheese Supplemented with Rosemary Extract (as Natural Antioxidant). *J. Am. Sci.* **2010**, *6,* 570–579.

El-Said, M. M.; Haggag, H. F.; Fakhr El-Din, H. M.; Gad, A. S.; Farahat, A. M. Antioxidant Activities and Physical Properties of Stirred Yoghurt Fortified with Pomegranate Peel Extracts. *Ann. Agr. Sci.* **2014**, *59,* 207–212.

El-Shourbagy, G. A.; El-Zahar, K. M. Oxidative Stability of Ghee as Affected by Natural Antioxidants Extracted from Food Processing Wastes. *Ann. Agr. Sci.* **2014**, *59,* 213–220.

Eriksson, C. E. Lipid Oxidation Catalysts and Inhibitors in Raw Materials and Processed Foods. *Food Chem.* **1982**, *9,* 3–19.

Farkye, N. Y. Significance of Milk Fat in Milk Powder. In *Advanced Dairy Chemistry Volume 2: Lipids;* 3rd ed.; Fox, P. F., McSweeney, P. L. H., Eds.; Springer: New York, 2006; pp 451–466.

Forss, D. A. Odor and Flavor Compounds from Lipids. *Prog. Chem. Fats Other Lipids.* **1972,** *13,* 177–258.

Foss, B.; Sliwka, H.; Partali, V; Cardounel, A.; Zweier, J.; Lockwood, S. Direct Superoxide Anion Scavenging by a Highly Water-Dispersible Carotenoid Phospholipid Evaluated by Electron Paramagnetic Resonance (EPR) Spectroscopy. *Bioorg. Med. Chem. Lett.* **2004,** *14,* 2807–2812.

Fox, P. F.; McSweeney, P. L. H.; Eds.; Milk Proteins. In *Dairy Chemistry and Biochemistry;* Blackie Academic and Professional: London, UK, 1998; pp 146–237.

Fox, P. F; McSweeney, P. L. H.; Eds.; Lactose: Chemistry and Properties. In *Advanced Dairy Chemistry. Volume 3: Lactose, Water, Salts and Minor Constituents*; 3rd ed.; Springer: New York, 2009; pp 1–16.

Frankel, E. N. Antioxidants. In *Lipid Oxidation*; The Oily Press Ltd.: Dundee, Scotland, 1998; pp 129–166.

FSSA. The Food Safety and Standards Act, Universals: New Delhi, India, 2006.

FSSAI. Food Safety Standards Authority of India, Notification, In *The Gazette of India*; Ministry of Health and Family Welfare: New Delhi, India, May 5, 2011.

Gad, A. S.; Abd El-Salam, M. H. The Antioxidant Properties of Skim Milk Supplemented with Rosemary and Green Tea Extracts in Response to Pasteurisation, Homogenisation and the Addition of Salts. *Int. J. Dairy Technol.* **2010,** *63,* 349–355.

Gad, A. S.; Sayd, A. F. Antioxidant Properties of Rosemary and Its Potential Uses as Natural Antioxidant in Dairy Products—A Review. *Food Nutr. Sci.* **2015,** *6,* 179–193.

Gandhi, K.; Pawar, N.; Kumar, A.; Arora, S. Effect of Vidarikand (Extracts) on Oxidative Stability of Ghee: A Comparative Study. *Res Rev: J. Dairy Sci. Technol.* **2013,** *2,* 1–10.

German, J. B. Butyric Acid: A Role in Cancer Prevention. *Nutr. Bull.* **1999,** *24,* 293–309.

Gordon, M.; Kourimska, L. Effect of Antioxidants on Losses of Tocopherols during Deep-Fat Frying. *Food Chem.* **1995,** *52,* 175–177.

Gülçin, L.; Oktay, M.; Kireçci, E.; Küfrevioğlu. Ö. İ. Screening of Antioxidant and Antimicrobial Activities of Anise (*Pimpinella anisum* L.) Seed Extracts. *Food Chem.* **2003,** *83,* 371–382.

Gunstone, F. D.; Hilditch, T. P. The Union of Gaseous Oxygen with Methyl Oleate, Linoleate, and Linolenate. *J. Chem. Soc.* **1945,** 836–841.

Gutteridge, J. M. C.; Halliwell, B. Antioxidants: Molecules, Medicines and Myths. *Biochem. Biophys. Res. Commun.* **2010,** *393,* 561–564.

Gutteridge, J. M. C.; Patterson, S. K.; Segal, A. W.; Halliwell, B. Inhibition of Lipid Peroxidation by the Iron-Binding Protein Lactoferrin. *Biochem. J.* **1981,** *199,* 259–261.

Halah M. F.; Mehanna, N. S. Use of Natural Plant Antioxidant and Probiotic in the Production of Novel Yogurt. *J. Evol. Biol. Res.* **2011,** *3,* 12–18.

Hall, G.; Andersson, J. Flavor Changes in Whole Milk Powder during Storage. III. Relationships between Flavor Properties and Volatile Compounds. *J. Food Qual.* **1985,** *7,* 237–253.

Hamm, D. L.; Hammond, E. G.; Parvanah, V.; Snyder, E. G. The Determination of Peroxides by the Stamm Method. *J. Am. Oil Chem. Soc.* **1965,** *42,* 920–922.

Hammond, E. G. Stabilizing Milk Fat with Antioxidants. *Am. Dairy Rev.* **1970,** *32,* 40–41, 76–77.

Han, J.; Britten, M.; St-Gelais, D.; Champagne, C. P.; Fustier, P.; Salmieri, S.; Lacroix, M. Polyphenolic Compounds as Functional Ingredients in Cheese. *Food Chem.* **2011,** *124,* 1589–1594.

Hansen, E.; Skibsted, L. H. Light-Induced Oxidative Changes in a Model Dairy Spread. Wavelength Dependence of Quantum Yields. *J. Agr. Food Chem.* **2000,** *48,* 3090–3094.

Hattori, M.; Yamaji-Tsukamoto, K.; Kumagai, H.; Feng, Y.; Takahashi, K. Antioxidative Activity of Soluble Elastin Peptides. *J. Agr. Food Chem.* **1998,** *46* (6), 2167–2170.

Haug, A.; Hostmark, A. T.; Harstad, O. M. Bovine Milk in Human Nutrition–A Review. *Lipids Health Dis.* **2007,** *6* (25), 1–16.

Hein, M.; Johnson, J.; Boess, F.; Bendik, I.; Weber, P.; Hunziker, W.; Fluhmann, B. Phytanic Acid, a Natural Peroxisome Proliferator Activated Receptor (PPAR) Agonist, Regulates Glucose Metabolism in Rat Primary Hepatocytes. *FASEB J.* **2002,** *16,*718–720.

Helal, F. R.; El-Bagoury; E., Rifaat. I. D.; Hofi, A. A.; El-Sokkary, A. The Effect of Some Antioxidants and Emulsifiers on the Keeping Quality of Buffalo Whole Milk Powder during Storage. *Egypt. J. Dairy Sci.* **1976,** *4,* 115–119.

Hellgren, L. I. Phytanic Acid—An Overlooked Bioactive Fatty Acid in Dairy Fat? *Ann. N Y Acad. Sci.* **2010,** *1190,* 42–49.

Hendricks, G.; Guo, M. Significance of Milk Fat in Infant Formulae. *In Advanced Dairy Chemistry. Volume 2 Lipids*; 3rd ed.; Fox, P. F., McSweeney, P. L. H., Eds.; Springer: New York. 2006; pp 467–480.

Henick, A. S.; Benca, M. F.; Mitchell Jr, J. H. Estimating Carbonyl Compounds in Rancid Fats and Foods. *J. Am. Oil. Chem. Soc.* **1954,** *31* (3), 88–91.

Henry, G. E.; Momin, R. A.; Nair, M. G.; Dewitt, D. L. Antioxidant and Cyclooxygenase Activities of Fatty Acids Found in Food. *J. Agr. Food Chem.* **2002,** *50,* 2231–2234.

Hoffman, J. R.; Falvo, M. J. Protein–Which is Best? *J. Sports Sci. Med.* **2004,** *3* (3), 118–130.

Howard, A. N.; Nigdikar, S. V.; Rajput-Williams, J.; Williams, N. R. Food Supplements. U. S. Patent, 6,086,910, 2000.

Hudson, B. J. F. *Food Antioxidants.* Elsevier applied science: London, UK, 1990.

Ip, C.; Banni, S.; Angioni, E.; Carta, G.; McGinley, J.; Thompson, H. J.; Barbano, D.; Bauman, D. Conjugated Linoleic Acid–Enriched Butter Fat Alters Mammary Gland Morphogenesis and Reduces Cancer Risk in Rats. *J. Nutr.* **1999,** *129* (12), 2135–2142.

Ito, N.; Hirose, M.; Fukushima, S.; Tsuda, H.; Shirai, T; Tatematsu, M. Studies on Antioxidants: Their Carcinogenic and Modifying Effects on Chemical Carcinogens. *Food Chem. Toxicol.* **1986,** *24* (10), 1071–1092.

Jung, J. E.; Yoon, H. J.; Yu, H. S.; Lee, N. K.; Jee, H. S.; Paik, H. D. Short Communication: Physicochemical and Antioxidant Properties of Milk Supplemented with Red Ginseng Extract. *J. Dairy Sci.* **2015,** *98,* 95–99.

Jung, M. Y.; Yoon, S. H.; Lee, H. O.; Min, D. B. Singlet Oxygen and Ascorbic Acid Effects on Dimethyl Disulfide and Off Favor in Skim Milk Exposed to Light. *J. Food Sci.* **1998,** *63,* 408–412.

Karaaslan, M.; Ozden, M.; Vardin, H.; Turkoglu, H. Phenolic Fortification of Yogurt Using Grape and Callus Extracts. *LWT - Food Sci. Technol.* **2011,** *44,* 1065–1072.

Karovicova, J.; Simko, P. Determination of Synthetic Phenolic Antioxidants in Food by High-Performance Liquid Chromatography. *J. Chromatogr. A.* **2000,** *882,* 271–281.

Keen, A. R.; Boon, P. M.; Walker, N. J. Off-Flavours in Stored Whole Milk Powder. 1. Isolation of Monocarbonyl Classes. *New Zealand J. Dairy Sci. Technol.* **1976,** *11,* 180–188.

Kendrick, J.; Watts, B. M. Acceleration and Inhibition of Lipid Oxidation by Heme Compounds. *Lipids.* **1969,** *4,* 454–458.

Keogh, M. K. Chemistry and Technology of Butter and Milk Fat Spreads. In *Advanced Dairy Chemistry. Volume 2: Lipids;* 3rd ed.; Fox, P. F, McSweeney, P. L. H., Eds.; 2006; pp 333–364.

Khokhar, S.; Owusu Apenten, R. K. Iron Binding Characteristics of Phenolic Compounds: Some Tentative Structure-Activity Relations. *Food Chem.* **2003,** *81,* 133–140.

Kim, Y. D.; Morr, C. V. Dynamic Headspace Analysis of Light Activated Flavor in Milk. *Int. Dairy J.* **1996,** *6* (2), 185–193.

King, R. L. Direct Addition of Tocopherol to Milk for Control of Oxidized Flavor. *J. Dairy Sci.* **1968,** *51,* 1705–1707.

Kinsella, J. E. What Makes Fat Important in Flavor. *Chem. Ind.* **1969b,** *2,* 36.

Kinsella, J. E. The Flavour Chemistry of Milk Lipids. *Chem. Ind.* **1969a,** *88,* 36–42.

Knipschildt, M. E. Drying of Milk and Milk Products. In *Modern Dairy Technology. Advances in Milk Processing*; Robinson, R. K., Ed.; Elsevier Applied Science Publishers: London, 1986; pp 131–234.

Koops, J. Antioxidant Activity of Ascorbyl Plamitate in Cold Stored Butter. *Neth. Milk Dairy J.* **1964,** *18,* 38–51.

Korycka-Dahl, M.; Richardson, T. Initiation of Oxidative Changes in Foods. *J. Dairy Sci.* **1980,** *63,* 1181–1198.

Kristensen, D.; Skibsted, L. H. Comparison of Three Methods based on Electron Spin Resonance Spectrometry for Evaluation of Oxidative Stability of Processed Cheese. *J. Agr. Food Chem.* **1999,** *47* (8), 3099–3104.

Kulisic, T.; Radonic, A.; Katalinic, V.; Milos, M. Use of Different Methods for Testing Activity of Oregano Essential Oil. *Food Chem.* **2004,** *85,* 633–640.

Labuza, T. P. Kinetics of Lipid Oxidation. *C R C Crit. Rev. Food Technol.* **1971,** *2,* 355–404.

Lee, J. H.; Min, D. B. Changes of Headspace Volatiles in Milk with Riboflavin Photosensitization. *J. Food Sci.* **2009,** *74,* 563–568.

Lee, K. G.; Shibamoto, T. Determination of Antioxidant Potential of Volatile Extracts Isolated from Various Herbs and Spices. *J. Agr. Food Chem.* **2002,** *50,* 4947–4952.

Loftus Hills, G.; Thiel, C. C. The Ferric Thiocyanate Method of Estimating Peroxide in the Fat of Butter, Milk and Dried Milk. *J. Dairy Res.* **1946,** *14* (03), 340–353.

López, M.; Martínez, F.; Del Valle, C.; Orte, C.; Miró, M. Analysis of Phenolic Constituents of Biological Interest in Red Wines by High-Performance Liquid Chromatography. *J. Chromatogr. A.* **2001,** *922,* 359–363.

Louli, V.; Ragoussis, N.; Magoulas, K. Recovery of Phenolic Antioxidants from Wine Industry by-Products. *Bioresour. Technol.* **2004,** *92,* 201–208.

Lwalokun, B. A.; Shittu, M. O. Effect of *Hibiscus sabdariffa* (Calyce) Extract on Biochemical and Organoleptic Properties of Yoghurt. *Pak. J. Nutr.* **2007,** *6,* 172–182.

Mariod, A. A.; Ali, R.T.; Ahmed, Y. A.; Abdel-Wahab, S. I.; Abdul, A. B. Effect of the Method of Processing on Quality and Oxidative Stability of Anhydrous Butter Fat (samn). *Afr. J. Biotechnol.* **2010,** *9,* 1046–1051.

Marshall, R. T.; Gov, H. D.; Hartel, R. W. *Ice Cream.* 6th ed.; Kluwer Academic/Plenum Publishers: New York, 2003.

McCarthy, T.; Kerry, J.; Kerry, J.; Lynch, P; Buckley, D. Evaluation of the Antioxidant Potential of Natural Food/Plant Extracts as Compared with Synthetic Antioxidants and Vitamin E in Raw and Cooked Pork Patties. *Meat Sci.* **2001,** *58,* 45–52.

Mehlenbacher, V. C. Analysis of Fats and Oils. Verlag The Garrard Press: Champaign, IL, 1960.

Mehta, B. Ragi (*Eleusine Coracana* L.)-a Natural Antioxidant for Ghee (butter Oil). *Int. J. Food Sci. Technol.* **2006,** *41,* 86–89.

Merai, M.; Boghra, V. R.; Sharma, R. S. Extraction of Antioxigenic Principles from Tulsi Leaves and Their Effects on Oxidative Stability of Ghee. *J. Food Sci. Technol.* **2003,** *40,* 52–57.

Min, D.; Boff, J. Lipid Oxidation of Edible Oil. In *Food Lipids*; Akoh C. C., Min D. B., Eds.; 2nd ed.; Marcel Dekker Inc.: New York, 2002; pp 335–364.

Molecular Probes. Product information, 2004, http://probes. invitrogen.com/media/pis/mp36002.pdf?id=mp36002

Morales, M. S.; Palmquist, D. L.; Weiss, W. P. Milk Fat Composition of Holstein and Jersey Cows with Control or Depleted Copper Status and Fed Whole Soybeans or Tallow. *J. Dairy Sci.* **2000,** *83,* 2112–2119.

Najgebauer-Lejko, D.; Grega, T.; Tabaszewska, M. Yoghurts with Addition of Selected Vegetables: Acidity, Antioxidant Properties and Sensory Quality. *Acta Sci. Pol. Technol.* **2014,** *13,* 35–42.

Najgebauer-Lejko, D.; Sady, M.; Grega, T.; Walczycka, M. The Impact of Tea Supplementation on Microflora, pH and Antioxidant Capacity of Yoghurt. *Int. Dairy J.* **2011,** *21,* 568–574.

Nielsen, B. R.; Stapelfeldt, H.; Skibsted, L. H. Early Prediction of the Shelf-Life of Medium-Heat Whole Milk Powders Using Stepwise Multiple Regression and Principal Component Analysis. *Int. Dairy J.* **1997,** *7* (5), 341–348.

O'Brien, J. Non-Enzymatic Degradation Pathways of Lactose and Their Significance in Dairy Products. In *Advanced Dairy Chemistry Volume 3: Lactose, Water, Salts and Minor Constituents*; McSweeney, P. L. H., Fox, P. F., Eds.; 3rd ed.; Springer: Verlag, NY, 2009; pp 231–294.

O'Brien, J.; Morrissey, P. A. Nutritional and Toxicological Aspects of the Maillard Browning Reaction in Foods. *Crit. Rev. Food Sci. Nutr.* **1989,** *28,* 211–248.

O'Connor, T. P.; O'Brien, N. M. Lipid Oxidation. In *Advanced Dairy Chemistry Volume 2 Lipids*; 3rd ed.; Fox, P. F., Mc Sweeney, Paul, L., H., Eds.; Springer: New York, 2006; pp 557–600.

Olmedo, R. H.; Asensio, C.; Nepote, V.; Mestrallet, M. G.; Grosso, N. R. Chemical and Sensory Stability of Fried-Salted Peanuts Flavored with Oregano Essential Oil and Olive Oil, *J. Sci. Food Agr.* **2009,** *89,* 2128–2136.

Olmedo, R. H.; Nepote, V.; Grosso N. R. Preservation of Sensory and Chemical Properties in Flavoured Cheese Prepared with Cream Cheese Base Using Oregano and Rosemary Essential Oils. *LWT - Food Sci. Technol.* **2013,** *53,* 409–417.

Olmedo, R. H.; Nepote, V.; Grosso, N. R. Aguaribay and Cedron Essential Oils as Natural Antioxidant in Oil-Roasted and Salted Peanuts. *J. Am. Oil Chem. Soc.* **2012,** *89,* 2195–2205.

Olmedo, R. H.; Nepote, V.; Mestrallet, M. G; Grosso, N. R. Effect of the Essential Oil Addition on the Oxidative Stability of Fried-Salted Peanuts. *Int. J. Food Sci. Technol.* **2008,** *43,* 1935–1944.

Østdal, H.; Bjerrum, M. J.; Pedersen, J. A.; Andersen H. J. Lactoperoxidase-Induced Protein Oxidation in Milk. *J. Agr. Food Chem.* **2000,** *48,* 3939–3944.

Palace, V. P.; Khaper, N.; Qin, Q.; Singal, P. K. Antioxidant Potentials of Vitamin A and Carotenoids and Their Relevance to Heart Disease. *Free Radical Biol. Med.* **1999,** *26,* 746–76.

Palmquist, D. L. Great Discoveries of Milk for a Healthy Diet and a Healthy Life. *R. Bras. Zootec.* **2010,** *39,* 465–477.

Pankaj, P.; Khamrui, K.; Devaraja, H. C.; Singh, R. R. B. The Effects of Alcoholic Extract of Arjuna (*Terminalia arjuna* Wight & Arn.) Bark on Stability of Clarified Butterfat. *J. Med. Plants Res.* **2013,** *7,* 2545–2550.

Pariza, M. W.; Park, Y.; Cook, M. E. Conjugated Linoleic Acid and the Control of Cancer and Obesity. *Toxicol. Sci.* **1999,** *52* (1), 107–110.

Parks, O. W.; Allen, C. Photodegradation of Riboflavin to Lumichrome in Milk Exposed to Sunlight. *J. Dairy Sci.* **1977,** *60* (7), 1038–1041.

Parodi, P. W. Anti-Cancer Agents in Milkfat. *Aust. J. Dairy Technol.* **2003,** 58 *(2),* 114.

Patton, S. The Mechanism of Sunlight Flavor Formation in Milk with Special Reference to Methionine and Riboflavin. *J. Dairy Sci.* **1954,** *37* (4), 446–452.

Pawar, N.; Arora, S.; Bijoy, R.; Wadhwa, B. The Effects of *Asparagus racemosus* (Shatavari) Extract on Oxidative Stability of Ghee, in Relation to Added Natural and Synthetic Antioxidants. *Int. J. Dairy Technol.* **2012,** *65,* 293–299.

Pawar, N.; Gandhi, K.; Purohit, A.; Arora, S.; Singh, R. R. B. Effect of Added Herb Extracts on Oxidative Stability of Ghee (Butter Oil) during Accelerated Oxidation Condition. *J. Food Sci. Technol.* **2014,** *51,* 2727–2733.

Pérez-Jiménez, J.; Aranz, S.; Tabernero, M.; Díaz-Rubio, M. E.; Serrano, J.; Goñi, I.; Saura-Calixto, F. Updated Methodology to Determine Antioxidant Capacity in Plant Foods, Oils and Beverages: Extraction, Measurement and Expression of Results. *Food Res. Int.* **2008,** *41,* 274–285.

Podsędek, A. Natural Antioxidants and Antioxidant Activity of Brassica Vegetables: A Review. *LWT-Food Sci. Technol.* **2007,** *40,* 1–11.

Pokorny, J.; Korczak, J. Preparation of Natural Antioxidants. In *Antioxidants in Food: Practical Application*; Pokorny, J., Yanishlieva, N., Gordon. M., Eds.; Woodhead Publishing Ltd.: Cambridge, UK, 2001; pp 311–330.

Pokorny, J. Major Factors Affecting the Auto-Oxidation of Lipids. In *Auto-oxidation of Unsaturated Lipids*; Chan, H. W. S., Ed.; Academic Press: London, 1987; pp 141–206.

Prasad, K. N.; Yang, B.; Yang, S.; Chen, Y.; Zhao, M.; Ashraf, M.; Jiang, Y. Identification of Phenolic Compounds and Appraisal of Antioxidant and Antityrosinase Activities from Litchi (*Litchi sinensis Sonn.*) Seeds. *Food Chem.* **2009,** *116,* 1–7.

Puranik, D. B.; Rao, H. G. R. Potentiality of Whey Protein as a Nutritional Ingredient. *Indian Dairyman.* **1996,** *48* (11), 17–21.

Rangappa, K. S.; Achaya, K. T. Rancidity in Ghee. In *Indian Dairy Products*; Rangappa, K. S., Achaya, K. T., Eds.; Asia Publishing House: Mumbai, India, **1975**; pp 327–345.

Reglero, R. G. J.; Tabera, G. J. J.; Ibáñez, E. M. E.; López-Sebastián, L. S.; Ramos, M. E.; Ballester, S. L.; Bueno, M. J. M. Proceso de Extracción con Fluidos Supercríticos Para la Producción de Antioxidantes Naturales y Antioxidante Obtenido Mediante Dicho Proceso. Patente española 2,128,996, 1999.

Ruggeri, S.; Straniero, R.; Pacifico, S.; Aguzzi, A.; Virgili, F. French Marine Bark Extract Pycnogenol as a Possible Enrichment Ingredient for Yoghurt. *J. Dairy Sci.* **2008,** *91,* 4484–4491.

Saad B.; Sing Y. Y.; Nawi M. A.; Hashim N.; Ali A. M.; Saleh M. I.; Ahmad, K. Determination of Synthetic Phenolic Antioxidants in Food Items using Reversed-Phase HPLC. *Food Chem.* **2007,** *105,* 389–394.

Saint-Eve, A.; Levy, C.; Martin, N.; Souchon, I. Influence of Proteins on the Perception of Flavored Stirred Yogurts. *J. Dairy Sci.* **2006,** *89,* 922–933.

Samuelsson, E. G. The Copper Content in Milk and the Distribution of Copper to Various Phases of Milk. *Milchwissenschaft.* **1966,** *21,* 335–341.

Sarwar, G. The Protein Digestibility-Corrected Amino Acid Score Method Overestimates Quality of Proteins Containing Antinutritional Factors and of Poorly Digestible Proteins Supplemented with Limiting Amino Acids in Rats. *J. Nutr.* **1997,** *127,* 758–764.

Sattar, A.; Alexander, J. C. Light-Induced Degradation of Vitamins I. Kinetic Studies on Riboflavin Decomposition in Solution. *Can. Inst. Food Sci. Technol. J.* **1977,** *10* (1), 61–64.

Schuster, G. S.; Dirksen, T. R.; Ciarlone, A. E.; Burnett, G. W.; Reynolds, M. T.; Lankford, M. T. Anticaries and Antiplaque Potential of Free-Fatty Acids in Vitro and in Vivo. *Pharmacol. Ther. Dent.* **1980**, *5*, 25–33.

Serrano, J.; Goňi, I; Saura-Calixto, F. Food Antioxidant Capacity Determined by Chemical Methods may Underestimate the Physiological Antioxidant Capacity. *Food Res. Int.* **2007**, *40*, 15–21.

Shahidi, F.; Zhong, Y. Novel Antioxidants in Food Quality Preservation and Health Promotion. *Eur. J. Lipid Sci. Technol.* **2010**, *112*, 930–940.

Shahidi, F.; Janitha, P. K.; Wanasundara, P. D. Phenolic Antioxidants. *Crit. Rev. Food Sci. Nutr.* **1992**, *32*, 67–103.

Shui, G.; Leong, L. P. Analysis of Polyphenolic Antioxidants in Star Fruit using Liquid Chromatography and Mass Spectrometry. *J. Chromatogr. A.* **2004**, *1022*, 67–75.

Singh, R.; Kumar, R.; Venkateshappa, R.; Mann, B.; Tomar, S. K. Studies on Physicochemical and Antioxidant Properties of Strawberry Polyphenol Extract–Fortified Stirred Dahi. *Int. J. Dairy Technol.* **2013**, *66*, 103–108.

Singh, S.; Immanuel, G. Extraction of Antioxidants from Fruit Peels and its Utilization in Paneer. *J. Food Process. Technol.* **2014**, *5*, 349.

Skrede, G.; Larsen, V. B.; Aaby, K.; Jorgensen, S.; Birkeland, S. E. Antioxidant Properties of Commercial Fruit Preparations and Stability of Bilberry and Black Currant Extracts in Milk Products. *J. Food Sci.* **2004**, *69*, S351–S356.

Smithers, G. W. Whey and Whey Proteins—from 'Gutter-to-Gold'. *Int. Dairy J.* **2008**, *18* (7), 695–704.

Spikes, J. D. Photosensitization. In *The Science of Photobiology*, 2nd ed.; Smith, K. C., Ed.; Plenum Press: New York, 1988; pp 79–110.

Stapelfeldt, H.; Nielsen, B. R.; Skibsted, L. H. Effect of Heat Treatment, Water Activity and Storage Temperature on the Oxidative Stability of Whole Milk Powder. *Int. Dairy J.* **1997**, *7*, 331–339.

Stapelfeldt, H.; Nielsen, K. N.; Jensen, S. K.; Skibsted, L. H. Free Radical Formation in Freeze-Dried Raw Milk in Relation to its α-tocopherol Level. *J. Dairy Res.* **1999**, *66*, 461–466.

Sun, C. Q.; O'Connor, C. J.; Roberton, A. M. The Antimicrobial Properties of Milkfat after Partial Hydrolysis by Calf Pregastric Lipase. *Chem. Biol. Interact.* **2002**, *140*, 185–198.

Sun-Waterhouse, D.; Zhou, J.; Wadhwa, S. S. Drinking Yoghurts with Berry Polyphenols Added before and after Fermentation. *Food Control.* **2013**, *32*, 450–460.

Sun-Waterhouse, D.; Zhou, J.; Wadhwa, S. S. Effects of Adding Apple Polyphenols before and after Fermentation on the Properties of Drinking Yoghurt. *Food Bioprod. Process.* **2012**, *5*, 2674–2686.

Swaisgood, H. E.; Abraham, P. Oxygen Activation by Sulfhydryl Oxidase and the Enzyme's Interaction with Peroxidase. *J. Dairy Sci.* **1980**, *63*, 1205–1210.

Tada, M.; Kobayashi, N.; Kobayashi, M. Studies on the Photosensitized Degradation of Food Products. Part II. Photosensitized Degradation of Methionine by Riboflavin. *J. Agr. Food Chem.* **1971**, *46*, 107–111.

Tappel, A. L. Catalysis of Linoleate Oxidation by Copper-Proteins. *J. Am. Oil Chem. Soc.* **1955**, *32* (5), 252–254.

Tauer, A.; Hasenkopf, K.; Kislinger, T.; Frey, I.; Pischetsrieder, M. Determination of Nε-carboxymethyllysine in Heated Milk Products by Immunochemical Methods. *Eur. Food Res. Technol.* **1999**, *209* (1), 72–76.

Taylor, M. J.; Richardson, T. Antioxidant Activity of Skim Milk: Effect of Heat and Resultant Sulfhydryl Groups. *J. Dairy Sci.* **1980,** *63,* 1783–1795.

Tena. M. T.; Valcarcel, M.; Hidalgo, P.; Ubera, J. L. Supercritical Fluid Extraction of Natural Antioxidants from Rosemary: Comparison with Liquid Solvent Sonication. *Anal. Chem.* **1997,** *69,* 521–526.

Terao, J.; Yamauchi, R.; Murakauri, H.; Matsushita, S. Inhibitory Effects of Tocopherols and β-carotene on Singlet Oxygen-Initiated Photooxidation of Methyl Linoleate and Soybean Oil. *J. Food Process. Pres.* **1980,** *4,* 79–93.

Thorat, I. D.; Jagtap, D. D.; Mohapatra, D.; Joshi, D. C.; Sutar, R. F.; Kapdi, S. S. Antioxidants, Their Properties, Uses in Food Products and Their Legal Implications. *Int. J. Food Stud.* **2013,** *2,* 81–104.

Thormar, H.; Isaacs, E. E.; Kim, K. S.; Brown, H. R. Interaction of Visna Virus and other Enveloped Viruses by Free Fatty Acids and Monoglycerides. *Ann. N. Y. Acad. Sci.* **1994,** *724,* 465–471.

Toran, A. A.; Barberá, R.; Farré, R.; Lagarda, M. J.; López, J. C. HPLC Method for Cyst (e) Ine and Methionine in Infant Formulas. *J. Food Sci.* **1996,** *61* (6), 1132–1136.

Trigueros, L.; Pérez-Alvarez, J. A.; Viuda-Martos, M.; Sendra, E. Production of Low-Fat Yoghurt with Quince (*Cydonia oblonga Mill.*) Scalding Water. *LWT - Food Sci. Technol.* **2011,** *44,* 1388–1395.

Truitt, A.; McNeill, G.; Vanderhoek, J. Y. Antiplatelet Effects of Conjugated Linoleic Acid Isomers. *BBA – Mol. Cell Biol. Lipids.* **1999,** *1438* (2), 239–246.

United States Dairy Export Council. Reference Manual for U.S. Whey Products, 2nd Edition, 1999.

Upadhyay, N.; Goyal, A.; Kumar, A.; Lal, D.; Singh, R. Preservation of Milk and Milk Products for Analytical Purposes: A Review. *Food Rev. Int.* **2014,** *30,* 203–224. DOI 10.1080/87559129.2014.913292.

van Aardt, M.; Duncan, S. E.; Marcy, J. E.; Long, T. E.; O'Keefe, S. R.; Nielsen-Sims, S. R. Aroma Analysis of Light-Exposed Milk Stored With and Without Natural and Synthetic Antioxidants. *J. Dairy Sci.* **2005,** *88,* 881–890.

Viljanen, K.; Sunberg, S.; Ohshima, T.; Heinonen, M. Carotenoids as Antioxidants to Prevent Photooxidation. *Eur. J. Lipid Sci. Technol.* **2002,** *104,* 353–359.

Vision, J. A.; Jang, J.; Yang, J.; Dabbagh, Y.; Liang, X.; Sery, M. Vitamins and Especially Flavonoids in Common Beverages are Powerful in Vitro Antioxidants Which Enrich Lower Density Lipoproteins and Increase Their Oxidative Resistance after ex Vivo Spiking in Human Plasma. *J. Agr. Food Chem.* **1999,** *47,* 2502–2504.

Wade, V.; Al-Tahiri, R.; Crawford, R. The Auto-Oxidative Stability of Anhydrous Milk Fat with and without Antioxidants. *Milchwisssenschaft.* **1986,** *41,* 479–482.

Wadhwa, B. K.; Jain, M. K. Studies on Lactone Profile of Ghee. III. Variations due to Method of Preparation. *Indian J. Dairy Sci.* **1985,** *38,* 31–35.

Walker, I. K.; Jackson, F. H. The Heat Balance in Spontaneous Ignition. 9. Influence of Thermal Conductivity for Reactions of Zero Order. *New Zealand J. Sci.* **1978,** *21,* 519–526.

Wanasundara, U. N.; Shahidi, F. Antioxidant and Pro-Oxidant Activity of Green Tea Extracts in Marine Oils. *Food Chem.* **1998,** *63,* 335–342.

Wojdyło, A.; Oszmiański, J.; Czemerys, R. Antioxidant Activity and Phenolic Compounds in 32 Selected Herbs. *Food Chem.* **2007,** *105,* 940–949.

Wold, J. P.; Bro, R.; Veberg, A.; Lundby, F.; Nilsen, A. N.; Moan, J. Active Photosensitizers in Butter Detected by Fluorescence Spectroscopy and Multivariate Curve Resolution. *J. Agr. Food Chem.* **2006,** *54* (26), 10197–10204.

Wold, J. P.; Veberg, A.; Nilsen, A.; Iani, V.; Juzenas, P.; Moan, J. The Role of Naturally Occurring Chlorophyll and Porphyrins in Light-Induced Oxidation of Dairy Products. A Study Based on Fluorescence Spectroscopy and Sensory Analysis. *Int. Dairy J.* **2005,** *14,* 343–353.

Wong, P. Y. Y.; Kitts, D. D. Chemistry of Buttermilk Solid Antioxidant Activity. *J. Dairy Sci.* **2003,** *86* (5), 1541–1547.

Wyatt, C. J.; Day, E. A. Evaluation of Antioxidants in Deodorized and Non-Deodorized Butter Oil Stored at 30°C. *J. Dairy Sci.* **1965,** *48,* 682–686.

Yamauchi, R.; Matsushita, S. Quenching Effect of Tocopherols on Methyl Linoleate Photo-oxidation and Their Oxidation Products. *Agr. Biol. Chem.* **1977,** *41,* 1425–1430.

Yanishlieva-Maslarova, N. Inhibition Oxidation. In *Antioxidants in Food Practical Applications*; Pokorny, J., Yanishlieva, N., Gordon, M., Eds.; CRC press, Woodhead publishing Ltd.: Cambridge, UK, 2001; pp 35–59.

Yee, J. J.; Shipe, W. F. Effects of Sulfhydryl Compounds on Lipid Oxidations Catalysed by Copper and Heme. *J. Dairy Sci.* **1982,** *65,* 1414–1420.

Yue, X.; Xu, Z.; Prinyawiwatkul, W.; Losso, J. N.; King, J. M; Godber, J. S. Comparison of Soybean Oils, Gum, and Defatted Soy Flour Extract in Stabilizing Menhaden Oil during Heating. *J. Food Sci.* **2008,** *73,* C19–C23.

CHAPTER 8

ANTIOXIDANT DIETARY FIBER: AN APPROACH TO DEVELOP HEALTHY AND STABLE MEAT PRODUCTS

ARUN K. VERMA[1,*], RITUPARNA BANERJEE[2], and V. RAJKUMAR[1]

[1]*Goat Products Technology Laboratory, ICAR-Central Institute for Research on Goats, Makhdoom, Farah, Mathura 281122, Uttar Pradesh, India*

[2]*ICAR-National Research Centre on Meat, Chengicherla, Hyderabad 500092, Telangana, India*

[*]*Corresponding author. E-mail: arun.lpt2003@gmail.com*

CONTENTS

ABSTRACT

The findings about association between healthy diet and consumer well-being have escalated the demands for the foods especially meat products supplying additional healthy nutrients such as dietary fiber (DF) and natural antioxidants. There are several reports in the literature linking the regular intake of DF and antioxidants help in preventing various diets related non-communicable diseases as well as degenerative problems. Researchers are also responding very well to the consumer's demands through the screening of various sources of antioxidant dietary fiber (ADF) such as fruits, vegetables, seeds and their by-products as well as other miscellaneous plant materials. These ADF ingredients are being added in different meat products and their effects on various qualities are monitored.

8.1   INTRODUCTION

Workload has considerably increased in modern society and very limited time is spared to take care of personal health. This time scarcity has forced us to opt for ready-to-eat and fast foods including meat products which lack adequate level of dietary fiber (DF) required in the daily diet. Inadequate DF in our meal can be a cause for several diet related problems. Several investigations have reported that one-third of all cancer cases and one-half of cardiovascular diseases and hypertension can be attributed to diet (Lee & Smith, 2000; Wolfe et al., 2003). The incidence of a number of non-infectious diseases in our modern societies, such as coronary heart disease, certain kind of cancer is partly associated to a low DF intake. Thus DF intake may be vital in reducing colonic cancer, in lowering serum cholesterol levels and in preventing hyperglycemias in diabetic patients (Nawirska & Kwasniewska, 2005). The main sources of DF are plant products like seeds, fruits, vegetables, and their products. It is widely accepted now that increased consumption of fruits and vegetables can reduce the risk of cancer, heart disease, and stroke (Liu, 2003). The possible beneficial health effects of diets rich in these plant products have been attributed to their DF and phytochemicals (Block et al., 1992; Lampe, 1999).

Meat and meat products are very much nutritious providing many key nutrients to our body. One of the several nutrients is the fat which is a dense source of energy and furnishes essential fatty acids. These fatty acids can possess various degrees of unsaturation. A high degree of unsaturation accelerates oxidative processes leading to deterioration in meat flavor, color, texture, and nutritional value (Mielnick et al., 2006). This oxidation is a highly complex process involving numerous reactions which produce a variety of chemical and physical changes (Sánchez-Alonso et al., 2008). The nutritional quality of meat and meat products is often being challenged due to absence of DF in them. However, this can be overcome by addition of DF rich ingredients while processing of meat products. There are several DF ingredients which are also gifted with many phytochemicals like polyphenols. Phenolic compounds present in fruits have been demonstrated to possess antioxidant, anti-inflammatory and anti-carcinogenic properties and the ability to prevent a variety of chronic diseases (Boyer & Liu, 2004). Adding these ingredients to meat products, which has almost negligible amount of DF, can be helpful in enhancing their nutritional and functional values, quality, and storage stability. Moreover, intake of natural antioxidant through these fiber rich meat products can help fighting stress of modern day hectic lifestyle. This chapter broadly deals with antioxidant dietary fiber (ADF), sources and effects of their incorporation on quality of meat products.

8.2   ANTIOXIDANT DIETARY FIBER

DFs are plentiful in plant products like fruits, vegetables, and grains. They are very well known as a beneficial component of healthy diet and their consumption is being linked with reductions in risks associated with cardio-vascular disease, cancer, and diabetes (Cho & Dreher, 2001). DF has hetero-geneous chemical structures and conventionally classified according to their solubility in water as soluble dietary fiber (SDF) and insoluble dietary fiber (IDF). In cereal bran, the IDF is largely predominant while fruit DFs are rich in SDF (Gorinstein et al., 2001; Grigelmo-Miguel & Martín-Belloso, 1998; Prosky et al., 1988). It is important to consume the DF in a balanced propor-tion, that is, the water-soluble fraction should represent between 30 and 50% of the total dietary fiber (TDF) (Eastwood, 1987; Spiller, 1986).

Interests in applying fruit processing wastes as functional food ingredi-ents is consistently increasing as they are rich source of DF, and several bene-ficial bioactive compounds (Balasundram et al., 2006) such as polyphenols. Such association of DF and polyphenols led to the proposal of the concept of ADF by Saura-Calixto (1998) with the criteria that 1 g of ADF should have 2,2-diphenyl-1-picrylhydrazyl (DPPH) free radical scavenging capacity equivalent to at least 50 mg vitamin E and DF content should be higher than 50% dry matter from the natural constituents of the material. Antioxidant DF is thus defined as a natural product that combines the beneficial effects of

DF and natural antioxidants, such as polyphenol compounds (Saura-Calixto, 1998).

The definition of DF according to the American Association of Cereals Chemists Expert Committee (De Vries, 2003) includes "cell wall polysaccharides, lignin and associated substances resistant to hydrolysis by the digestive enzymes of humans." In this definition, "associated compounds" are included for the first time in an official definition of DF. Polyphenols linked with DFs could be one of the associated compounds. Later the concept of the antioxidant activity of insoluble material was introduced (Serpen et al., 2007) as it is able to exert a marked antioxidant activity also by a solid–liquid interaction. The insoluble materials, particularly ADFs survive in the gastrointestinal tract for a long time and quench the soluble radicals that are continuously formed in the intestinal tract (Babbs, 1990).

The most abundant phenolic compounds in cereals belong to the chemical class of hydroxycinnamic acids (HCA). Ferulic acid (FA) is the main, followed by diferulic acids and by sinapic acid, p-coumaric acid, and caffeic acid; benzoic acid derivatives have also been described (Vitaglione et al., 2008). About 95% of grain polyphenols are linked to cell wall polysaccharides through covalent ester bonds. FA is bound to the arabinoxylans via the acid group acetylating the primary hydroxyl at the C_5 position of α-arabinofuranosyl residues (Hatfield et al., 1999). In case of fruits and vegetable, phenolic compounds could also be linked to pectin and to other polysaccharide structures (Ishii, 1997; Jiménez-Escrig et al., 2001a; Saura-Calixto & Díaz-Rubio, 2007).

The antioxidant capacity of DF linked polyphenols has been largely underestimated because of their low water and organic solvent solubility. In order to get actual value, it is necessary to perform multiple step extraction and an appropriate chemical hydrolysis to release phenolics and to permit them to exert antioxidant activity in the *in vitro* assays. In the view of constituents present in the ADF, it can be used on one hand as a dietary supplement to improve gastrointestinal health and to prevent cardiovascular diseases (Pérez-Jiménez et al., 2008), and on the other as an ingredient in meat products to improve technological qualities and prevent lipid oxidation (Sánchez-Alonso et al., 2007).

8.3 POLYPHENOLS AND DIETARY FIBER QUALITY

The antioxidant moiety associated with DF fundamentally determines its structure and consequently physical properties. The presence of FA linked to polysaccharide chains represents a suitable way to cross-link them through the formation of diferulates (Bunzel et al., 2001). The principal mechanism for crosslinking cell wall polysaccharides is ferulate dehydrodimerization via radical coupling reactions producing a number of different diferulates (Bunzel et al., 2004). They form bridge structures between chains of poly-saccharides. Moreover, ferulates have a significant role in cross-linking polysaccharides to lignin; thus they deeply influence physical parameters of DF determining its reticulation, molecular weight, and water solubility. The amount of diferulates found in the SDF of different cereals is more than 100-fold lower than the amount present in the corresponding IDF (Bunzel et al., 2001). The extent of diferulates severely affects the biological signifi-cance of DF and it is believed that the amount of DF diferulates is inversely correlated to the fermentability by intestinal microflora (Kroon, 2000; Wang et al., 2004). The bacterial β-glucosidases and esterases cannot attack the highly cross-linked IDF while their action is easier when the DF is less struc-tured such as in SDF (Kroon et al., 1997; Zhao et al., 2003). Although it is also reported that low to moderate levels of diferulates do not interfere with hydrolysis of non-lignified cell walls by human gut microbiota (Funk et al., 2007). It is generally accepted that higher the SDF/IDF ratio, the higher is DF polyphenols bioaccessibility.

8.4 DIETARY FIBER PROCESSING AND QUALITY

The by-products from cereals as well as fruits and vegetables are subjected different processing steps and conditions to prepare the DF or ADF. The extent and intensity of these processing steps and conditions may affect various properties and activity of this bioactive ingredient. The treatment like extent and intensity of blanching and drying, nature of solvent used should be taken into consideration while processing of ADF.

DF production typically involves pre-treatment methods, such as blanching or chemical treatments depending on the type of raw material, prior to drying, to inactivate enzymes responsible for degradation of many active compounds (Wolfe & Liu, 2003). However, number of sensitive compounds may degrade during blanching, depending on the type (steam- or water-blanching) and conditions (Zhang & Hamauzu, 2004). Loss of SDF and solubilization of structural polymers such as protopectin may happen while blanching of high fiber products (Maté et al., 1998). Drying can result in oxidation, thermal degradation and other events such as collapse of microstructure that lead, directly or indirectly, to lower levels of antioxidants or their bioaccessibility.

In a study on evolution of antioxidant compounds from lime residue during drying Kuljarachanan et al. (2009) found that blanching decreased both the antioxidant contents and activities of the residues due to thermal degradation and loss with the blanched water. During drying initially nomilin and limolin increased followed by sharp decrease due to thermal degradation. In this case the product temperature was found to be a major factor controlling the changes of limonoids. The amounts of vitamin C and phenolic compounds decreased as the product temperature increased and the moisture content decreased during drying.

Effect of air-drying temperature on physicochemical properties of DF and antioxidant capacity of orange (*Citrus aurantium* v. Canoneta) by-products (peel and pulp) was investigated (Garau et al., 2007). It was observed that dehydration promoted important modifications affecting both the physicochemical properties of DF and the antioxidant capacity of orange by-products. The major modifications on the DF components were observed when either extended drying periods, or elevated drying temperatures were applied. Dehydration at around 50–60 °C apparently promoted the minor disruption of cell wall polymers, in particular of pectic substances. However, significant decreases in water retention capacity, fat adsorption capacity, and solubility values were detected for both by-products with increased drying temperature. The by-products studied here were quite resistant to the different heat treatments applied within the range of 40–70 °C. It was suggested that in order to preserve the fiber quality and/or the antioxidant capacity, air drying temperature should be controlled.

8.5 SOURCES OF ANTIOXIDANT DIETARY FIBER

Almost all the plant materials contain ample amount of DF. These materials are also enriched with various secondary metabolites acting as defense systems which are either present in free form or associated with DF. The plant materials as a source of ADF can be numerous; however, in the present chapter only those sources are mentioned which have been investigated as a source of antioxidant and DF. These sources are categorized here into four groups (Fig. 8.1).

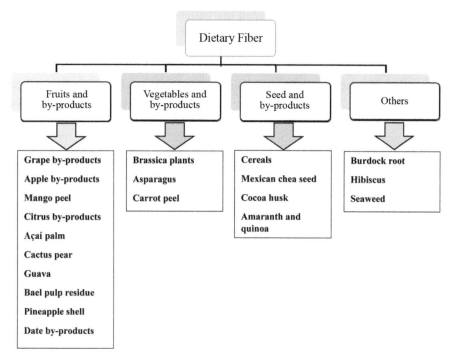

FIGURE 8.1   Sources of antioxidant dietary fiber.

8.5.1   FRUITS AND BY-PRODUCTS

8.5.1.1   GRAPE BY-PRODUCTS

Grape pomace is a by-product which constitutes ~20% of the harvested grapes (Laufenberg et al., 2003). At present, only least amounts of these wastes are upgraded or recycled. Investigation on grape pomace is limited, but it is undoubtedly rich in polyphenols (Amico et al., 2004; Kammerer et al., 2004). The major constituents of grape pomace peels and seeds, has been reported by several authors, with high polyphenols such as anthocyanins, catechins, flavonoids, phenolic acids as well as DF contents (Bravo & Saura-Calixto, 1998; Valiente et al., 1995; Larrauri et al., 1999; Mazza, 1995; Mazza & Miniati, 1993). The phenolic compounds in grape pomace include catechins, namely monomeric and oligomeric flavan-3-ols (proanthocyanidins) and glycosylated flavonols. Catechins, together with other polyphenols, are potent free radical-scavengers (Ruberto et al., 2007).

Several studies have highlighted the beneficial effects of grape or wine polyphenols on human health (Ho et al., 1994; Rice-Evans et al., 1997; Lurton, 2003; Frankel, 1999). More specifically, the antioxidative properties of many natural polyphenols may exert a chemopreventive role toward cardiovascular and degenerative diseases (Halpern et al., 1998; Renaud & De Lorgeril, 1992), including neurodegenerative pathologies (Esposito et al., 2002) and cancer (Bozidar, 1995). Grape skins are rich in anthocyanins, a group of polyphenols well known for their beneficial properties (Katsube et al., 2003; Wang et al., 1997; Ghiselli et al., 1998; Kong et al., 2003; Kähkönen & Heinonen, 2003). Pérez-Jiménez et al. (2008) found that fibers from grapes show higher reducing efficacy in lipid profile and blood pressure with respect to oat fiber or psyllium due to combined effect of DF and antioxidants. Wine grape pomace as ADF not only retarded human low-density lipoprotein oxidation *in vitro* (Meyer et al., 1998) but also helped to enhance the gastrointestinal health of the host by promoting a beneficial microbiota profile (Pozuelo et al., 2012).

The general composition of two by-products of the Manto Negro red grape (*Vitis vinifera*) variety, namely pomace and stem, were determined (Llobera & Cañellas, 2007). Both by-products had high contents of TDF, comprising three fourths of the total dry matter. Notable were the high percentage of soluble fiber (15%) in relation to the TDF for the pomace, as well as the high content of Klason lignin (KL) in both by-products, especially in the stem (31.6%). The free radical scavenging capacity (EC_{50}) of the by-products was found as 0.46 mg DM/mg DPPH (stem) and 1.41 mg DM/mg DPPH (pomace). Thus both by-products of the vinification process, particularly the stem, present excellent antioxidant properties as free radical-scavengers. The methanolic extracts (MeOH) obtained from de-stemmed grape pomace samples of five Sicilian red grape cultivars were evaluated for their DPPH· and ABTS· radical scavenging capacity (Ruberto et al., 2007). All the MeOH extracts showed significant antioxidant activity, with some differences between the two methods employed. There was large variability in the total anthocyanin (TA) and flavonol (TF) contents of the extracts, as well as in the quantitative distribution of the single anthocyanins and flavonols.

Grape skins, comprising on average, 82% of the wet weight of wine grape pomace (Jiang et al., 2011), contain multiple types of polyphenols, including 39 types of anthocyanins, HCA, catechins, and flavonols (Kammerer et al., 2004). The skins of two white wine grape pomace (WWGP) and three red wine grape pomace (RWGP) were analyzed for their DF and phenolic composition (Deng et al., 2011). IDF represented 95.5% of TDF, composed of KL (7.9–36.1% DM), neutral sugars (4.9–14.6% DM), and uronic acid (3.6–8.5% DM) in all five WGP varieties. WWGP was significantly lower in DF (17.3–28.0% DM) than those of RWGP (51.1–56.3%), but extremely higher in soluble sugar (55.8–77.5% DM vs 1.3–1.7% DM). Compared with WWGP, RWGP had higher values in total phenolic content (TPC) (21.4–26.7 mg GAE/g DM vs 11.6–15.8 mg GAE/g DM) and DPPH radical scavenging activity (32.2–40.2 mg AAE/g DM vs 20.5–25.6 mg AAE/g DM). The total flavanol and proanthocyanidin contents were ranged from 31.0 to 61.2 mg CE/g DM and 8.0 to 24.1 mg/g DM, respectively, for the five WGP varieties.

8.5.1.2   APPLE BY-PRODUCTS

Apples are well known and widespread fruit of the genus Malus belonging to the family Rosaceae and play significant part in our diet. They are important source of bioavailable polyphenolic compounds such as flavonols, monomeric and oligomeric flavanols, dihydrochalcones, anthocyanidins, p-hydroxycinnamic, and p-hydroxybenzoic acids (Escarpa & Gonzalez, 1998). The phenolic contents in apple are variable among different varieties, and between the peel and the flesh; apple peels contain a higher concentration of phenolic compounds (Escarpa & Gonzalez, 1998; Vrhovsek et al., 2004).

Apple pomace is a by-product of the apple juice processing, which is a rich source of polyphenols, minerals, and DF (Boyer & Liu, 2004; Fernando et al., 2005; Schieber et al., 2001; Sudha et al., 2007). Disposal of apple pomace may present an added cost to beverage industry. Fernando et al. (2005) evaluated some functional properties of apple pomace and reported that fiber concentrates from apple pomace can be considered as a potential source for fiber enrichment. According to Lu and Foo (2000), polyphenols present in apple pomace could be a cheap and readily available source of dietary antioxidants. The antioxidant capacity of apple pomace is related to its phenolic profile. Procyanidins have long been recognized as the major contributors to antioxidant activity of apples (Chinnici et al., 2004; Tsao et al., 2005) and derivatives (Oszmianski et al., 2008), and the capacity depends on their polymerization degree and substituents (Lotito et al., 2000). Additionally, the antioxidant activity of hydroxycinnamic and benzoic acids and flavonols has been ascertained (Kim et al., 2002; Tsao et al., 2005). Apple pomace procured from fruit juice industry, contained 10.8% moisture, 0.5% ash, and 51.1% of DF. The total phenol content in apple pomace was

7.16 mg/g (Sudha et al., 2007). Apple extracts have been shown to have potent antioxidant activity and anti-proliferative activity against human cancer cells (Boyer & Liu, 2004; Leontowicz et al., 2002).

Apple pomaces were subjected to evaluation as potential sources of anti-oxidant phytochemicals on the basis of their total content of phenolics (from 4.22 to 8.67 mg/g), total flavonoids (from 0.45 to 1.19 mg/g) and total flavan-3-ols (from 2.27 to 9.51 mg/g), and *in vitro* antiradical activities (Ćetković et al., 2008). Some individual phenolic compounds including caffeic and chlorogenic acids (+)-catechin, and (-)-epicatechin, rutin, quercetin glyco-sides, and phloridzin were identified and quantified by high performance liquid chromatography (HPLC). The antiradical activity of apple pomaces was tested by measuring their ability to scavenge DPPH and hydroxyl radicals. Eleven different cider apple pomaces (six single-cultivar and five from the cider-making industry) were analyzed for low molecular phenolic profiles and antioxidant capacity (García et al., 2009). The Folin index ranged between 2.3 and 15.1 g gallic acid per kg of dry matter. Major phenols were flavanols, dihydrochalcones (phloridzin and phloretin-20-xyloglucoside), flavonols, and cinnamic acids (chlorogenic and caffeic acids). The group of single-cultivar pomaces had higher contents of chlorogenic acid (-)-epicat-echin, procyanidin B_2 and dihydrochalcones, whereas the industrial samples presented higher amounts of up to four unknown compounds, with absorption maxima between 256 and 284nm. The antioxidant capacity of apple pomace, as determined by the DPPH and ferric reducing antioxidant power (FRAP) assays, was between 4.4 and 16.0 g ascorbic acid per kg of dry matter.

Methanolic and acetonic extracts of apple pomace were evaluated for phenolic profiles, antioxidant properties, and antiviral effects against herpes simplex virus type 1 (HSV-1) and 2 (HSV-2) (Suárez et al., 2010). Acetone extraction yielded the higher amounts of phenolic compounds. The extrac-tion method influenced the phenolic composition although antioxidant activity correlated weakly with phenols concentration. Among the poly-phenols analyzed, quercetin glycosides were the most important family, followed by dihydrochalcones. It was observed that apple pomace extracts were able to inhibit both HSV-1 and HSV-2 replication in Vero cells by more than 50%, at non-cytotoxic concentrations. Selectivity indexes (SI) ranged from 9.5 to 12.2.

Apple skin is rich in many health-enhancing phytonutrients including flavonoids and phenolic acids (Boyer & Liu, 2004). Apple skin has three to six folds more flavonoids than apple flesh and has unique flavonoids, such as quercetin glycosides, not found in the flesh (Wolfe et al., 2003; Wolfe & Liu, 2003). Apple fruit skin is rich source of DF and phenolics. The blanched, dehydrated, and ground apple skin powder contained ~41% TDF and oxygen radical absorption capacity (ORAC) of 52 mg Trolox equivalents/g dry weight (Rupasinghe et al., 2008).

8.5.1.3   MANGO PEEL

Mango (*Mangifera indica* L. Anacardiaceae) is one of the most important tropical fruits. It is a seasonal fruit thus processed into various popular products such as puree, nectar, leather, pickles, canned slices, and so forth (Loelillet, 1994). Processing of mango mainly for pulp and amchur powder, peel is a by-product which is not utilized for any commercial purpose and discarded as a waste. This waste should be treated as a specialized residue due to high levels of residual phenolics as well as DF (Larrauri et al., 1996a). Peel constitutes about 20% of the whole fruit and its disposal has become a great problem. According to Ajila et al. (2007) the polyphenol contents in these peels ranged from 55 to 110 mg/g dry peel. DF content ranged from 45 to 78 % of peel and was found at a higher level in ripe peels. Similarly, carotenoid content was higher in ripe fruit peels. Vitamins C and E contents ranged from 188 to 392 and 205 to 509 μg/gm dry peels, respectively, and these were also found at a higher level in ripe peels. The study conducted by Ajila et al. (2008) indicated that mango peel contained 51.2% of TDF, 96 mg GAE/g of polyphenols, and 3092 mg/g of carotenoids.

8.5.1.4   CITRUS BY-PRODUCTS

The amount of residue obtained from the citrus fruits after juice and essential oil extraction accounts for 50% of the original whole fruit mass (Cohn & Cohn, 1997). They consist of peels (albedo and flavedo), which are almost one-fourth of the whole fruit mass, seeds, and fruit pulp (Braddock, 1999). Citrus peel has been reported to be a good source of pectin and DF in general, with an equilibrated proportion of soluble and insoluble fractions (Baker, 1994; Larrauri et al., 1994). Approaches to the development of products with increased dietary benefits from citrus peel have placed emphasis not only on the recovery of carbohydrates and pectin (Baker, 1994) but also on the production of potentially important secondary metabolites, such as polyphenols (Manthey & Grohmann, 1996). These fiber-associated polyphenols are known to exert important health promoting effects (Middleton & Kadaswami, 1994). Thus, citrus fruit by-products could be interesting not only for its important fiber content but also because of its antioxidant capacity (Kang et al., 2006; Rehman, 2006). They have a high fiber and vitamin contents as well as other associated bioactive compounds such as flavonoids and terpenes possessing antioxidant properties (Lario et al., 2004). According to Saura-Calixto (1998) DF from citrus residues has good functional properties and low caloric content. Moreover, the residues contain more natural anti-oxidants than do the flesh or juice (Larrauri et al., 1996b).

Limonoids, the major cause of bitterness in citrus juice, have been reported to possess substantial antioxidant and anti-cancer activities. The anti-carcinogenic activity of limonoids has been tested successfully in labo-ratory animals (Lam et al., 1989; Miller et al., 1989). They appear in many forms, including nomilin, obacunone, ichangin, and limonin. Limonin and nomilin are the most prevalent forms of the citrus limonoids. Among the classes of phytochemicals with anti-cancer properties designated by the National Cancer Institute of USA, carotenoids, coumarins, flavonoids, gluca-rates, monoterpenes, and phenolic acids are present in citrus fruits (Nagy & Attaway, 1992). They were reported in the highest concentration in citrus peel and were mainly composed of ferulic, sinapic, coumaric, and caffeic acids as well as hesperidin and naringin, among others (Peleg et al., 1991; Nagy & Attaway, 1992). Citrus flavonoids have been extensively investi-gated because of their health-promoting properties (Middleton & Kandas-wami, 1994). Nogata et al. (1996) reported that albedo tissue extracts from lemon, inhibited both cyclooxygenase and lipoxygenase activities more than those in orange, which could be related to the prevention of thrombosis, atherosclerosis, and carcinogenesis.

The chemical components of citrus fiber (pectin, lignin, cellulose, and hemicellulose), together with other compounds, such as flavonoids, were analyzed in nine different industrial sources (Marín et al., 2007). Final fiber composition was found to be more dependent on the industrial process than on the type of citrus. The chemical changes gone by citrus fiber showed losses of functional values; that is, SDF and ascorbic acid content decreased when waste products were transformed into fibers. The water holding and lipid holding capacities of analyzed citrus fibers suggested a non-linear behavior of these properties.

High DF powders from Valencia orange and Persa lime peels were prepared and their composition and antioxidant capacity were studied (Larrauri et al., 1996b). Fibers from both peels had high TDF content (61–69%) with an appreciable amount of soluble fiber (19–22%). The concentration of anti-oxidant (AA_{50}) required to achieve a 50% inhibition of oxidation of linoleic acid at 40 °C was measured using the ferric-thiocyanate method. Lime peel fiber (AA_{50}) had half the value of DL-α tocopherol and 23 times lower than orange peel fiber; the AA_{50} of commercial butylated hydroxyanisole (BHA) was half the value of lime fiber. The HPLC analyses of the polyphenols extracted from orange and lime peels fibers showed the presence of caffeic and FAs, as well as naringin, hesperidin, and myricetin in both fruit fibers. The different antioxidant power of these fibers could be in part explained by the presence of ellagic acid, quercetin and kaempferol in lime peel fiber which are strong antioxidant polyphenols.

Lemon (*Citrus limon cv* Fino) possesses the highest antioxidant potential among citrus fruits and it is the most suitable fiber for dietary prevention of cardiovascular and other diseases (Gorinstein et al., 2001). Lario et al. (2004) used lemon juice industry by-products to obtain high DF powder. The effect of processing variables (direct drying, and washing previous to drying) on functional properties, fiber content and type, microbial quality and physicochemical properties of the fiber were evaluated. The fiber had good functional and microbial qualities as well as favorable physicochemical characteristics to be used in food formulations. It was observed that processing conditions affected fiber composition and properties. Water holding capacity was enhanced by washing and slightly decreased by the reduction in fiber particle size. Oil holding capacity was not affected by those factors. Acid detergent and neutral detergent fibers were highest in powder from washed lemon residue. Washing prevented fiber browning during drying as reflected in color parameters. Washing water rinsed green components.

High DF powders from Persian and Mexican lime peels were prepared and their fiber composition and antioxidant capacities were determined (Ubando-Rivera et al., 2005). The TDF contents of both varieties were high; 70.4 and 66.7%, respectively. Both lime peel varieties had an appropriate ratio of soluble/insoluble fractions. The water-holding capacities of fiber concentrates were high (6.96–12.8 g/g) which was related to the SDF which was higher in the DF concentrate of Mexican lime. Fiber concentrates of Persian lime peel had greater polyphenol contents than those of Mexican lime peel. The polyphenols associated with the DF in both lime peel varieties showed a good antioxidant activity. It was suggested that from a nutritional standpoint, DF lime concentrates may be suitable as food additives.

8.5.1.5   AÇAÍ PALM

Açaí (*Euterpe oleracea*), also known as cabbage palm, is a tropical species which bears a dark purple, berry-like fruit, clustered into bunches. Recently, much attention has been paid to its antioxidant capacity and its possible role as a functional food or food ingredient (Pozo-Insfran et al., 2006; Ribeiro et al., 2010; Schreckinger et al., 2010). Anthocyanins, proanthocyanidins, and other flavonoids were found to be the major phytochemicals in freeze-dried açaí (Schauss et al., 2006) and some works have also been carried out on antioxidant capacity of açaí pulp (de Souza et al., 2009; Rufino et al., 2009a; Rufino et al., 2009b; Rufino et al., 2010). Rufino et al. (2011) reported the concentrations of DF and antioxidant capacity in fruits (pulp and oil) of a new açaí cultivar—"BRS-Pará." The result showed that "BRS-Pará" açaí fruit has a high content of DF (71% dry matter) and oil (20.82%) as well as a high antioxidant capacity in both defatted matter and oil. These features provide açaí "BRS-Pará" fruits with considerable potential for nutritional and health applications.

8.5.1.6 CACTUS PEAR

Opuntia ficus-indica (cactus pear) is a cactus well adapted to extreme climate and edaphic conditions. The genus Opuntia embraces about 1500 species of cactus and many of them produce edible tender stems and fruits (Hegwood, 1994). The tender young part of the cactus stem, or cladode, is frequently consumed as a vegetable in salads, while the cactus pear fruit is consumed as a fresh fruit. Studies on the chemical composition of the edible portion of cladodes and fruits from *O. ficus-indica* showed that these foods have a high nutritional value, mainly due to their mineral, protein, DF, and phytochemical contents (Bensadón, 2010). Interestingly, antioxidant activity has also been reported (Corral-Aguayo, 2008).

The by-products are the outer coating of these plants, which is removed before food preparation and contains spines and a large quantity of glochids and pulp. Around 20 and 45% of the fresh weight of cladodes and fruits, respectively, are by-products (Muñoz de Chávez & Ledesma-Solano, 2002). These by-products are rich in DF, minerals, and antioxidant bioactive compounds. Bensadón et al. (2010) determined the nutritional value of by-products obtained from cladodes and fruits from two varieties of *O. ficus-indica*, examining their DF and natural antioxidant compound contents. They found that the materials studied were rich in good quality DF and natural antioxidants, especially Milpa Alta and Alfajayucan cultivars. It was concluded that by-products from cladodes and fruits of *Opuntia sp.* could be attractive for use as functional food ingredients.

8.5.1.7   GUAVA

Guava (*Psidium guajava* L.), now being recognized as *"super food"* is getting very much attention in the agro-food business attributed to presence of health promoting bioactive components and functional elements. The fruit is considered as highly nutritious due to presence of high level of ascorbic acid (50–300 mg/100 g fresh weight) and has several carotenoids such as phytofluene, β-carotene, β-cryptoxanthin, γ-carotene, lycopene, rubixan-thin, cryptoflavin, lutein, and neochrome (Mercadante et al., 1999). Phenolic compounds such as myricetin and apigenin (Miean & Mohamed, 2001), ellagic acid, and anthocyanins are also at high levels in guava fruits. Jiménez-Escrig et al. (2001a) reported IDF, SDF, and TDF content in dried guava as 46.72–47.65, 1.77–1.83, and 48.55–49.42%, respectively. According to researchers, peel and pulp of *P. guajava* fruit has high levels of DF, indigest-ible fraction, and phenolic compounds. Nahar et al. (1990) found a similar relative value for IDF (91% of TDF) in the edible portion of *P. guajava.*

The total phenolics (on fresh mass basis) was 344.9 mg GAE/100 g in "Allahabad Safeda" and ranged from 170.0 to 300.8 mg GAE/100 g in the pink pulp clones (Thaipong et al., 2006). According to Corrêa et al. (2011) total phenolics in guava varied from 158 to 447 mg GAE/100 g. Luximon-Ramma et al. (2003) have reported that white pulp guava had higher anti-oxidant activity and total phenolics than pink pulp guava in which the antioxidant activity was 142.6 and 72.2 mg/100 g in white and pink pulp, respectively, and the total phenolics was 247.3 and 126.4 mg GAE/100 g in white and pink pulp, respectively.

8.5.1.8   BAEL PULP RESIDUE

Bael fruit pulp is endowed with many functional and bioactive compounds such as DF, carotenoids, phenolics, alkaloids, coumarins, flavonoids, terpe-noids, and other antioxidants (Suvimol & Pranee, 2008). Major antioxidants in bael fruit are phenolics, flavonoids, carotenoids, and vitamin C (Morton, 1987; Roy & Khurdiya, 1995). Quantitative analyses have indicated that the bael fruit is rich in carbohydrates and fibers and also a good source of protein, vitamins, and minerals (Ramulu & Rao, 2003).

TPC (mg of GAE/100 g of decoction) in crude aqueous extract of bael fruit powder was reported as 336.1 (Gheisari et al., 2011). According to Suvimol and Pranee (2008), bael fruit pulps had TPC of 87.34 mg GAE/g dry weight while Jain et al. (2011) reported the total polyphenols (mgGAE/g) in bael fruit extract as 95.33. Abdullakasim et al. (2007) reported that the bael fruit drink possess high quantities of total phenolic compounds (37.6–83.89 mg GAE/100 ml). The aqueous extract of the bael fruit pulp possesses potent antioxidant effect (Kamalakkannan & Prince, 2003a; 2003b). The hydro-alcoholic extract of bael pulp is also shown to possess nitric oxide scavenging activities *in vitro* (Jagetia & Baliga, 2004). Suvimol and Pranee (2008) found that bael fruit pulps had TDF, SDF, and IDF contents of 19.84, 11.22, and 8.62 g/100 g dry weight, respectively. According to these workers the bael fruit is relatively rich in DF and it was in range of fruits which are defined as high DF fruits. According to Parichha (2004) fresh bael pulp without seed contains 31.8% TDF.

8.5.1.9 PINEAPPLE SHELL

Pineapple is one of the most important fruits in the world, and most of its production is used in processing. It is consumed as canned slices, chunks, dice, or fruit salads and in the preparation of juices, concentrates, and jams (Salvi & Rajput, 1995). By-products obtained from industrial processing represent 25–35% of the fruit, and the shell is the major constituent. The shell has been used to produce alcohol, citric acid, vinegar, bromelain, wine, sugar syrup, wax, sterols, and cattle feed (Joseph & Mahadeviah, 1988; Salvi & Rajput, 1995). The DF content and composition of pineapple flesh has been reported (Bartolomé & Rupérez, 1995).

Properties of a high DF powder prepared from pineapple fruit shell were evaluated and compared to those of several commercial fruit fibers (Larrauri et al., 1997). TDF content (70.6%) was similar to some commercial DFs from apple and citrus fruits; however, its sensory properties were better than those from commercial fibers. The IDF fraction accounted 99% of the TDF. Major neutral sugars in SDF and IDF were xylose and glucose, respectively. Total uronic acids and KL were 5.1 and 11.2%, respectively. At the concentration of 0.5 g of powdered sample/100 mL in the assay mixture, pineapple fiber showed a higher antioxidant activity (86.7%) than orange peel fiber (34.6%). Myricetin was the major identified polyphenol in pineapple fiber.

8.5.1.10 DATE BY-PRODUCTS

Dates of date palm tree (*Phoenix dactylifera* L.) are popular among the population of the Middle Eastern countries. The fruit is composed of a fleshy pericarp and seed which constitutes between 10 and 15% of date fruit weight (Hussein et al., 1998). The date seeds considered as a waste product of many date processing plants producing pitted dates, date syrup, and date confectionery. Generally, seeds are used as an animal feed; however, could be a valuable source of DF and phenolics.

Three native sun dried date varieties from Oman (Mabseeli, Um-sellah, and Shahal) their syrups and by-products (press cake and seed) were examined for their proximate composition, DF, total phenolics, and total antioxidant activity (Al-Farsi et al., 2007). Carbohydrate was the predominant component in all date varieties, syrups, and their by-products, followed by moisture, along with small amounts of protein, fat, and ash. The DF content in seeds and press cakes were found to be 77.75–80.15% fresh weight and 25.39–33.81% fresh weight, respectively. Among dates, syrups, and their by-products, seeds had the highest contents of total phenolics (3102–4430 mg of GAE/100 g fresh weight) and antioxidant activity (580–929 μmol of Trolox equivalents/g fresh weight). The researcher concluded that date by-products, particularly seeds serve as a good source of natural antioxidants and could potentially be considered as a functional food or functional food ingredient.

Al-Farsi and Lee (2008) conducted the work to optimize extraction conditions of phenolics and DF from date seeds. The effects of solvent to sample ratio, temperature, extraction time, number of extractions, and solvent type on phenolic extraction efficiency were observed. Two stage extractions, each stage lasting for 1 h duration at 45 °C with a solvent to sample ratio of 60:1, was considered optimum. Acetone (50%), and butanone were the most efficient solvents for extraction and purification, increasing the yield and phenolic contents of seed concentrate to 18.10 and 36.26%, respectively. The TDF of seeds increased after water and acetone extractions. Nine phenolic acids were detected in seeds with p-hydroxybenzoic, protocatechuic, and m-coumaric acids found to be among the highest. Protocatechuic, caffeic, and FAs were the major phenolic acids found in the concentrates. It was suggested that date seed concentrates could potentially be economical source of natural DF and antioxidants.

8.5.2 VEGETABLE BY-PRODUCTS

8.5.2.1 BRASSICA PLANTS

Epidemiological studies have shown that high consumption of Brassica vegetables, including cauliflower, cabbages, and broccoli, is associated with reduced risk of certain cancers such as lung cancer, colorectal cancer, breast cancer, and prostate cancer (Ciska & Pathak, 2004; Higdon et al., 2007). Brassica vegetables have been reported to contain high amount of DF and various bioactive agents with high antioxidant activity. Phenolic compounds and vitamin C are the major antioxidants of brassica vegetables (Podsedek, 2007). Lipid-soluble antioxidants (carotenoids and vitamin E) are responsible for up to 20% of the brassica total antioxidant activity.

Cauliflower has a very high waste index (Kulkarni et al., 2001) and is an excellent source of protein, cellulose, and hemicellulose (Wadhwa et al., 2006). It is considered as a rich source of DF and possesses both antioxidant and anticarcinogenic properties. The level of non-starch polysaccharide (NPS) in the upper cauliflower stem remains similar to that of the floret and both are rich in pectic polysaccharides, while the cauliflower lower stem NPS is rich in cellulose and xylan (Femenia et al., 1998).

White cabbages (*Brassica oleracea* var. capitata) have been reported to contain high amount of DF and bioactive agents with high antioxidant activity as well as glucosinolates which are claimed to possess anticarcinogenic activity. About 40% of cabbage leaves, which are processed into many products, including salads and ready-to-eat vegetables, are lost and regarded as waste which contains high amount of DF and various phytochemical compounds (Nilnakara et al., 2009). Processing of these residues could therefore add much value to the products. The main constituents in white cabbage are carbohydrates, comprising nearly 90% of the dry weight, where approximately one-third is DF and two-thirds are low-molecular-weight carbohydrates (Wennberg et al., 2004). Additionally, white cabbage also possesses significant amounts of antioxidants such as ascorbic acid, phenolic compounds, and tocopherols (Kim et al., 2004; Wennberg et al., 2004; Singh et al., 2006).

The cabbage glucosinolates are the most interesting compounds. Glucosinolates are a group of sulfur-containing plant secondary metabolites and can be hydrolyzed by myrosinase to form different products, for example, thiocyanates, isothiocyanates, epithionitrile, nitrile, and oxazolidine-thione (Wennberg et al., 2006). The breakdown products (especially isothiocyanates) possess anticancer activity via modulation of phase II enzymes, including glutathione S-transferase and quinine reductase. These enzymes are reported to help inactivate cancer by blocking normal cells from DNA damage (Verkerk et al., 2001). Production of DF powder from cabbage outer leaves involves mechanical and thermal processes which may affect the amount of glucosinolates. Several reports have indeed shown that glucosinolates in Brassica vegetables decreased upon blanching because of enzymatic breakdown, thermal breakdown and leaching into blanching water (Verkerk et al., 2001; Wachtel-Galor et al., 2008; Cartea & Velasco, 2008). Drying is also reported to affect glucosinolates evolution. Mrkic et al. (2010) studied the effect of temperature (50–100 °C) of the air that was used to dry broccoli and reported that glucosinolates content decreased upon drying, especially at higher drying temperatures.

Tanongkankit et al. (2012) investigated the effects of processing steps, that is, slicing, blanching, and drying, on the changes of total glucosinolates in cabbage outer leaves, changes in DF composition and color. They noted that the preparation steps did not lead to any significant changes of the DF powder compositions. On the other hand, steam blanching was noted to better preserve glucosinolates than water blanching. Drying methods and conditions did not lead to any significant effect on the powder compositions; however, vacuum drying led to better retention of glucosinolates. Color of the DF powder was not affected by the drying methods and conditions.

Jongaroontaprangsee et al. (2007) produced high DF powder from outer leaves of cabbage and reported that the powder contained ~41–43% TDF (dry basis). Moreover, the powder possessed high water holding capacity and swelling capacity. The production of DF powder associated with antioxidant activity from cabbage outer leaves was studied by Nilnakara et al. (2009). The effects of hot water blanching and hot air drying temperature (70–90 °C) on the quality of DF powder produced from cabbage outer leaves were also investigated. Parameters like proximate composition, visual color, phenolic content, vitamin C as well as total antioxidant activity were evaluated.

8.5.2.2   ASPARAGUS

Besides their culinary quality, green asparagus spears are known for their composition of bioactive compounds. Eastern civilizations have been using asparagus extracts as stimulants, laxatives, antitussives, diuretics, and so forth, for hundreds of years. Asparagus has been reported as rich in the quality and quantity of its antioxidants (Pellegrini et al., 2003; Vinson et al., 1998). During industrial processing, around half of the total length of each spear is discarded, which creates significant waste for producers. It is expected that the by-products have similar composition to the edible part of the spears and could be a promising source of phytochemicals and fiber (Nindo et al., 2003).

The asparagus extracts possess number of biological activities, including anti-tumor and antioxidant activities and participate in the prevention of cardiovascular diseases (Cushine & Lamb, 2005; Nijveldt et al., 2001). Among all the bioactive compounds present in asparagus spears, saponins, flavonoids, and hydroxycinnamates are the main compounds responsible for the characteristics cited above. Rutin is the most abundant flavonoid in asparagus spears, in addition to others that have been recently described (Fuentes-Alventosa et al., 2007, 2008).

According to Fuentes-Alventosa et al. (2009a) the method by which asparagus by-products are treated affects the phytochemical composition and antioxidant activity of the fiber rich powders. They studied factors such as the treatment intensity, the solvent used, and the drying system. Among the asparagus phytochemicals, HCA, saponins, flavonoids, sterols, and fructans were quantified. HCA varied from 2.31 to 4.91 mg/g of fiber, the content being affected by the drying system and, in some cases, the solvent. Treatment intensity while isolation was found to affect the saponin content in fibers. Saponin content ranged from 2.14 to 3.64 mg/g of fiber. Flavonoids were most affected by processing conditions, being present (0.6–1.8 mg/g of fiber) only in three of the samples analyzed. Sterols and fructans were present in minor amounts, 0.63–1.03 mg/g of fiber and 0.2–1.4 mg/g of fiber, respectively.

Fuentes-Alventosa et al. (2009b) investigated the effect of extraction method on chemical composition and functional characteristics of high DF powders obtained from asparagus by-products. The by-products represented around 50% of the processed vegetable. All the fiber rich powders had high concentrations of TDF (62–77%). The proportion of insoluble fiber to soluble fiber decreased with the severity of treatment, in this way increasing the physiological quality of the fiber. Functional properties, namely water holding capacity, oil holding capacity, solubility, and glucose dialysis retardation index (GDRI), varied according to the preparation procedure. These properties make fiber rich powders from asparagus by-products a valuable source of DF to be included in the formulation of fiber-enriched foods.

8.5.2.3 CARROT PEEL

Carrot (*Daucus carota* L.) is a good source of natural antioxidants, especially carotenoids and phenolic compounds (Prakash et al., 2004; Zhang & Hamauzu, 2004). After processing, carrot residues such as peels, pomace, are usually discarded or used as animal feed. These by-products contain high contents of beneficial substances, especially bioactive compounds with antioxidant activities (Zhang & Hamauzu, 2004). The feasibility study of using carrot peels as a starting raw material to produce ADF powder was investigated (Chantaro et al., 2008). The effects of blanching and hot air drying (60–80 °C) on the drying kinetics and physicochemical properties of DF powder were evaluated. The results showed that blanching had a significant effect on the fiber contents and compositions, water retention and swelling capacities of the fiber powder. In contrast, drying temperature in the selected range did not affect the hydration properties. As far as antioxidant activity is concerned, thermal degradation during both blanching and drying caused a decrease in the contents of β-carotene and phenolic compounds, hence leading to the loss of antioxidant activity of the final product.

8.5.3 SEED AND BY-PRODUCTS

8.5.3.1 CEREALS

Cereals and legumes containing wide range of phenolics are good sources of natural antioxidants (Krings et al., 2000). It has been reported that phenolic compounds are concentrated in the bran portion of cereal kernels and may contribute to the total antioxidant activities, suggesting bran a potent source of antioxidants (Onyeneho & Hettiarachy, 1992). However, it is not clear at which extent both free and carbohydrate-bound compound are measured by a given assay (Zhou et al., 2004). Wheat is one of the popular cereal grains, and its bran represents not only a good source of DFs (Alabaster et al., 1997), but also of phenolic acids (Baublis et al., 2000). Significant levels of antioxidant activities have been detected in wheat (Yu et al., 2003; Zielinski & Kozlowska, 2000), and wheat-based food products (Baublis et al., 2000), suggesting that wheat may serve as an excellent dietary source of natural antioxidants for disease prevention and health promotion. Yu et al. (2002) reported a significant level of TPC, free radical scavenging capacities, chelating activity, and inhibitory effect on lipid peroxidation of the three-wheat grain extracts with significant differences among the varieties. Antioxidant activity of wheat bran and flour extracts varies with cultivar and location (Yu et al., 2002).

Wheat bran, a by-product generated in large amounts during wheat processing, consists of 36.5–52.4% TDF (Vitaglione et al., 2008), which makes it a good source of DF. Additionally, wheat and wheat bran has shown strong antioxidative activities (Li et al., 2005). Several phenolic acids, including vanillic acid, p-coumaric acid, and, largely, FA have been found in wheat bran extracts (Kähkönen et al., 1999). These compounds, particularly FA, are not evenly distributed in the wheat; most are found in the bran (Baublis et al., 2002). According to Onyeneho and Hettiarachchy (1992) extract of wheat bran, having high concentration of phenolic acids and have stronger antioxidant activity than other fractions of wheat. Wheat bran has been reported to be able to inhibit lipid oxidation catalyzed by either iron or peroxyl radicals (Baublis et al., 2000). Zhou et al. (2004) reported that wheat grain, bran, and fractions had different antioxidant activities and TPCs. Their study also showed that FA was a major contributor to the antioxidant activity.

Antioxidant activity of bran extracts from five wheat varieties indigenous to Pakistan, that is, Punjab-96, Bhakkar-2002, Uqab-2000, SH- 2002, and Pasban-90, was evaluated (Iqbal et al., 2007). All the bran extracts exhibited appreciable TPC (2.12–3.37 mg GAE/g bran), total flavonoid content (epicatechin equivalent 262–304 mg/g bran), chelating activity (EDTA equivalent 597–716 mg/g bran), DPPH radical scavenging activity (51–79%), ABTS radical cation scavenging activity (Trolox equivalent 27–36 mmol/g), oxygen radical absorbance capacity (ORAC) (97–123 mmol/g), and TA content (30–38 mg/kg bran). Tocopherol (22–26 ppm) and tocotrienol content (59–74 ppm) were determined by reversed phase HPLC (RP-HPLC). It was observed that all the varieties exhibited appreciable antioxidant potential and significant differences were observed among the varieties in different systems of antioxidant activity evaluation.

Four different types of wheat bran were extracted and analyzed for phenolic acids using the Folin–Ciocalteu method and HPLC (Kim et al., 2006). The extracts and their hydrolysis products were also evaluated for antioxidant activities. The TPC of the red wheat bran was found higher than that of the bran from white wheat. The majority of the phenolic acids existed in a bound form in wheat bran. These phenolic acids can be released by hydrolyzing the bran under alkaline or acidic conditions; however, the former was more efficient. Ferulic, vanillic, and syringic acids were the major individual phenolic acids in the studied wheat bran. The major portion of the total FA was from alkaline hydrolysis. The alkaline hydrolysable fractions had greater antioxidant activities, while the acid hydrolysable fractions showed lower activities in both the red and white bran. The antioxidant activity of bran extract was stronger than that of free phenolic acids.

The impact of thermal processing on antioxidant activity of purple wheat bran, heat-treated purple wheat bran was evaluated to assess potential health benefits (Li et al., 2007). TPC and ORAC values of sample extracts were significantly affected by various extracting solvents. The conditions selected for heat treatment did not markedly change antioxidant activity of purple wheat bran. Though, there was a significant reduction in TPCs, ORAC values, and TAs during processing of purple wheat bran. Esposito et al. (2005) selected the fractions of durum wheat bran having different functional and nutritional characteristics. Wheat bran by-products were obtained by an industrial milling process. Beside the single fractions, two commercial products B&B 50 and B&B 70, obtained by blending some of the durum wheat fractions were also studied. The soluble fiber content of the durum wheat by-product ranged between 0.9 and 4.1%; while that of insoluble fiber between 21 and 64%. The B&B 70 had a TDF content of 61%, while B&B 50 has 42%. These workers observed that water-holding capacity of each fraction is strictly related to the amount of insoluble fiber and to the granulometry of the by-products. The antioxidant activity was found higher for the internal bran fraction and it increases in fractions having reduced granulometry.

Wheat bran DF powders was prepared by ultrafine grinding, whose effects were investigated on the composition, hydration, and antioxidant properties of the wheat bran DF products (Zhu et al., 2010). The results showed that as particle size decrease, the hydration properties (water holding capacity, water retention capacity, and swelling capacity) of wheat bran DF were significantly decreased and a redistribution of fiber components from insoluble to soluble fractions was observed. Compared with DF before and after grinding, micronized IDF showed increased chelating activity, reducing power and total phenolic compounds yet decreased DPPH radical scavenging activity.

8.5.3.2   MEXICAN CHIA SEED

The seeds of the species *Salvia hispanica* L. commonly known as "chia," "chia sage" and "Spanish sage," were an important staple food, oil source and medicine for Mesoamericans in pre-Columbian times. The curative properties of the seeds were also appreciated, for example, for treating eye obstructions, infections, and respiratory malaises. The seeds soaked in water or fruit juice were and still are consumed in some regions as a refreshing drink (Cahill, 2003). The presence of cinnamic, chlorogenic, and caffeic acids together with the flavonoids, myricetin, quercetin, and kaempferol in methanolic hydrolyzed extracts has also been reported (Taga et al., 1984).

Chia seeds from two different regions in the states of Jalisco and Sinaloa were analyzed for soluble and insoluble fiber and antioxidant activity of phenolic compounds (Reyes-Caudillo et al., 2008). The soluble and insoluble fiber content of the Sinaloa and Jalisco seeds was similar. The major compounds identified in hydrolyzed and crude extracts were quercetin and kaempferol, while caffeic and chlorogenic acids were present in low concentrations. The crude extract of the Jalisco seed has an antioxidant activity comparable to the commercial antioxidant Trolox used as a reference. Different concentrations of the hydrolyzed and crude extracts of the seeds from both regions showed antioxidant effect when tested in a model water-in-oil food emulsion.

8.5.3.3 COCOA HUSK

Besides flavonoids, cocoa is rich in other component of significant nutritional interest such as DF. Polyphenolic compounds usually accumulate in the outer parts of plants such as shells, skins, and so forth (Bravo, 1998). Though, information on the polyphenolic content of cocoa husks is very limited. It has been suggested that cocoa hull may be a good source of DF, with reported values ranging from 38 to 44% of TDF as NPSs plus KL (Martín-Cabrejas et al., 1994; Serra-Bonvehí & Aragay-Benería, 1998). Considering the health benefits associated to the consumption of DF and polyphenols in the diet, the presence of both bioactive components in cocoa bean husks could highlight the interest of this product as a potential ingredient for the functional food industry.

Cocoa polyphenols have been suggested to positively influence cardiovascular health through inhibition of lipid peroxidation, platelet activation or cyclo-oxygenase, and lipoxygenase activities, and enhancing levels of the endothelial-derived relaxing factor, nitric oxide (Karim et al., 2000; Rein et al., 2000; Schewe et al., 2002; Steinberg et al., 2003; Wan et al., 2001; Wiswedel et al., 2004). Moreover, cocoa polyphenols have exhibited antimutagenic activity (Yamagishi et al., 2000). A decreased levels of 8-hydroxy-20-deoxyguanosine, a biomarker of oxidative damage to DNA, have been reported in rats after consumption of cocoa suggesting a potential role in cancer (Orozco et al., 2003).

The proximate composition and DF content of a fiber-rich product obtained from cocoa were studied (Lecumberri et al., 2007). This product contained 60.54% (dry matter basis) of DF, composed of mainly insoluble fiber although with appreciable amounts of SDF (10.09%). The presence of associated polyphenolic compounds (1.32 and 4.46% of soluble polyphenols and condensed tannins, respectively) provides this fiber material with intrinsic antioxidant capacity as determined by the FRAP and trolox equivalent antioxidant capacity (TEAC) methods.

8.5.3.4   AMARANTH AND QUINOA

Amaranth (*Amaranthus hypochondriacus*) is an ancient and very nutritious food crop cultivated mainly in South America and Mexico, but grows also very well across the world. Of late, the use of amaranth and quinoa has broadened not only in the common diet, but also in diet of people with celiac disease or allergies to typical cereals (Berti et al., 2005). These pseudocereals seeds have high nutritional and functional values which are associated with the quality and quantity of their proteins, fats, and antioxidant potential (Gorinstein et al., 2002; Gorinstein et al., 2007; Paśko et al., 2007). Amaranth is a crop naturally resistant to water deficit and is a good source of nutritious seeds. Barba de la Rosa et al. (2009) analyzed physical and proximal-nutritional properties of amaranth seeds obtained from different varieties (Tulyehualco, Nutrisol DGETI, and Gabriela) and characterized their phenolic acids and flavonoids. Polyphenols as rutin (4.0–10.2 mg/g flour) and nicotiflorin (7.2–4.8 mg/g flour) were detected.

Two varieties (Centenario and Oscar Blanco) of Andean native grain, *kiwicha (Amaranthus caudatus)*, were evaluated as sources of DF and of some bioactive compounds (Repo-Carrasco-Valencia et al., 2009). The impact of low-cost extrusion on the content of these components was studied for technological applications. The content of TDF in Centenario was higher (16.4%) than in Oscar Blanco (13.8%). The extrusion process decreased the total and IDF contents in both varieties while in Centenario, the content of SDF increased, from 2.5 to 3.1%. The content of phytic acid in raw *kiwicha* was 0.3% for both varieties, and the content of total phenolic compounds was 98.7 and 112.9 mg GAE/100 g of sample, for Centenario and Oscar Blanco, respectively. Antioxidant activity of both the varieties of raw *kiwicha* was evaluated through DPPH and ABTS methods. The content of total phenolics, phytic acid and the antioxidant activity decreased in both varieties during the extrusion process.

The consumption of sprouts—the atypical vegetable, is becoming a new trend which has received attention as functional foods, because of their nutritive value including amino acid, fiber, trace elements, vitamins as well as flavonoids, and phenolic acids (Paśko et al., 2008). Its intake has become very popular among people interested in improving and maintaining their health status by changing dietary habits. The sprouts of amaranth and quinoa are new vegetables, which can be used as a source of nutrition. Total antioxidant capacity, TPCs and anthocyanins contents were determined in *Amaranthus cruentus* and *Chenopodium quinoa* seeds and sprouts (Paśko et al., 2009). Sprouts activity depended on the length of their growth, and the peak values were reached on the fourth day in the case of amaranth and on the sixth day in the case of quinoa. It was suggested that amaranth and quinoa seeds and sprouts can be used in food, because it is a good source of anthocyanins and total phenolics with high antioxidant activity.

8.5.4 OTHER SOURCES

8.5.4.1 BURDOCK ROOT

It is a source of inulin and popular vegetable in Japan. Burdock root has been extensively analyzed for its components due to their antioxidant properties (Chow et al., 1997) as well as for its extractable components having antimicrobial activity (Duh, 1998; Lin et al., 1996). The simultaneous ultrasonic/microwave assisted extraction (UMAE) of inulin and production of phenols rich DF powder from burdock root was studied (Lou et al., 2009). The DF powder prepared from the residue of burdock root after inulin extraction was rich in phenols (302.62 mg GAE/100 g powder). It was seen that drying temperature in the selected range did not significantly affect the hydration properties.

8.5.4.2 HIBISCUS

Hibiscus sabdariffa L., commonly known as roselle, red sorrel, or karkadè, is widely grown in Africa, South East Asia, and some tropical countries of America. Its fleshy flowers provide a soft drink consumed as a cold or hot beverage. Pharmacological actions have been identified in *H. sabdariffa* L. flowers, petals, and seeds (Ali-Bradeldin et al., 2005). Roselle is an important source of vitamins, minerals, and bioactive compounds, such as organic acids, phytosterols, and polyphenols. The phenolic content in the plant consists mainly of anthocyanins like delphinidin-3-glucoside, sambubioside, and cyanidin-3- sambubioside; other flavonoids like gossypetin, hibiscetin, and their respective glycosides; protocatechuic acid, eugenol, and sterols like—sitosterol and ergosterol (Ali-Bradeldin et al., 2005). The health effects include cardioprotective action; reduction of urinary concentrations of creatinine, uric acid, citrate, tartrate, calcium, sodium, potassium, phosphate; antihypertensive action; effectiveness against low-density lipoprotein oxidation and hyperlipidemia (Herrera-Arellano et al., 2004; Jonadet et al., 1990; Chen et al., 2003; Mojiminiyi et al., 2000; Odigie et al., 2003). In a study on roselle flower and beverage prepared from it, Sáyago-Ayerdi et al. (2007) quantified the DF, associated polyphenols, and antioxidant capacity. It was reported that roselle flower contained DF as the largest component (33.9%) and was rich in phenolic compounds (6.13%).

8.5.4.3 SEAWEED

Fucus vesiculosus L. is a brown-colored, perennial, *dioecious* edible seaweed forming dense belts in cold rocky littoral habitats (Jiménez-Escrig et al., 2001b; Jormalainen & Honkanen, 2004), covering large areas from a few decimeters below the water surface to a depth of several meters (Nilsson et al., 2004). It contains protein, minerals, iodine, vitamins, monounsaturated and polyunsaturated fatty acids (Herbreteau et al., 1997; Rupérez & Saura-Calixto, 2001; Rupérez et al., 2002; Morel et al., 2005). However, the main components that make Fucus nutritionally significant are non-digestible polysaccharides (DF) and polyphenols. DF from seaweeds has proven to have a positive effect on cholesterol metabolism and blood pressure (Jiménez-Escrig & Sánchez-Muñiz, 2000). DF of *F. vesiculosus* is composed of fucans, alginates, laminaranes, and cellulose, with fucoidan as the predominant polysaccharide (Rioux et al., 2007). Fucoidan as observed in several studies in rats and humans, showing beneficial effects as an anti-coagulant, antithrombotic, antiviral, and anti-cancer agent (Béress et al., 1993; Aisa et al., 2005), as well as in the treatment of chronic renal failure (Zhang et al., 2003). *F. vesiculosus* has shown high antioxidant capacity by several methods (Morel et al., 2005) due to the synergic effect of phloro-tannins, vitamin E, and certain carotenoids (Le Tutour et al., 1998; Toth & Pavia, 2001).

The presence of the functional components DF and antioxidants in *F. vesiculosus* in a higher proportion than in other edible seaweeds (Jiménez-Escrig et al., 2001b; Rupérez & Toledano, 2003) has led to the development of a large number of functional ingredients and dietary supplements derived from this like fucoidan powders, *F. vesiculosus* capsules, or *F. vesiculosus* antioxidant extracts. Díaz-Rubio et al. (2009) compared the antioxidant capacity and polysaccharide composition of raw Fucus with those of some common commercial nutraceuticals. All tested products contained a high percentage of DF (45–59%), raw Fucus powder being the sample with the highest content. Moreover, Fucus powder exhibited significantly higher antioxidant capacity as determined by FRAP, ABTS, and ORAC assays than the commercial fucoidans and commercial antioxidant extracts. Polyphenols (phlorotannins) seem to be the main contributors to antioxidant capacity in both raw powder and commercial fucoidans.

8.6   APPLICATION IN MEAT PRODUCTS

In addition to promoting consumers health, application of ADF in different meat foods can play an important role in maintaining their quality and extension of storage stability. Incorporation of ADF in meat products is a quite new concept in the area of developing functional meat products thus the available literatures on ADFs in meat system are limited. The positive effects of ADF on meat and meat products are depicted in Figure 8.2. The effects of ADFs in meat products may vary according to their source, physicochemical characteristics, amount of phenolic compounds and TDF as well as ratio of soluble to IDF.

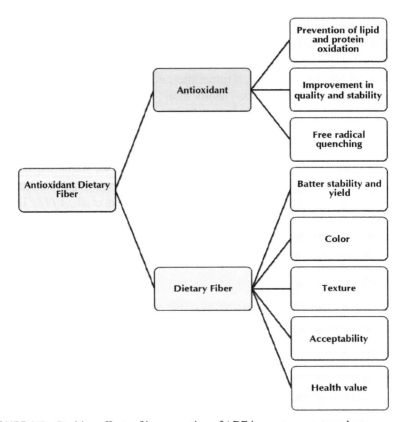

FIGURE 8.2   Positive effects of incorporation of ADF in meat or meat products.

The effect of grape antioxidant dietary fiber (GADF) addition (0, 2, and 4%) to minced fish muscle (MFM) on lipid stability during frozen storage (6 months) was studied (Sánchez-Alonso et al., 2007). GADF was characterized in terms of DF, total polyphenols and antioxidant capacity, and multifunctional antioxidant assays were carried out on all the MFM samples. The addition of red grape fiber considerably delayed lipid oxidation in minced horse mackerel (*Trachurus trachurus*) muscle during the first 3 months of frozen storage. In another study, white grape antioxidant dietary fiber (WGDF) obtained from white grape (*V. vinifera,* var. Airén) pomace from wine production was evaluated for antioxidant capacity in MFM during frozen storage at three different levels viz., 0, 2, and 4% (Sánchez-Alonso et al., 2008). WGDF was evaluated for DF (insoluble and soluble), total polyphenols and antioxidant capacity, and multifunctional antioxidant assays were done on all the MFM samples. The addition of WGDF considerably delayed lipid oxidation in minced horse mackerel muscle during the frozen storage. Vacuum-packing the sample with 2% WGDF significantly enhanced the antioxidant properties of WGDF.

Sánchez-Alonso and Borderías (2008) again investigated the effect of adding GADF at same levels to horse mackerel minced muscle as a technological ingredient in MFM for over six months of frozen storage (−20 °C). Protein solubility, water retention, color, mechanical properties, lipid oxidation, and sensory analyses were carried out immediately after preparation of samples, and during and after storage. Bound water after thawing and cooking minced samples was proportional to the amount of GADF used. Mechanical properties indicated softness and loss of cohesiveness depending on the amount of GADF. There was inhibition of lipid oxidation during frozen storage when GADF was added. Based on chemical, physical, and sensory analyses it was suggested that GADF is a highly active technological ingredient in frozen dark minced fish. Grape pomace concentrate (GPC) is a natural source of phenolic compounds with high antioxidant capacity. The effect of a diet containing GPC on lipid peroxidation levels via measuring thiobarbituric reactive substances (TBARS) and antioxidant capacity (ABTS method) of raw and cooked chicken breast meat patties stored in chilled conditions (4 °C) for 0, 3, 6, 13, and 20 days, and frozen storage (six months) was investigated (Sáyago-Ayerdi et al., 2009a). Chickens were fed GPC at levels of 0, 30, and 60 mg/kg from three to six weeks of age. Dietary GPC significantly exerted an inhibitory effect on lipid oxidation of raw and cooked breast chicken patties compared with samples obtained from birds fed the control diet at chilling and frozen storage. Radical scavenging capacity was significantly increased at 20 days in cooked samples and significantly reduced at six months of storage in raw and cooked samples. The higher concentration of dietary GPC increased the ABTS values only in the raw samples. The researchers stated that dietary GPC could be effective in inhibiting lipid oxidation of chilled and frozen stored chicken patties.

Efficiency of four concentrations (0.5, 1, 1.5, and 2%) of GADF on susceptibility of raw and cooked chicken breast hamburger to lipid oxidation was investigated after 0, 3, 5, and 13 days of refrigerated storage at 4 °C (Sáyago-Ayerdi et al., 2009b). A significant reduction in lightness and yellowness and increase in redness as a result of GADF addition were observed in raw and cooked chicken hamburgers. Addition of GADF significantly improved the oxidative stability and the radical scavenging activity in raw and cooked chicken hamburgers. The ability of GADF to prevent lipid oxidation was concentration-dependent. Acceptability of chicken meat was not affected by the addition of GADF.

Guava powder (0.5 and 1 %) was used as a source of ADF in sheep meat nuggets and its effect was evaluated against control (Verma et al., 2013). Guava powder was found rich in DF (43.21%), phenolics (44.04 mgGAE/g) and possessed good radical scavenging activity as well as reducing power. Total phenolics and TDF content significantly increased in nuggets with added guava powder. Product redness value was significantly improved due to guava powder. Guava powder was found to retard lipid peroxidation of cooked sheep meat nuggets as measured by TBARS number during refrigerated storage. Acceptability of the product remained unchanged due to addition of guava powder.

The antioxidant potential of bael (*Aegle marmelos* L.) pulp residue (BPR) and its influence at two different levels (0.25 and 0.5%) as an ADF on the quality of goat meat nuggets was investigated (Das et al., 2015). The antioxidant potential (total phenolics, radical-scavenging activity, and ferric reducing antioxidant power) and DF content of BPR were evaluated. BPR contained good amount of total phenolics (15.16 mgGAE/g dry weight) and DF (56.91%). BPR significantly improved the emulsion stability, cooking yield, ash, total phenolics, DF, and color characteristics of the meat products. On another side, BPR decreased the hardness, gumminess, and chewiness. Sensory evaluation of the products revealed significant improvement in the appearance score and non-significant increase in the score of other attributes. BPR decreased lipid peroxidation and microbial counts in meat products during 21 days of refrigerated storage (4±1 °C) period. It was concluded that BPR being a rich in bioactive components such as phenolic compounds and DF, could be used as an ADF in muscle food products without affecting its quality and acceptability.

8.7 CONCLUSION

The threats of non-communicable diseases which are associated with the life-style and diet have forced modern consumers to look toward functional foods especially meat products which can be the source of DF and antioxidants. Almost every plant-based materials are rich in natural antioxidants which may be associated with the DF leading to the concept of ADF way back in 1998. Inclusion of these materials into meat products can help in improving their technological and functional characteristics as well as stability while storage. It is also supposed that regular consumption of such functional ADF rich meat products may improve the physiological status of consumers. The incorporation of such functional ingredients in meat products is a newer concept and very scant works are available in the literature. The impacts of adding ADF on the quality, acceptability, and stability of the meat product have been reported to be very much interesting. However, assessment of the effects on population consuming these products is somewhat missing till now. In the near future it is hoped that a lots of ADF with defined amount of natural antioxidants and DF would be explored. Additionally these ingredients would be attempted in the meat products and their impact on meat products quality as well as consumer health would assessed.

KEYWORDS

- **antioxidant dietary fiber**
- **dietary fiber**
- **natural antioxidants**
- **meat products**

REFERENCES

Abdullakasim, P.; Songchitsomboon, S.; Techagumpuch, M.; Balee, N.; Swatsitang, P.; Sungpuag, P. Antioxidant Capacity, Total Phenolics and Sugar Content of Selected Thai Health Beverages. *Int. J. Food Sci. Nutri.* **2007,** *58,* 77–85.

Aisa, Y.; Miyakawa, Y.; Nakazato, T.; Shibata, H.; Saito, K.; Ikeda, Y.; Kizaki, M. Fucoidan Induces Apoptosis of Human HS-Sultan Cells Accompanied by Activation of Caspase-3 and Down-Regulation of ERK Pathways. *Am. J. Hematol.* **2005,** *78* (1), 7–14.

Ajila, C. M.; Leelavathi K.; Prasada Rao, U. J. S. Improvement of Dietary Fiber Content and Antioxidant Properties in Soft Dough Biscuits with the Incorporation of Mango Peel Powder. *J. Cereal Sci.* **2008,** *48,* 319–326.

Ajila, C. M.; Bhat, S. G.; Prasada Rao, U. J. S. Valuable Components of Raw and Ripe Peels from Two Indian Mango Varieties. *Food Chem.* **2007,** *102,* 1006–1011.

Alabaster, O.; Tang, Z.; Shivapurkar, N. Inhibition by Wheat Bran Cereals of the Development of Aberrant Crypt Foci and Colon Tumours. *Food Chem. Toxicol.* **1997,** *35,* 517–522.

Al-Farsi, M.; Lee, C. Y. Optimization of Phenolics and Dietary Fibre Extraction from Date Seeds. *Food Chem.* **2008,** *108,* 977–985.

Al-Farsi, M.; Alasalvar, C.; Al-Abid, M.; Al-Shoaily, K.; Al-Amry, M.; Al-Rawahy F. Compositional and Functional Characteristics of Dates, Syrups, and Their By-products. *Food Chem.* **2007,** *104,* 943–947.

Ali, B. H.; Wabel, N. A.; Blunden, G. Phytochemical, Pharmacological and Toxicological Aspects of *Hibiscus sabdariffa* L.: A Review. *Phytother. Res.* **2005,** *19* (5), 369–375.

Amico, V.; Napoli, E. M.; Renda, A.; Ruberto, G.; Spatafora, C.; Tringali, C. Constituents of Grape Pomace from the Sicilian Cultivar Nerello Mascalese. *Food Chem.* **2004,** *88* (4), 599–607.

Babbs, C. F. Free Radicals and the Etiology of Colon Cancer. *Free Radic. Biol. Med.* **1990,** *8,* 191–200.

Baker, R. A. Potential Dietary Benefits of Citrus Pectin and Fiber. *Food Technol.* **1994,** *48* (11), 133–139.

Balasundram, N.; Sundram, K.; Samman, S. Phenolic Compounds in Plants and Agri-Industrial By-products: Antioxidant Activity, Occurrence, and Potential Uses. *Food Chem.* **2006,** *99* (1), 191–203.

Barba de la Rosa, A. P.; Fomsgaard, I. S.; Laursen, B.; Mortensen, A. G.; Olvera-Martínez, L.; Silva-Sánchez, C.; Mendoza-Herrera, A.; González-Castañeda, J.; DeLeón-Rodríguez, A. Amaranth (*Amaranthus hypochondriacus*) as an Alternative Crop for Sustainable Food Production: Phenolic Acids and Favonoids with Potential Impact on its Nutraceutical Quality. *J. Cereal Sci.* **2009,** *49,* 117–121.

Bartolomé, A. P.; Rupérez, P. Dietary Fibre in Pineapple Fruit. *J. Clin. Nutr.* **1995,** *49,* S261–S263.

Baublis, A. J.; Decker, E. A.; Clydesdale, F. M. Antioxidant Effect of Aqueous Extracts from Wheat Based Ready-to-Eat Breakfast Cereals. *Food Chem.* **2000,** *68,* 1–6.

Baublis, A. J.; Lu, C.; Clydesdale, F. M.; Decker, E. A. Potential of Wheat-Based Breakfast Cereals as a Source of Dietary Antioxidants. *J. Am. Coll. Nutr.* **2002,** *19,* 308S–311S.

Bensadón, S.; Hervert-Hernández, D.; Sáyago-Ayerdi, S. G.; Goñi, I. By-products of Opuntia Ficus-Indica as a Source of Antioxidant Dietary Fiber. *Plant Foods Human Nutr.* **2010,** *65* (3), 210–216.

Béress, A.; Wassermann, O.; Bruhn, T.; Béress, L.; Kraiselburd, E. N.; Gonzalez, L. V.; de Motta, G. E.; Chavez, P. I. A New Procedure for the Isolation of Anti-HIV Compounds (Polysaccharides and Polyphenols) from the Marine Alga *Fucus vesiculosus*. *J. Nat. Prod.* **1993,** *56* (4), 478–488.

Berti, C.; Riso, P.; Brusamolino, A.; Porrini, M. Effect on Appetite Control of Minor Cereal and Pseudocereal Products. *Brit. J. Nutr.* **2005,** *94,* 850–858.

Block, G.; Patterson, B.; Subar, A. Fruit, Vegetables, and Cancer Prevention: A Review of the Epidemiological Evidence. *Nutr. Cancer.* **1992,** *18,* 1–29.

Boyer, J.; Liu, R. H. Apple Phytochemicals and their Health Benefits. *Nutr. J.* **2004,** *3,* 5–19.

Bozidar, S. The Role of Polyphenols as Chemopreventers. *Polyphén. Actual.* **1995**, *13,* 24–25.

Braddock, R. J. *Handbook of Citrus By-products and Processing Technology;* John Wiley & Sons Inc.: New York, 1999.

Bravo, L. Polyphenols: Chemistry, Dietary Sources, Metabolism, and Nutritional Significance. *Nutr. Rev.* **1998**, *56,* 317–333.

Bravo, L.; Saura-Calixto, F. Characterization of Dietary fiber and the *in vitro* Indigestible Fraction of Grape Pomace. *Am. J. Enol Vitic.* **1998**, *49,* 135–141.

Bunzel, M.; Ralph, J.; Lu, F.; Hatfield, R. D.; Steinhart, H. Lignins and Ferulate-Coniferyl Alcohol Cross-Coupling Products in Cereal Grains. *J. Agric. Food Chem.* **2004**, *52,* 6496–6502.

Bunzel, M.; Ralph, J.; Marita, J. M.; Hatfeld, R. D.; Steinhart, H. Diferulates as Structural Components in Soluble and Insoluble Cereal Dietary Fibre. *J. Sci. Food Agric.* **2001**, *81,* 653–660.

Cahill, J. P. Ethnobotany of Chia, *Salvia hispanica* L. (Lamiaceae). *Econ. Bot.* **2003**, *57,* 604–618.

Cartea, M. E.; Velasco, P. Glucosinolates in Brassica Foods: Bioavailability in Food and Significance for Human Health. *Phytochem. Rev.* **2008**, *7,* 213–229.

Ćetković, G.; Ćanadanović-Brunet, J.; Djilas, S.; Savatović, S.; Mandić, A.; Tumbas, V. Assessment of Polyphenolic Content and *in vitro* Antiradical Characteristics of Apple Pomace. *Food Chem.* **2008**, *109* (2), 340–347.

Chantaro, P.; Devahastin, S.; Chiewchan N. Production of Antioxidant High Dietary Fiber Powder from Carrot Peels. *LWT-Food Sci.Technol.* **2008**, *41,* 1987–1994.

Chen, C. C.; Hsu, J. D.; Wang, S. F.; Chiang, H. C.; Yang, M. Y.; Kao, E. S.; Ho, Y. C.; Wang, C. J. *Hibiscus sabdariffa* Extract Inhibits the Development of Atherosclerosis in Cholesterol-Fed Rabbits. *J. Agric. Food Chem.* **2003**, *51* (18), 5472–5477.

Chinnici, F.; Bendini, A.; Gaiani, A.; Riponi, C. Radical Scavenging Activities of Peels and Pulps from Cv. Golden Delicious Apples as Related to Their Phenolic Composition. *J. Agric. Food Chem.* **2004**, *52,* 4684–4689.

Cho, S. S.; Dreher, M. L. Handbook of Dietary Fiber. Mercel Dekker: New York, 2001.

Chow, L. W.; Wang, S. J.; Duh, P. D. Antibacterial Activity of Burdock. *Food Sci.* **1997**, *24* (2), 195–202.

Ciska, E.; Pathak, D. R. Glucosinolate Derivatives in Stored Fermented Cabbage. *J. Agric. Food Chem.* **2004**, *52,* 7938–7943.

Cohn, R.; Cohn, A. L. Subproductos del Procesado de Las Frutas. In *Procesado de Frutas;* Arthey, D., Ashurst, P. R., Eds.; Acribia: Zaragoza, Spain, 1997; p 288.

Corral-Aguayo, R. D.; Yahia, E. M.; Carrillo-Lopez, A.; González-Aguilar, G. Correlation between Some Nutritional Components and the Total Antioxidant Capacity Measured with Six Different Assays in Eight Horticultural Crops. *J. Agric. Food Chem.* **2008**, *56,* 10498–10504.

Corrêa, L. C.; Santos, C. A. F.; Vianello, F.; Lima, G. P. P. Antioxidant Content in Guava (*Psidium guajava*) and Araca (*Psidium spp.*) Germplasm from Different Brazilian Regions. *Plant Genet. Resour.* **2011**, *9* (03), 384–391.

Cushine, T.; Lamb, A. J. Antimicrobial Activity of Flavonoids. *Int. J. Antimicrob. Agents.* **2005**, *26,* 343–356.

Das, A. K.; Rajkumar, V.; Verma, A. K. Bael Pulp Residue as a New Source of Antioxidant Dietary Fiber in Goat Meat Nuggets. *J. Food Process. Preserv.* **2015**, *39,* 1626–1635.

De Souza, M. C.; Figueiredo, R. W.; Maia, G. A.; Alves, R. E.; Brito, E. S.; Moura, C. F. H.; Rufino, M. S. M. Bioactive Compounds and Antioxidant Activity on Fruits from Different Açaí (*Euterpe oleracea Mart*) Progenies. *Acta Hortic.* **2009,** *841,* 455–458.

De Vries, J. W. On Defining Dietary Fibre. *Proc. Nutr. Soc.* **2003,** *62,* 37–43.

Deng, Q.; Penner, M. H.; Zhao, Y. Chemical Composition of Dietary Fiber and Polyphenols of Five Different Varieties of Wine Grape Pomace Skins. *Food Res. Int.* **2011,** *44* (9), 2712–2720.

Díaz-Rubio, M. E.; Pérez-Jiménez, J.; Saura-Calixto, F. Dietary Fiber and Antioxidant Capacity in *Fucus vesiculosus* Products. *Int. J. Food Sci. Nutr.* **2009,** *60* (Sup2), 23–34.

Duh, P. D. Antioxidant Activity of Burdock (*Arctium lappa Linne*): Its Scavenging Effect on Free-Radical and Active Oxygen. *J. Am. Oil Chem. Soc.* **1998,** *75* (4), 455–461.

Eastwood, M. A. Dietary Fibre and Risk of Cancer. *Nutr. Rev.* **1987,** *7,* 193–202.

Escarpa, A.; González, M. C. High Performance Liquid Chromatography with Diode-Array Detection for the Determination of Phenolic Compounds in Peel and Pulp from Different Apple Varieties. *J. Chromat.* **1998,** *823,* 331–337.

Esposito, E.; Rotilio, D.; Di Matteo, V.; Di Giulio, C.; Cacchio, M.; Algeri, S. A Review of Specific Dietary Antioxidants and the Effects on Biochemical Mechanisms Related to Neurodegenerative Processes. *Neurobiol. Aging.* **2002,** *23* (5), 719–735.

Esposito, F.; Arlotti G.; Bonifati, A. M.; Napolitano A.; Vitale D.; Fogliano, V. Antioxidant Activity and Dietary Fibre in Durum Wheat Bran By-products. *Food Res. Int.* **2005,** *38,* 1167–1173.

Femenia, A.; Robertson, J. A.; Waldron, K. W. Cauliflower (*Brassica oleracea* L), Globe Artichoke (*Cynara scolymus*) and Chicory Witloof (*Cichorium intybus*) Processing By-products as Sources of Dietary Fibre. *J. Sci. Food. Agric.* **1998,** *77,* 511–518.

Fernando, F.; Maria, L. H.; Ana Maria, E.; Chiffelle, I.; Fernando, A. Fibre Concentrates from Apple Pomace and Citrus Peel as Potential Source for Food Enrichment. *Food Chem.* **2005,** *91,* 395–401.

Frankel, E. N. Food Antioxidants and Phytochemicals. Present and Future Perspectives. *Fett-Lipid.* **1999,** *101* (12), 450–455.

Fuentes-Alventosa, J. M.; Jaramillo, S.; Rodríguez-Gutiérrez, G.; Cermeño, P.; Espejo, J. A.; Jiménez-Araujo, A.; Rodríguez-Arcos, R. Flavonoid Profile of Green Asparagus Genotypes. *J. Agric. Food Chem.* **2008,** *56,* 6977–6984.

Fuentes-Alventosa, J. M.; Rodríguez, G.; Cermeño, P.; Jiménez, A.; Guillén, R.; Fernandez-Bolaños, J.; Rodríguez-Arcos, R. Identification of Flavonoid Diglycosides in Several Genotypes of Asparagus from the Huétor-Tájar Population Variety. *J. Agric. Food Chem.* **2007,** *55,* 10028–10035.

Fuentes-Alventosa, J. M.; Jaramillo-Carmona S.; Rodríguez-Gutiérrez G.; Rodríguez-Arcos, R.; Fernández-Bolaños, J.; Guillén-Bejarano, R.; Espejo-Calvo, J. A.; Jiménez-Araujo, A. Effect of the Extraction Method on Phytochemical Composition and Antioxidant Activity of High Dietary Fibre Powders Obtained from Asparagus By-products. *Food Chem.* **2009a,** *116,* 484–490.

Fuentes-Alventosa, J. M.; Rodríguez-Gutiérrez, G.; Jaramillo-Carmona, S.; Espejo-Calvo, J. A.; Rodríguez-Arcos, R.; Fernández-Bolaños, J.; Guillén-Bejarano, R.; Jiménez-Araujo, A. Effect of Extraction Method on Chemical Composition and Functional Characteristics of High Dietary Fibre Powders Obtained from Asparagus By-products. *Food Chem.* **2009b,** *113,* 665–671.

Funk, C.; Braune, A.; Grabber, J. H.; Steinhart, H.; Bunzel, M. Moderate Ferulate and Diferulate Levels do not Impede Maize Cell Wall Degradation by Human Intestinal Microbiota. *J. Agric. Food Chem.* **2007,** *55,* 2418–2423.

Garau, M. C.; Simal, S.; Rossello, C.; Femenia, A. Effect of Air-Drying Temperature on Physico-Chemical Properties of Dietary Fibre and Antioxidant Capacity of Orange (*Citrus aurantium* v. Canoneta) By-products. *Food Chem.* **2007,** *104* (3), 1014–1024.

García, Y. D.; Valles, B. S.; Lobo, A. P. Phenolic and Antioxidant Composition of By-products from the Cider Industry: Apple Pomace. *Food Chem.* **2009,** *117* (4), 731–738.

Gheisari, H. R.; Amiri, F.; Zolghadri, Y. Antioxidant and Antimicrobial Activity of Iranian Bael (*Aegle Marmelos*) Fruit against Some Food Pathogens. *Int. J. Curr. Pharm. Res.* **2011,** *3,* 85–88.

Ghiselli, A.; Nardini, M.; Baldi, A.; Scaccini, C. Antioxidant Activity of Different Phenolic Fractions Separated from Italian Red Wine. *J. Agric. Food Chem.* **1998,** *46,* 361–367.

Gorinstein, S.; Martín-Belloso, O.; Park, Y.; Haruenkit, R.; Lojek, A.; Ciz, M.; Caspi, A.; Libman, I.; Trakhtenberg, S. Comparison of Some Biochemical Characteristics of Different Citrus Fruits. *Food Chem.* **2001,** *74,* 309–315.

Gorinstein, S.; Pawelzik, E.; Delgado-Licon, E.; Haruenkit, R.; Weisz, M.; Trakhtenberg, S. Characterisation of Pseudocereal and Cereal Proteins by Protein and Amino Acid Analyses. *J. Sci. Food Agric.* **2002,** *82,* 886–891.

Gorinstein, S.; Vargas, O. J. M.; Jaramillo, N. O.; Salas, I. A.; Ayala, A. L. M.; Arancibia-Avila, P.; Toledo, F.; Katrich, E.; Trakhtenberg, S. The Total Polyphenols and the Antioxidant Potentials of Some Selected Cereals and Pseudocereals. *Eur. Food Res. Technol.* **2007,** *225,* 321–328.

Grigelmo-Miguel, N.; Martín-Belloso, O. Characterization of Dietary Fiber from Orange Juice Extraction. *Food Res. Int.* **1998,** *31* (5), 355–361.

Halpern, M. J.; Dahlgren, A. L.; Laakso, I.; Seppanen-Laakso, T.; Dahlgren, J.; McAnulty, P. A. Red-Wine Polyphenols and Inhibition of Platelet Aggregation: Possible Mechanisms, and Potential Use in Health Promotion and Disease Prevention. *J. Int. Med. Res.* **1998,** *26* (4), 171–180.

Hatfield, R. D.; Ralph, J.; Grabber, J. H. Cell Wall Crosslinking by Ferulates and Diferulates in Grasses. *J. Sci. Food Agric.* **1999,** *79,* 403–407.

Hegwood, D. A. Human Health Discoveries with *Opuntia sp.* (Prickly Pear). *Hortic. Sci.* **1994,** *25,* 1515–1516.

Herbreteau, F.; Coiffard, L.; Derrien, A.; Roeck-Holtzhauer, D. The Fatty Acid Composition of Five Species of Macroalgae. *Bot. Marina.* **1997,** *40* (1–6), 25–28.

Herrera-Arellano, A.; Flores-Romero, S.; Chavez-Soto, M.; Tortoriello, J. Effectiveness and Tolerability of a Standardized Extract from *Hibiscus sabdariffa* in Patients with Mild to Moderate Hypertension: A Controlled and Randomized Clinical Trial. *Phytomedicine.* **2004,** *11* (5), 375–382.

Higdon, J. V.; Delage, B.; Williams, D. E.; Dashwood, R. H. Cruciferous Vegetables and Human Cancer Risk: Epidemiological Evidence and Mechanistic Basis. *Pharmacol. Res.* **2007,** *55,* 224–236.

Ho, C. T.; Osawa, T.; Huang, M. T.; Rosen, R. T. Eds.; *Food Phytochemicals for Cancer Prevention II – Teas, Spices, and Herbs;* ACS Symposium Series 547, American Chemical Society: Washington, DC, 1994.

Hussein, A. S.; Alhadrami, G. A.; Khalil, Y. H. The Use of Dates and Date Pits in Broiler Starter and Finisher Diets. *Bioresour. Technol.* **1998,** *66,* 219–223.

Iqbal, S.; Bhanger M. I.; Anwar F. Antioxidant Properties and Components of Bran Extracts from Selected Wheat Varieties Commercially Available in Pakistan. *LWT.* **2007,** *40,* 361–367.

Ishii, T. Structure and Functions of Feruloylated Polysaccharides. *Plant Sci.* **1997,** *127,* 111–127.

Jagetia, G. C.; Baliga, M. S. The Evaluation of Nitric Oxide Scavenging Activity of Certain Indian Medicinal Plants *in Vitro*: A Preliminary Study. *J. Medi. Food.* **2004,** *7,* 343–348.

Jain, N.; Goyal, S.; Ramawat, K. Evaluation of Antioxidant Properties and Total Phenolic Content of Medicinal Plants Used in Diet Therapy during Postpartum Healthcare in Rajasthan. *Int. J. Pharm. Pharma. Sci.* **2011,** *3,* 248–253.

Jiang, Y.; Simonsen, J.; Zhao, Y. Compression-Molded Biocomposite Boards from Red and White Wine Grape Pomaces. *J. Appl. Polym. Sci.* **2011,** *119,* 2834–2846.

Jiménez-Escrig, A.; Jiménez-Jiménez, I.; Pulido, R.; Saura-Calixto, F. Antioxidant Activity of Fresh and Processed Edible Seaweeds. *J. Sci. Food Agric.* **2001b,** *81* (5), 530–534.

Jiménez-Escrig, A.; Rincón, M.; Pulido, R.; Saura-Calixto, F. Guava Fruit (*Psidium guajava* L.) as a New Source of Antioxidant Dietary Fiber. *J. Agric. Food Chem.* **2001a,** *49* (11), 5489–5493.

Jiménez-Escrig, A.; Sánchez-Muniz, F. Dietary Fibre from Edible Seaweeds: Chemical Structure, Physicochemical Properties and Effects on Cholesterol Metabolism. *Nutr. Res.* **2000,** *20* (4), 585–598.

Jonadet, M.; Bastide, J.; Bastide, P. Activités Inhibitrices Enzymatiques *in vitro* Déxtraits de Karkadé (*Hibiscus sabdariffa* L.). *J. Pharm. Belg.* **1990,** *45,* 120–124.

Jongaroontaprangsee, S.; Tritrong, W.; Chokanaporn, W. Effects of Drying Temperature and Particle Size on Hydration Properties of Dietary Fiber Powder from Lime and Cabbage By-products. *Int. J. Food Prop.* **2007,** *10,* 887–897.

Jormalainen, V.; Honkanen T. Variation in Natural Selection for Growth and Phlorotannins in the Brown Alga *Fucus vesiculosus*. *J. Evol. Biol.* **2004,** *17,* 807–820.

Joseph, G.; Mahadeviah, M. Utilization of Waste from Pineapple Processing Industry. *Ind. Food Pack.* **1988,** *42* (1), 46–58.

Kähkönen, M. P.; Heinonen, M. Antioxidant Activity of Anthocyanins and Their Aglycons. *J. Agric. Food Chem.* **2003,** *51,* 628–633.

Kähkönen, M. P.; Hopia, A. I.; Vuorela, H. J.; Rauha, J. P.; Pihlaja, K.; Kujala, T. S.; Heinonen, M. Antioxidant Activity of Plant Extracts Containing Phenolic Compounds. *J. Agric. Food Chem.* **1999,** *47,* 3954–3962.

Kamalakkannan, N.; Prince, P. Effect of *Aegle marmelos Correa* (Bael) Fruit Extract on Tissue Antioxidants in Streptozotocin Diabetic Rats. *Ind. J. Experi. Biol.* **2003b,** *41,* 1285–1288.

Kamalakkannan, N.; Prince, P. Hypoglycaemic Effect of Water Extracts of *Aegle marmelos* Fruits in Streptozotocin Diabetic Rats. *J. Ethnopharm.* **2003a,** *87,* 207–210.

Kammerer, D.; Claus, A.; Carle, R.; Schieber, A. Polyphenol Screening of Pomace from Red and White Grape Varieties (*Vitis vinifera* L.) by HPLC–DAD–MS/MS. *J. Agric. Food Chem.* **2004,** *52,* 4360–4367.

Kang, H. J.; Chawla, S. P.; Jo, C.; Kwon, J. H.; Byun, M. W. Studies on the Development of Functional Powder from Citrus Peel. *Bioresour. Technol.* **2006,** *97,* 614–620.

Karim, M.; McCormick, K.; Kappagoda, C. T. Effects of Cocoa Extracts on Endothelium-Dependent Relaxation. *J. Nutr.* **2000,** *130,* 2108S–2119S.

Katsube, N.; Iwashita, K.; Tsushida, T.; Yamaki, K.; Kobori, M. Induction of Apoptosis in Cancer Cells by Bilberry (*Vaccinium myrtillus*) and the Anthocyanins. *J. Agric. Food Chem.* **2003,** *51,* 68–75.

Kim, D. O.; Lee, K. W.; Lee, H. J.; Lee, C. Y. Vitamin C Equivalent Antioxidant Capacity (VCEAC) of Phenolic Phytochemicals. *J. Agric. Food Chem.* **2002**, *50,* 3713–3717.

Kim, D. O.; Padilla-Zakour, O. I.; Griffiths, P. D. Flavonoids and Antioxidant Capacity of Various Cabbage Genotypes at Juvenile Stage. *J. Food Sci.* **2004**, *69,* 685–689.

Kim, K. H.; Tsao, R.; Yang, R.; Cui, S. W. Phenolic Acid Profiles and Antioxidant Activities of Wheat Bran Extracts and the Effect of Hydrolysis Conditions. *Food Chem.* **2006**, *95,* 466–473.

Kong, J. M.; Chia, L. S.; Goh, N. K.; Chia, T. F.; Brouillard, R. Analysis and Biological Activities of Anthocyanins. *Phytochemistry.* **2003**, *64* (5), 923–933.

Krings, U.; El-saharty, Y. S.; El-Zeany, B. A.; Pabel, B.; Berger, R. G. Antioxidant Activity of Extracts from Roasted Wheat Germ. *Food Chem.* **2000**, *71,* 91–95.

Kroon, P. A. What Role for Feruloyl Esterases Today? *Polyphén. Actual.* **2000**, *19,* 4–5.

Kroon, P. A.; Faulds, C. B.; Ryden, P.; Robertson, J. A.; Williamson, G. Release of Covalently Bound Ferulic Acid from Fibre in Human Colon. *J. Agric. Food Chem.* **1997**, *45,* 661–667.

Kuljarachanan, T.; Devahastin, S.; Chiewchan, N. Evolution of Antioxidant Compounds in Lime Residues during Drying. *Food Chem.* **2009**, *113* (4), 944–949.

Kulkarni, M.; Mootey, R.; Lele, S. S. In *Biotechnology in Agriculture, Industry and Environment,* Proceedings of the International Conference of SAARC Countries, Karad, India, Dec 28–30, 2001; Microbiologist Society: Karad, India, 2001.

Lam, L. K. T.; Li, Y.; Hasegawa, S. Effects of Citrus Limonoids on Glutathione S-Transferase Activity in Mice. *J. Agric. Food Chem.* **1989**, *37,* 878–880.

Lampe, J. W. Health Effects of Vegetables and Fruit: Assessing Mechanisms of Action in Human Experimental Studies. *Am. J. Clin. Nutr.* **1999**, *70,* 475S–490S.

Lario, Y.; Sendra, E.; García-Pérez, J.; Fuentes, C.; Sayas-Barberá, E.; Fernández-López, J.; Pérez-Alvarez, J. A. Preparation of High Dietary Fiber Powder from Lemon Juice By-products. *Innov. Food Sci. Emerg. Technol.* **2004**, *5,* 113–117.

Larrauri, J. A.; Rodruiguez, J. L.; FernPndez, M.; Borroto, B. NOTA. Fibra Dietética Obtenida a Partir de Hollejos Ciltricos y Cáscaras de Piñas. *Rev. Esp. Cienc. Tecnol. Aliment.* **1994**, *34,* 102–107.

Larrauri, J. A.; Ruperez, P.; Borroto, B.; Saura-calixto, F. Mango Peel as a New Tropical Fiber: Preparation and Characterization. *Lebensm. Wiss. Technol.* **1996a**, *29,* 729–733.

Larrauri, J. A.; Rupéréz, P.; Bravo, L.; Saura-Calixto F. High Dietary Fiber from Orange and Lime Peels: Associated Polyphenols and Antioxidant Capacity. *Food Res. Int.* **1996b**, *29* (8), 757–762.

Larrauri, J. A.; Ruperez, P.; Saura-calixto, F. New Approaches in the Preparation of High Dietary Fibre from Fruit By-products. *Trends Food Sci.Technol.* **1999**, *29,* 729–733.

Larrauri, J. A.; Rupérez, P.; Saura-Calixto F. Pineapple Shell as a Source of Dietary Fiber with Associated Polyphenols. *J. Agric. Food Chem.* **1997**, *45,* 4028–4031.

Laufenberg, G.; Kunz, B.; Nystroem, M. Transformation of Vegetable Waste into Value Added Products: (A) the Upgrading Concept, (B) Practical Implementations. *Bioresour. Technol.* **2003**, *87* (2), 167–198.

Le Tutour, B.; Benslimane, F.; Gouleau, M.; Gouygou, J.; Saadan, B.; Quemeneur, F. Antioxidant and Pro-Oxidant Activities of the Brown Algae, *Laminaria digitata, Himanthalia elongata, Fucus vesiculosus, Fucus serratus* and *Ascophyllum nodosum. J. Appl. Phycol.* **1998**, *10* (2), 121–129.

Lecumberri, E.; Mateos, R.; Izquierdo-Pulido, M.; Rupérez, P.; Goya, L.; Bravo L. Dietary Fibre Composition, Antioxidant Capacity and Physico-Chemical Properties of a Fibre-Rich Product from Cocoa (*Theobroma cacao* L.). *Food Chem.* **2007**, *104,* 948–954.

Lee, C. Y.; Smith, N. L. Apples: An Important Source of Antioxidants in the American Diet. *New York Fruit Quart.* **2000**, *8,* 8–10.

Leontowicz, H.; Gorinstein, S.; Lojek, A.; Leontowicz, M.; Číž, M.; Soliva-Fortuny, R.; Park, Y. S.; Jung, S. T.; Trakhtenberg, S.; Martin-Belloso, O. Comparative Content of Some Bioactive Compounds in Apples, Peaches and Pears and Their Influence on Lipids and Antioxidant Capacity in Rats. *J. Nutr. Biochem.* **2002**, *13* (10), 603–610.

Li, W. D.; Shan, F.; Sun, S. C.; Corke, H.; Beta, H. Free Radical Scavenging Properties and Phenolic Content of Chinese Black-Grained Wheat. *J. Agric. Food Chem.* **2005**, *53,* 8533–8536.

Li, W. D.; Pickard, M. D.; Beta, T. Effect of Thermal Processing on Antioxidant Properties of Purple Wheat Bran. *Food Chem.* **2007**, *104,* 1080–1086.

Lin, C. C.; Lin, J. M.; Yang, J. J.; Chuang, S. C.; Ujiie, T. Anti-Inflammatory and Radical Scavenging Effect of *Arctium lappa*. *Am. J. Chin. Med.* **1996**, *24* (2), 127–137.

Liu, R. H. Health Benefits of Fruit and Vegetables are from Additive and Synergistic Combinations of Phytochemicals. *Am. J. Clin. Nutr.* **2003**, *78,* 517S–520S.

Llobera, A.; Cañellas, J. Dietary Fibre Content and Antioxidant Activity of Manto Negro Red Grape (*Vitis vinifera*): Pomace and Stem. *Food Chem.* **2007**, *101* (2), 659–666.

Loelillet, D. The European Mango Market: A Promising Tropical Fruit. *Fruit.* **1994**, *49,* 332–334.

Lotito, S.; Actis-Goretta, L.; Renart, M. L.; Caligiuri, M.; Rein, D.; Schmitz, H. H.; Steinberg, F. M., Keen, C. L.; Fraga, C. G. Influence of the Oligomer Chain Length on the Antioxidant Activity of Procyanidins. *Biochem. Biophys. Res. Commun.* **2000**, *276,* 945–951.

Lou, Z.; Wang, H.; Wang, D.; Zhang, Y. Preparation of Inulin and Phenols-Rich Dietary Fibre Powder from Burdock Root. *Carbohydr. Polym.* **2009**, *78* (4), 666–671.

Lu, Y.; Foo, L. Y. Antioxidant and Radical Scavenging Activities of Polyphenols from Apple Pomace. *Food Chem.* **2000**, *68,* 81–85.

Lurton, L. Grape Polyphenols: A New Powerful Health Ingredients. *Innov. Food Technol.* **2003**, *18,* 28–30.

Luximon-Ramma, A.; Bahorun, T.; Crozier, A. Antioxidant Actions and Phenolic and Vitamin C Contents of Common Mauritian Exotic Fruits. *J. Sci. Food Agric.* **2003**, *83* (5), 496–502.

Manthey, J. A.; Grohmann, K. Concentration of Hesperidin and Other Orange Peel Flavonoids in Citrus Processing Byproducts. *J. Agric. Food Chem.* **1996**, *44,* 811–814.

Marín, F. R.; Cristina, S. R.; Benavente-García, O.; Castillo, J.; Pérez-Alvarez, J. A. By-products from Different Citrus Processes as a Source of Customized Functional Fibres. *Food Chem.* **2007**, *100,* 736–741.

Martín-Cabrejas, M. A.; Valiente, C.; Esteban, R. M.; Mollá, E.; Waldron, K. Cocoa Hull: A Potential Source of Dietary Fibre. *J. Sci. Food Agric.* **1994**, *66,* 307–311.

Maté, J. I.; Quartaert, C.; Meerdink, G.; Riet, K. V. Effect of Blanching on Structural Quality of Dried Potato Slices. *J. Agric. Food Chem.* **1998**, *46,* 676–681.

Mazza, G. Anthocyanins in Grapes and Grape Products. *Crit. Rev. Food Sci. Nutr.* **1995**, *35* (4), 341–371.

Mazza, G.; Miniati, E. *Anthocyanins in Fruits Vegetables and Grains;* CRC Press: Boca Raton, Ann Harbor, London, 1993.

Mercadante, A. Z.; Steck, A.; Pfander, H. Carotenoids from Guava (*Psidium guajava* L): Isolation and Structure Elucidation. *J. Agric. Food Chem.* **1999**, *47,* 145–151.

Meyer, A. S.; Jepsen, S. M.; Sorensen, N. S. Enzymatic Release of Antioxidants for Human Low-Density Lipoprotein from Grape Pomace. *J. Agric. Food Chem.* **1998**, *46* (7), 2439–2446.

Middleton, E.; Kandaswami, C. Potential Health Promoting Properties of Citrus Flavonoids. *Food Technol.* **1994,** *8* (11), 115–119.

Miean, K. H.; Mohamed, S. Flavonoid (*Myricetin, Quercetin, Kaempferol, Luteolin,* and *Apigenin*) Content of Edible Tropical Plants. *J. Agric. Food Chem.* **2001,** *49,* 3106–3112.

Mielnick, M. B.; Olsen, E.; Vogt, G.; Adeline, D.; Skrede, G. Grape Seed Extract as Antioxidant in Cooked, Cold Stored Turkey Meat. *LWT-Food Sci. and Technol.* **2006,** *39,* 191–198.

Miller, E. G.; Fanous, R.; Rivera-Hidalgo, F.; Binnie, W. H.; Hasegawa, S. The Effect of Citrus Limonoids on Hamster Buccal Pouch Carcinogenesis. *Carcinogenesis.* **1989,** *10,* 1535–1537.

Mojiminiyi, F. B. O.; Adegunloye, B. J.; Egbeniyi, Y. A.; Okolo, R. U. An Investigation of the Diuretic Effect of an Aqueous Extract of the Petal of *Hibiscus sabdariffa. J. Med. Sci.* **2000,** *2,* 77–80.

Morel, J. M.; Perrey, F.; Lejeune, R.; Goetz P. *Fucus vesiculosus* L. *Phytothérapie.* **2005,** *5,* 218–221.

Morton, J. F. *Fruits of Warm Climates;* Creative Resource Systems: Winterville, NC, 1987.

Mrkic, V.; Redovnikovic, I. R.; Jolic, S. M.; Delonga, K.; Dragovic-Uzelac, V. Effect of Drying Conditions on Indole Glucosinolates Level in Broccoli. *Acta Aliment.* **2010,** *39,* 167–174.

Muñoz de Chávez, M.; Ledesma-Solano, J. A. Los *Alimentos y Sus Nutrientes. Tablas de Valor Nutritive;* McGraw-Hill Interamericana: Mexico, 2002.

Nagy, S.; Attaway, J. A. Anticarcinogenic Activity of Phytochemicals in Citrus Fruit and Their Juice Products. *Proc. Florida State Hortic. Soc.* **1992,** *105,* 162–168.

Nahar, N.; Rahman, S.; Mosihuzzaman, M. Analysis of Carbohydrates in Seven Edible Fruits of Bangladesh. *J. Sci. Food Agric.* **1990,** *51* (2), 185–192.

Nawirska, A.; Kwasniewska, M. Dietary Fibre Fractions from Fruit and Vegetable Processing Waste. *Food Chem.* **2005,** *91,* 221–225.

Nijveldt, R. J.; van Nood, E.; van Hoorn, D. E.; Boelens, P.; van Norren, K.; van Leeuwen, P. Flavonoids: A Review of Probable Mechanisms of Action and Potential Applications. *Am. J. Clin. Nutr.* **2001,** *74,* 418–425.

Nilnakara, S.; Chiewchan, N.; Devahastin, S. Production of Antioxidant Dietary Fiber Powder from Cabbage Outer Leaves. *Food Bioprod. Process.* **2009,** *87,* 301–307.

Nilsson, J.; Engkvist, R.; Persson, L. E. Long-Term Decline and Recent Recovery of Fucus Populations along the Rocky Shores of Southeast Sweden, Baltic Sea. *Aquat. Ecol.* **2005,** *38* (4), 587–598.

Nindo, C. I.; Sun, T.; Wang, S. W.; Tang, J.; Powers, J. R. Evaluation of Drying Technologies for Retention of Physical Quality and Antioxidants in Asparagus. *Eur. Food Res. Technol.* **2003,** *36,* 507–516.

Nogata, Y.; Yoza, K. I.; Kusumoto, K. I.; Kohyama, N.; Sekiya, K.; Ohta, H. Screening for Inhibitory Activity of Citrus Fruit Extracts against Platelet Cyclooxygenase and Lipoxygenase. *J. Agric. Food Chem.* **1996,** *44,* 725–729.

Odigie, I.; Ettarh, R.; Adigun, S. Chronic Administration of Aqueous Extract of *Hibiscus sabdariffa* Attenuates Hypertension and Reverses Cardiac Hypertrophy in 2K-1C Hypertensive Rats. *J. Ethnopharmacol.* **2003,** *86* (2), 181–185.

Onyeneho, S. N.; Hettiarachy, N. S. Antioxidant Activity of Durum Wheat Bran. *J. Agric. Food Chem.* **1992,** *40,* 1496–1500.

Orozco, T. J.; Wang, J. F.; Keen, C. L. Chronic Consumption of a Flavonols- and Procyanidin-Rich Diet is Associated with Reduced Levels of 8-Hydroxy-2'-Deoxyguanosine in Rat Testes. *J. Nutr. Biochem.* **2003,** *14,* 104–110.

Oszmianski, J.; Wolniak, M.; Wojdylo, A.; Wawer, I. Influence of Apple Purée Preparation and Storage on Polyphenol Contents and Antioxidant Activity. *Food Chem.* **2008,** *107,* 1473–1484.

Parichha, S. Bael (*Aegle marmelos*) Nature's Most Natural Medicinal Fruit. *Orissa Rev.* **2004,** *9,* 16–17.

Paśko, P.; Bartoń, H.; Folta, M.; Gwizdz, J. Evaluation of Antioxidant Activity of Amaranth (*Amaranthus cruentus*) Grain and By-products (Flour, Popping, Cereal). *Rocz. Państwowego Zakł. Hig.* **2007,** *58,* 35–40.

Paśko, P.; Bartoń, H.; Zagrodzki, P.; Gorinstein, S.; Folta, M.; Zachwieja, Z. Anthocyanins, Total Polyphenols and Antioxidant Activity in Amaranth and Quinoa Seeds and Sprouts during Their Growth. *Food Chem.* **2009,** *115* (3), 994–998.

Paśko, P.; Sajewicz, M.; Gorinstein, S.; Zachwieja, Z. Analysis of the Selected Phenolic Acids and flavonoids in *Amaranthus cruentus* and *Chenopodium quinoa* Seeds and Sprouts by HPLC Method. *Acta Chromatogr.* **2008,** *20* (4), 661–672.

Peleg, H.; Naim, M.; Rouseff, R. L.; Zehavi, U. Distribution of Bound and Free Phenolic Acids in Oranges (*Citrus sinensis*) and Grapefruits (*Citrus paradise*). *J. Sci. Food Agric.* **1991,** *57,* 417–426.

Pellegrini, N.; Serafini, M.; Colombi, B.; Del Rio, D.; Salvatore, S.; Bianchi, M.; Brighenti, F. Total Antioxidant Capacity of Plant Foods, Beverages and Oils Consumed in Italy Assessed by Three Different *in vitro* Assays. *J. Nutr.* **2003,** *133,* 2812–2819.

Pérez-Jiménez, J.; Serrano, J.; Tabernero, M.; Arranz, S.; Díaz-Rubio, M. E.; García-Diz, L.; Goñi, I.; Saura-Calixto, F. Effects of Grape Antioxidant Dietary Fiber in Cardiovascular Disease Risk Factors. *Nutrition.* **2008,** *24,* 646–653.

Podsedek, A. Natural Antioxidant Capacity of Brassica Vegetables: A Review. *LWT- Food Sci. Technol.* **2007,** *40,* 1–11.

Pozo-Insfran, D. D.; Percival, S. S.; Talcott, S. T. Açaí (*Euterpe oleracea* Mart.) Polyphenolics in Their Glycoside and Aglycone forms Induce Apoptosis of HL-60 Leukemia Cells. *J. Agric. Food Chem.* **2006,** *54,* 1222–1229.

Pozuelo, M. J.; Agis-Torres, A.; Hervert-Hernández, D.; Elvira López-Oliva, M.; Muñoz-Martínez, E.; Rotger, R.; Goñi, I. Grape Antioxidant Dietary Fiber Stimulates Lactobacillus Growth in Rat Cecum. *J. Food Sci.* **2012,** *77* (2), H59–H62.

Prakash, S.; Jha, S. K.; Datta, N. Performance Evaluation of Blanched Carrots Dried by Three Different Driers. *J. Food Eng.* **2004,** *62,* 305–313.

Prosky, L.; Asp, N. G.; Schweizer, T. F.; De Vries, J. W.; Furda, I. Determination of Insoluble, Soluble and Total Dietary Fibre in Foods and Food Products: Interlaboratory Study. *J. AOAC Int.* **1988,** *71,* 1017–1023.

Ramulu, P.; Rao, U. P. Total, Insoluble and Soluble Dietary Fiber Contents of Indian Fruits. *J. Food Compo. Anal.* **2003,** *16,* 677–685.

Rehman, Z. Citrus Peel Extract – A Natural Source of Antioxidant. *Food Chem.* **2006,** *99,* 450–454.

Rein, D.; Paglieroni, T. G.; Wun, T.; Pearson, D. A.; Schmitz, H. H.; Gosselin, R.; Keen, C. L. Cocoa Inhibits Platelet Activation and Function. *Am. J. Clin. Nutr.* **2000,** *72,* 30–35.

Renaud, S.; De Lorgeril, M. Wine, Alcohol, Platelets, and the French Paradox for Coronary Heart. *Lancet.* **1992,** *339* (8808), 1523–1526.

Repo-Carrasco-Valencia, R.; Peña, J.; Kallio, H.; Salminen, S. Dietary Fiber and Other Functional Components in Two Varieties of Crude and Extruded *kiwicha* (*Amaranthus caudatus*). *J. Cereal Sci.* **2009,** *49* (2), 219–224.

Reyes-Caudillo, E.; Tecante, A.; Valdivia-López, M. A. Dietary Fiber Content and Antioxidant Activity of Phenolic Compounds Present in Mexican Chia (*Salvia hispanica* L.) Seeds. *Food Chem.* **2008,** *107,* 656–663.

Ribeiro, J. C.; Antunes, L. M. G.; Aissa, A. F.; Darin, J. D. C.; Veridiana Rosso, V.; Mercadante, A. Z.; Bianchi, M. d. L. P. Evaluation of the Genotoxic and Antigenotoxic Effects after Acute and Subacute Treatments with Açaí Pulp (*Euterpe oleracea* Mart.) on Mice Using the Erythrocytes Micronucleus Test and the Comet Assay. *Mutat. Res.* **2010,** *695,* 22–28.

Rice-Evans, C. A.; Miller, N. J.; Paganga, G. Antioxidant Properties of Phenolic Compounds. *Trends Plant Sci.* **1997,** *2* (4), 152–159.

Rioux, L. E.; Turgeon, S. L.; Beaulieu, M. Characterization of Polysaccharides Extracted from Brown Seaweeds. *Carbohydr. Polym.* **2007,** *69* (3), 530–537.

Roy, S. K.; Khurdiya, D. S. Other Subtropical Fruit. In *Handbook of Fruit Science and Technology: Production, Composition, Storage and Processing;* Salunkhe, D. K., Kadam S. S., Eds.; CRC Press: New York, 1995; p 539.

Ruberto, G.; Renda, A.; Daquino, C.; Amico, V.; Spatafora, C.; Tringali, C.; De Tommasi, N. Polyphenol Constituents and Antioxidant Activity of Grape Pomace Extracts from Five Sicilian Red Grape Cultivars. *Food Chem.* **2007,** *100* (1), 203–210.

Rufino, M. d. S. M.; Alves, R. E.; Brito, E. S.; Pérez-Jiménez, J.; Saura-Calixto, F.; Mancini-Filho, J. Bioactive Compounds and Antioxidant Capacities of 18 Non-Traditional Tropical Fruits from Brazil. *Food Chem.* **2010,** *121,* 996–1002.

Rufino, M. d. S. M.; Alves, R. E.; Brito, E. S.; Pérez-Jiménez, J.; Saura-Calixto, F. D. Total Phenolic Content and Antioxidant Activity in Acerola, Açaí, Mangaba and Uvaia Fruits by DPPH Method. *Acta Hort.* **2009a,** *841,* 459–462.

Rufino, M. d. S. M.; Fernandes, F. A. N.; Alves, R. E.; Brito, E. S. Free Radical Scavenging Behaviour of Some North-East Brazilian Fruits in a DPPH System. *Food Chem.* **2009b,** *114,* 693–695.

Rufino, M. d. S. M.; Pérez-Jiménez, J.; Arranz, S.; Alves, R.; Brito, E.; Oliveira, M. S.; Calixto, S. Açaí (*Euterpe oleraceae*)'BRS Pará': A Tropical Fruit Source of Antioxidant Dietary Fiber and High Antioxidant Capacity Oil. *Food Res. Int.* **2011,** *44* (7), 2100–2106.

Rupasinghe, H. V.; Wang, L.; Huber, G. M.; Pitts, N. L. Effect of Baking on Dietary Fibre and Phenolics of Muffins Incorporated with Apple Skin Powder. *Food Chem.* **2008,** *107* (3), 1217–1224.

Rupérez, P.; Ahrazem, O.; Leal, J. A. Potential Antioxidant Capacity of Sulfated Polysaccharides from the Edible Marine Brown Seaweed *Fucus vesiculosus. J. Agric. Food Chem.* **2002,** *50* (4), 840–845.

Ruperez, P.; Saura-Calixto, F. Dietary Fibre and Physicochemical Properties of Edible Spanish Seaweeds. *Eur. Food Res. Technol.* **2001,** *212* (3), 349–354.

Rupérez, P.; Toledano, G. Indigestible Fraction of Edible Marine Seaweeds. *J. Sci. Food Agric.* **2003,** *83* (12), 1267–1272.

Salvi, M. J.; Rajput, C. Pineapple. In *Handbook of Fruit Science and Technology. Production, Composition, Storage, and Processing*; Salunkhe, D. K., Kadam S. S., Eds.; CRC Press: New York, 1995; p 171.

Sánchez-Alonso, I.; Borderías, A. J. Technological Effect of Red Grape Antioxidant Dietary Fibre Added to Minced Fish Muscle. *Int. J. Food Sci. Technol.* **2008,** *43* (6), 1009–1018.

Sánchez-Alonso, I.; Jiménez-Escrig, A.; Saura-Calixto, F.; Borderías, A. J. Effect of Grape Antioxidant Dietary Fibre on the Prevention of Lipid Oxidation in Minced Fish: Evaluation by Different Methodologies. *Food Chem.* **2007,** *101* (1), 372–378.

Sánchez-Alonso, I.; Jiménez-Escrig, A.; Saura-Calixto, F.; Borderías, A. J. Antioxidant Protection of White Grape Pomace on Restructured Fish Products during Frozen Storage. *LWT-Food Sci. Technol.* **2008**, *41* (1), 42–50.

Saura-Calixto, F. Antioxidant Dietary Fiber Product: A New Concept and a Potential Food Ingredient. *J. Agric. Food Chem.* **1998**, *46* (10), 4303–4306.

Saura-Calixto, F.; Díaz-Rubio, M. E. Polyphenols Associated with Dietary Fibre in Wine: A Wine Polyphenols Gap? *Food Res. Int.* **2007**, *40* (5), 613–619.

Sáyago-Ayerdi, S. G.; Arranz, S.; Serrano, J.; Goñi, I. Dietary Fiber Content and Associated Antioxidant Compounds in Roselle Flower (*Hibiscus sabdariffa* L.) Beverage. *J. Agric. Food Chem.* **2007**, *55* (19), 7886–7890.

Sáyago-Ayerdi, S.; Brenes, A.; Goñi, I. Effect of Grape Antioxidant Dietary Fiber on the Lipid Oxidation of Raw and Cooked Chicken Hamburgers. *LWT-Food Sci. Technol.* **2009b**, *42* (5), 971–976.

Sáyago-Ayerdi, S.; Brenes, A.; Viveros, A.; Goñi, I. Antioxidative Effect of Dietary Grape Pomace Concentrate on Lipid Oxidation of Chilled and Long-Term Frozen Stored Chicken Patties. *Meat Sci.* **2009a**, *83* (3), 528–533.

Schauss, A. G.; Wu, X.; Prior, R. L.; Ou, B.; Patel, D.; Huang, D.; Kababick, J. P. Phytochemical and Nutrient Composition of the Freeze-Dried Amazonian Palm Berry, *Euterpe oleraceae* Mart. (Açaí). *J. Agric. Food Chem.* **2006**, *54*, 8598–8603.

Schewe, T.; Kühn, H.; Sies, H. Flavonoids of Cocoa Inhibit Recombinant Human 5-Lipoxygenase. *J. Nutr.* **2002**, *132*, 1825–1829.

Schieber, A.; Keller, P.; Carle, R. Determination of Phenolic Acids and flavonoids of Apple and Pear by High-Performance Liquid Chromatography. *J. Chromatogr A.* **2001**, *910* (2), 265–273.

Schreckinger, M. E.; Lotton, J.; Lila, M. A.; Mejia, E. G. Berries from South America: A Comprehensive Review on Chemistry, Health Potential, and Commercialization. *J. Med. Food.* **2010**, *13*, 233–246.

Serpen, A.; Capuano, E.; Fogliano, V.; Gökmen, V. A New Procedure to Measure the Antioxidant Activity of Insoluble Food Components. *J. Agric. Food Chem.* **2007**, *55* (19), 7676–7681.

Serra-Bonvehí, J.; Aragay-Benería, M. Composition of Dietary Fibre in Cocoa Husk. *Z. Lebensm-Unters. Forsch. A.* **1998**, *207*, 105–109.

Singh, J.; Upadhyay, A. K.; Bahadur, A.; Singh, B.; Singh, K. P.; Rai, M. Antioxidant Phytochemicals in Cabbage (*Brassica oleracae* L. var. capitata). *Sci. Hortic. Amst.* **2006**, *108*, 233–237.

Spiller, G. A. Suggestions for a Basis on which to Determine Adesirable Intake of Dietary-fibre. In *CRC Handbook of Dietary Fibre in Human Nutrition*; Spiller G. A., Ed.; CRC Press: Boca Raton, FL, 1986.

Steinberg, F. M.; Bearden, M. M.; Keen, C. L. Cocoa and Chocolate Flavonoids: Implications for Cardiovascular Health. *J. Am. Diet. Assoc.* **2003**, *103*, 215–223.

Suárez, B.; Álvarez, Á. L.; García, Y. D.; del Barrio, G.; Lobo, A. P.; Parra, F. Phenolic Profiles, Antioxidant Activity and *in vitro* Antiviral Properties of Apple Pomace. *Food Chem.* **2010**, *120* (1), 339–342.

Sudha, M. L.; Baskaran, V.; Leelavathi, K. Apple Pomace as a Source of Dietary Fiber and Polyphenols and Its Effect of the Rheological Characteristics and Cake Making. *Food Chem.* **2007**, *104*, 686–692.

Suvimol, C.; Pranee A. Bioactive Compounds and Volatile Compounds of Thai Bael Fruit (*Aegle Marmelos* L. Correa) as a Valuable Source for Functional Food Ingredients. *Int. Food Res. J.* **2008**, *15*, 287–295.

Taga, M. S.; Miller, E. E.; Pratt, D. E. Chia Seeds as a Source of Natural Lipid Antioxidants. *J. Am. Oil Chem. Soc.* **1984**, *61*, 928–932.

Tanongkankit, Y.; Chiewchan N.; Devahastin, S. Physicochemical Property Changes Of Cabbage Outer Leaves Upon Preparation into Functional Dietary Fiber Powder. *Food Bioprod. Process.* **2012**, *90*, 541–548.

Thaipong, K.; Boonprakob, U.; Crosby, K.; Cisneros-Zevallos, L.; Byrne, D. H. Comparison of ABTS, DPPH, FRAP, and ORAC assays for Estimating Antioxidant Activity from Guava Fruit Extracts. *J. Food Compos. Anal.* 2006 *19* (6), 669–675.

Toth, G. B.; Pavia, H. Removal of Dissolved Brown Algal Phlorotannins Using Insoluble Polyvinylpolypyrrolidone (PVPP). *J. Chem. Ecol.* **2001**, *27* (9), 1899–1910.

Tsao, R.; Yang, R.; Xie, S.; Sockovie, E.; Khanizadeh, S. Which Polyphenolic Compounds Contribute to the Total Antioxidant Activities of Apple? *J. Agric. Food Chem.* **2005**, *53*, 4989–4995.

Ubando-Rivera, J.; Navarro-Ocaña, A.; Valdivia-López, M. A. Mexican Lime Peel: Comparative Study on Contents of Dietary Fibre and Associated Antioxidant Activity. *Food Chem.* **2005**, *89*, 57–61.

Valiente, C.; Arrigoni, E.; Esteban, R. M.; Amado, R. Grape Pomace as a Potential Food Fiber. *J. Food Sci.* **1995**, *60*, 818–820.

Verkerk, R.; Dekker, M.; Jongen, W. M. F. Post-Harvest Increase of Indolyl Glucosinolates in Response to Chopping and Storage of Brassica Vegetables. *J. Sci. Food Agric.* **2001**, *81*, 953–958.

Verma, A. K.; Rajkumar, V.; Banerjee, R.; Biswas, S.; Das, A. K. Guava (*Psidium guajava* L.) Powder as an Antioxidant Dietary Fibre in Sheep Meat Nuggets. *Asian-Austr. J. Anim. Sci.* **2013**, *26* (6), 886–895.

Vinson, J. A.; Hao, Y.; Su, X.; Zubik, L. Phenol Antioxidant Quantity and Quality in Foods: Vegetables. *J. Agric. Food Chem.* **1998**, *46*, 3630–3634.

Vitaglione, P.; Napolitano, A.; Fogliano, V. Cereal Dietary Fibre: A Natural Functional Ingredient to Deliver Phenolic Compounds into the Gut. *Trends Food Sci. Technol.* **2008**, *19* (9), 451–463.

Vrhovsek, U.; Rigo, A.; Tonon, D.; Mattivi, F. Quantitation of Polyphenols in Different Apple Varieties. *J. Agric. Food Chem.* **2004**, *52*, 6532–6538.

Wachtel-Galor, S.; Wong, K. W.; Benzie, I. F. F. The Effect of Cooking on Brassica Vegetables. *Food Chem.* **2008**, *110*, 706–710.

Wadhwa, M.; Kaushal, S.; Bakshi, M. P. S. Nutritive Evaluation of Vegetable Wastes as Complete Feed for Goat Bucks. *Small Rumin. Res.* **2006**, *64*, 279–284.

Wan, Y.; Vinson, J. A.; Etherton, T. D.; Proch, J.; Lazarus, S. A.; Kris-Etherton, P. M. Effects of Cocoa Powder and Dark Chocolate on LDL Oxidative Susceptibility and Prostaglandin Concentrations in Humans. *Am. J. Clin. Nutr.* **2001**, *74*, 596–602.

Wang, H.; Cao, G.; Prior, R. L. Oxygen Radical Absorbing Capacity of Anthocyanins. *J. Agric. Food Chem.* **1997**, *45*, 304–309.

Wang, X.; Geng, X.; Egashira, Y.; Sanada, H. Purification and Characterization of a Feruloyl Esterase from the Intestinal Bacterium *Lactobacillus acidophilus*. *Appl. Environ. Microbiol.* **2004**, *70*, 2367–2372.

Wennberg, M.; Ekvall, J.; Olsson, K.; Nyman, M. Changes in Carbohydrate and Glucosinolate Composition in White Cabbage (*Brassica oleracea* var. capitata) during Blanching and Treatment with Acetic Acid. *Food Chem.* **2006,** *96,* 226–236.

Wennberg, M.; Engqvist, G.; Nyman, E. Effects of Boiling on Dietary Fibre Components in Fresh and Stored White Cabbage (*Brassica oleracea* var. capitata). *J. Food Sci.* **2004,** *68,* 1615–1621.

Wiswedel, I.; Hirsch, D.; Kropf, S.; Gruening, M.; Pfister, E.; Schewe, T.; Sies, H. Flavanol-Rich Cocoa Drink Lowers Plasma F-Isoprostane Concentrations in Humans. *Free Radic. Biol. Med.* **2004,** *37,* 411–421.

Wolfe, K. E.; Liu, R. H. Apple Peels as a Value-Added Food Ingredient. *J. Agric. Food Chem.* **2003,** *51,* 1676–1683.

Wolfe, K.; Wu, X.; Liu, R. H. Antioxidant Activity of Apple Peels. *J Agric. Food Chem.* **2003,** *51,* 609–614.

Yamagishi, M.; Natsume, M.; Magaki, A.; Adachi, T.; Osakabe, N.; Takizawa, T.; Kumon, H.; Osawa, T. Antimutagenic Activity of Cacao: Inhibitory Effect of Cacao Liquor Polyphenols on the Mutagenic Action of Heterocyclic Amines. *J. Agric. Food Chem.* **2000,** *48,* 5074–5078.

Yu, L.; Haley, S.; Perret, J.; Harris, M.; Wilson, J.; Qian, M. Free Radical Scavenging Properties of Wheat Extracts. *J. Agric. Food Chem.* **2002,** *50,* 1619–1624.

Yu, L.; Perret, J.; Harris, M.; Wilson, J.; Haley, S. Antioxidant Properties of Bran Extracts from *"Akron"* Wheat Grown at Different Locations. *J. Agric. Food Chem.* **2003,** *51,* 1566–1570.

Zhang, D.; Hamauzu, Y. Phenolic Compounds and their Antioxidant Properties in Different Tissues of Carrots. *Food Agric. Environ.* **2004,** *2,* 95–100.

Zhang, Q.; Li, Z.; Xu, Z.; Niu, X.; Zhang, H. Effects of Fucoidan on Chronic Renal Failure in Rats. *Planta Med.* **2003,** *69* (6), 537–541.

Zhao, Z.; Egashira, Y.; Sanada, H. Digestion and Absorption of Ferulic Acid Sugar Esters in Rat Gastrointestinal Tract. *J. Agric. Food Chem.* **2003,** *51,* 5534–5539.

Zhou, K.; Laux, J. J.; Yu, L. Comparison of Swiss Red Wheat Grain and Fractions for their Antioxidant Properties. *J. Agric. Food Chem.* **2004,** *52,* 1118–1123.

Zhu, K. X.; Huang, S.; Peng, W.; Qian, H.F.; Zhou, H.M. Effect of Ultrafine Grinding on Hydration and Antioxidant Properties of Wheat Bran Dietary Fiber. *Food Res. Int.* **2010,** *43,* 943–948.

Zieliński, H.; Kozłowska, H. Antioxidant Activity and Total Phenolics in Selected Cereal Grains and Their Different Morphological Fractions. *J. Agric. Food Chem.* **2000,** *48,* 2008–2016.

CHAPTER 9

CONTROL OF LIPID OXIDATION IN MUSCLE FOOD BY ACTIVE PACKAGING TECHNOLOGY

JOSÉ M. LORENZO[1,*], RUBEN DOMÍNGUEZ[1], and JAVIER CARBALLO[2]

[1]*Centro Tecnológico de la Carne de Galicia, Rua Galicia No. 4, Parque Tecnológico de Galicia, San Cibrao das Viñas, Ourense 32900, Spain*

[2]*Área de Tecnología de los Alimentos, Facultad de Ciencias de Ourense, Universidad de Vigo, Ourense 32004, Spain*

Corresponding author. E-mail: jmlorenzo@ceteca.net

CONTENTS

ABSTRACT

Lipid oxidation is a major cause of deterioration in meat, and preventing this alteration process during storage is actually a major challenge for the food technology. The function of food packaging has evolved from a simple physical barrier to include more specific aspects of convenience for approaching to specific concerns, and active packaging seems to be an effective tool to prevent lipid oxidation in meat and increase shelf life.

In this chapter, after a brief reminder on the lipid oxidation in meat, the natural and artificial antioxidants used in foods, and the modified atmosphere packaging (MAP), and its effect on the oxidation processes in meat, the active packaging as a solution against oxidative processes during meat storage is treated. The modes of action for antioxidant packages, the criteria of selection of the antioxidant compounds used, the methodologies for producing antioxidant packaging systems, and the materials for food active antioxidant packaging are addressed, and information in literature on the use of antioxidant packaging in the preventing lipid oxidation during meat storage are reviewed.

9.1 INTRODUCTION

Lipid oxidation is a major cause of quality deterioration in meat, which leads to off-odors and off-flavors, which are usually described as rancid (Gray & Pearson, 1994) as well as discoloration or texture changes of muscle foods during refrigerated storage (Kanner, 1994; Shahidi, 2002; Gong et al., 2010). In addition, oxidation causes loss of nutritional values, and generates and accumulates compounds that may pose continual risks to human health (Kanner, 1994; Min & Ahn, 2005).

Nowadays, consumers are finding less time to prepare meals. Food industry is responding to this by increasing the availability of pre-cooked meats. However, as in fresh meat, the major problem with precooked meats is the development of an objectionable warmed-over flavor via lipid oxidation (Ang & Lyon, 1990). Therefore, lipid oxidation needs to be controlled during storage in order to prevent the formation of off-odors and off-flavors in foods (Richards, 2006). The first step is know the specific changes that these foods undergo to select appropriate packaging material and package format options, and so minimizes quality loss (Krotcha, 2006).

One method is to reduce the concentration of oxygen in the fat by packing the products under vacuum or nitrogen (Chu & Hwang, 2002).

Nevertheless the fresh meat must be packaged with an atmosphere rich in O_2 to maintain the red color of the meat. In fact, red meat is usually packaged in modified atmosphere packaging (MAP) with 70–80% O_2. The drawback to high O_2 MAP is that although it maintains redness during storage, rancidity often develops in the meat while color is still desirable (Jayasingh et al., 2002).

Other method is the use of synthetic antioxidants such as butylated hydroxy anisole (BHA). However, consumers' concerns about the use of artificial preservatives in meat products have been increased because of their possible toxicity to human health. Consequently, attention has focused on the use of natural antioxidants to replace synthetic antioxidants (Min & Ahn, 2012). Different antioxidant agents, such as rosemary extract, tocopherol, ascorbic acid, and different plant extracts may be successfully included in bio-based films, to decrease oxidative reactions in meat products (Coma & Kerry, 2012).

Research and development in the area of active packaging systems for meat products has received much attention recently and will continue to do so in the near future (Walsh & Kerry, 2002). Packaging materials with antioxidant properties could be particularly efficient (Nabrzyski, 2002). In fact, a number of active packaging technologies for meat-based products have been extensively reviewed (Kerry et al., 2006; Hogan & Kerry, 2008; O'Grady & Kerry, 2008).

Therefore, the simultaneous application of both natural antioxidant and MAP not only meets consumer demands for replacement of synthetic preservatives, but also provides stronger protective effects on lipid oxidation in fresh meat (Min & Ahn, 2012).

9.2 LIPID OXIDATION

9.2.1 FACTORS AFFECTING THE DEVELOPMENT OF LIPID OXIDATION IN MEAT

There are many factors that affect the development of oxidative rancidity in meat, some of them are intrinsic, such as species, muscle type, amount and type of fat in the diet, enzymes, differences in fat content and fatty acid composition, endogenous antioxidants (carnosine and related dipeptides), and others extrinsic such as storage conditions, O_2 concentration, and processing treatments (heat, mincing, irradiation, etc.).

9.2.1.1 MEAT COMPOSITION

Two factors greatly influence lipid oxidation in raw meat are fat content and fatty acid composition. According to Min et al. (2008), the composition of fat is more important than the amount of fat in meat, because the suscepti- bility of muscle lipid to lipid peroxidation depends upon the degree of poly- unsaturation in fatty acids. Unsaturated lipids are generally more susceptible to lipid oxidation because hydrogen atoms can be more easily abstracted from polyunsaturated fats than saturated fats (Kanner et al., 1987; Gong et al., 2010). In fact, fats containing high proportions of linoleic or linolenic acids are more prone to oxidation than oils high in oleic acid. Thus one nutri- tional effect of oxidation is to reduce the essential fatty acid content of fats.

It is thought that the polyunsaturated fatty acids from polar phospholipids rather than triglycerides are responsible for the initial development of lipid oxidation in muscle foods (Renerre & Ladabie, 1993). During the course of oxidation, the total unsaturated fatty acid content of lipids decreases with a concurrent increase in the amount of primary and secondary oxidation prod- ucts such as lipid hydroperoxides, aldehydes, ketones, hydrocarbons, and alcohols. Therefore, rancidity in food occurs when unsaturated fatty acids decompose into volatile compounds. Increasing levels of unsaturated fatty acids in meat increase lipid oxidation rates (and rancidity) and thus decrease shelf life of the muscle foods. The autoxidation rate greatly depends on the rate of fatty acid or acylglycerol alkyl radical formation, and the radical formation rate depends mainly on the types of fatty acid or acylglycerol.

Consequently, the susceptibility of meat to lipid peroxidation varies among meats from different animal species and muscles from the same animal (Min et al., 2008). There are numerous studies showing that both the species and fat location significantly affects fatty acid composition of the meat. Cava et al. (2003) found that muscles with higher proportions of phospholipids also presented higher amounts of polyunsaturated fatty acids. In the same way, Domínguez et al. (2015) also observed that *Psoas major* muscle (oxidative muscle) had significantly higher amounts of polyunsatu- rated fatty acids than *Longissimus dorsi* (glycolytic muscle). Differences in fatty acid composition between oxidative and glycolytic muscles might be due to a higher number of cellular and sub-cellular membranes, and the difference in the ratio of mitochondria to other membranes between oxida- tive and glycolytic muscles. Therefore, the different polyunsaturated fatty acid amounts making lipids from oxidative muscles more susceptible to oxidative processes than those from the glycolytic muscles.

Regarding to the animal species, poultry meats which contain high levels of polyunsaturated fatty acids are most susceptible to lipid oxidation, followed by pork, beef, and lamb (Cross et al., 1987). In fact, a higher proportion of unsaturated fatty acids in the triglycerides of pork and chicken, compared with beef or lamb, produce more unsaturated volatile aldehydes in these meats and these compounds may be important in determining the specific aromas of meat species (Mottram, 1991).

9.2.1.2   PROCESSING AND STORAGE CONDITIONS

However, not only the meat composition affects lipid oxidation. Other factors, such as processing and storage conditions have a great impact on meat oxidation.

a) Irradiation

Irradiation is a preservation method that has been more extensively investigated for preservation of poultry than red meats (Morehouse, 2002; Argyri et al., 2012). According to Kanatt et al. (2005), who investigated the effect of irradiation processing on the quality of chilled meat products, concluded that irradiated samples showed significantly higher thiobarbituric reactive substances (TBARS) values than non-irradiated. In addition, the increase in TBARS values was dose-dependent. Katusin-Razem et al. (1992) also reported that irradiation of pork and poultry meat accelerates lipid oxidation. This fact is due to that when ionization radiation is absorbed by matter, ions, and excited molecules are produced. These ions and excited molecules can dissociate to form free radicals (Richards, 2006).

Moreover, not only the dose increases lipid oxidation, but the type of packaging also has great importance in meat oxidation. To this regard, Nam and Ahn (2003a, 2003b) studied the effects of combining aerobic and anaerobic packaging and the oxidant combinations on color, lipid oxidation, and volatile production to establish a modified packaging method to control quality changes in irradiated raw turkey meat. These authors reported that lipid oxidation is the major problem with aerobically packaged irradiated turkey breast, and concluded that the combination of double packaging and antioxidants was more effective in reducing sulfur volatiles and lipid oxidation, when compared with aerobic packaging.

b) Cooking

A typical treatment that greatly affects lipid oxidation is cooking. Heating accelerates lipid peroxidation and volatile production in meat (Broncano et al., 2009; Alfaia et al., 2010; Domínguez et al., 2014a, 2014b) by disrupting muscle cell structure, inactivating antioxidant enzymes, and other antioxidant compounds. In addition, cooking has the ability to increase iron concentrations in biological systems (which act as pro-oxidant) by stimulating the release of iron from heme-proteins (Decker & Welch, 1990; Smiddy et al., 2002; Richards, 2006). The resultant breakup of cell compartments permits the interaction of pro-oxidants with unsaturated fatty acids and oxygen, the generation of free radicals and propagation of the oxidative reaction (Asghar et al., 1988). High temperature causes reduction of activation energy for lipid peroxidation and decomposes preformed hydroperoxides to free radicals, which stimulates autoxidation process and off-flavor development further (Min et al., 2008).

In lipid oxidation, the most important parameters are the conditions of heat treatment (temperature and time of cooking) (Byrne et al., 2002; Domínguez et al., 2014a). The use of high temperature during cooking causes an increase of the oxidation processes in meat (Broncano et al., 2009). However, according to Domínguez et al. (2014a) oxidation processes during cooking are more affected by cooking time than temperature. Therefore, the application of heat during a long time produces higher oxidation compared to the changes caused by the use of a higher temperature during a shorter time.

The type of heating method used also has a great effect on the lipid oxidation. While it is widely accepted that the microwave oven has greatly contributed to the daily lives of modern society, the use of microwave increase the lipid oxidation. Domínguez et al. (2014a) found that meat cooked with microwave oven showed the highest values of TBARS. The fact that samples cooked by microwave had high levels of oxidation compounds suggests some interaction between microwave and meat fat which causes oxidation of polyunsaturated fatty acids (Broncano et al., 2009).

c) High pressure

High-pressure processing is a non-thermal technology, which applies pressures up to 1000 MPa for a variable time. It has been reported as a preservation method as it is able to extend the shelf life of food without modifying its sensory properties or nutrient content (Cheftel & Culioli, 1997; Hendrickx et al., 1998). However, there are several studies that relate the treatment of meat with high pressures and increased oxidation. Regarding to this, Cheah and Ledward (1996) showed that high pressure (800 MPa, 20 min) treated pork mince samples revealed faster oxidation than control samples, and that pressure treatment at greater than 300–400 MPa caused conversion of reduced myoglobin/oxymyoglobin to the denatured ferric form. According to Orlien and Hansen (2000), 500 MPa is a critical pressure for lipid oxidation and development of rancidity in chicken breast muscle. Therefore it appears that the iron released from metal complexes during pressure treatment catalyzed lipid oxidation in meat (Cheah & Ledward, 1997) but it also be related to membrane damage.

d) Mincing

A typical way of finding the meat is like minced meat. However, this treatment has a great effect on rancidity development. It is well known that compartmentation of cellular and extracellular reactants should be critical in controlling rates of lipid oxidation. Therefore, mincing can cause significant disruption of the cellular compartmentalization structure which facilitates the meeting of pro-oxidants with unsaturated fatty acids resulting in the generation of free radicals and propagation of the oxidative reaction (Buckley et al., 1995; Walsh & Kerry, 2002). According to Takama et al. (1974), minced flesh was susceptible to rancidity due to the dispersed blood pigments in the meat caused by the mechanical destruction of the tissue. In addition, other study concluded that TBARS values increase most rapidly with decreasing particle sizes, as the latter are related to greater cell disruption (Ladikos & Lougovois, 1990).

e) Light

Usually the meat is exposed in a supermarket to be attractive to consumers, and therefore it is directly exposed to light. This fact increases the oxidation of fatty acids. In addition, photo-oxidation is much faster than autoxidation. To this regard, small amounts of O_2 (for example in MAP packaging), when combined with exposure to light, cause significant oxidative deterioration of products (Jakobsen et al., 2005). This is due to ultraviolet radiation decomposes existing hydroperoxides, peroxides, and carbonyl and other oxygen-containing compounds, producing radicals that initiate autoxidation (Frankel, 1998).

9.2.1.3   PRO-OXIDANT FACTORS

a) Metals

Muscle contains notable amounts of iron, a known pro-oxidant, and trace amounts of copper, which are potent catalysts of lipid oxidation (Richards, 2006). Iron is a part of the active site of lipoxygenase, which may participate in lipid oxidation (Nabrzyski, 2002). These metals are believed to be pivotal in the generation of species capable of abstracting a proton from an unsaturated fatty acid (Gutteridge & Halliwell, 1990; Kanner, 1994).

The reaction between ferrous ion and oxygen produce hydrogen peroxide:

$$Fe^{2+} + O_2 \rightarrow Fe^{3+} + O_2^-\bullet$$

$$O_2^-\bullet + 2H^+ \rightarrow H_2O_2$$

Ferrous iron can also then react with H_2O_2 or preformed lipid hydroperoxides to produce hydroxyl or alkoxyl, and hydroxyl radicals, respectively:

$$Fe^{2+} + H_2O_2 \rightarrow Fe^{3+} + OH^- + \bullet OH^-$$

$$Fe^{2+} + ROOH \rightarrow Fe^{3+} + \bullet OH^- + RO^-\bullet$$

b) Heme-proteins

Hemoglobin and myoglobin are the predominant heme-proteins in muscle foods. Therefore, it is easy to imagine that meat with higher proportion of heme-proteins (such as pork, beef, or horse) is more susceptible to lipid oxidation than meat with lower amounts of heme-proteins (such as chicken or turkey). Similarly, to the above in the section of metals, heme-proteins can react with lipid hydroperoxides to produce alkoxyl and hydroxyl radicals:

$$ROOH + Fe^{2+}\text{-complex} \rightarrow Fe^{3+}\text{-complex} \rightarrow RO\bullet + OH^-$$

$$ROOH + Fe^{3+}\text{-complex} \rightarrow ROO\bullet + H^+ + Fe^{2+}\text{-complex}$$

The ability of heme pigments to accelerate the propagation step of the free-radical chain mechanism can explain the rapid rate of oxidation in cooked meats (O'sullivan & Kerry, 2012).

c) Enzymes

Various endogenous enzymes are of great importance in the development of rancidity. The oxidation of fatty acids may occur either directly or indirectly through the action of enzyme systems, of which three major groups are involved: microsomal enzymes, peroxidases, and dioxygenases (Erickson, 2002).

Lipoxygenases is capable of the hydrogen abstraction from a polyunsaturated fatty acid even in polar lipids bound to membrane to generate lipid hydroperoxides. Therefore, lipoxygenase can be involved in the initiation of lipid peroxidation of meat (Min et al., 2008).

The off-flavor is due to the volatiles that are produced from breakdown of the lipoxygenase-derived lipid hydroperoxides (Richards, 2006). These enzymes may also be responsible for formation of rancid odors by providing critical amounts of lipid hydroperoxides that can be broken down by metals or heme-proteins to produce rancid odor.

9.2.1.4 ANTIOXIDANT FACTORS

The addition of antioxidants is the most commonly used method of retarding lipid oxidation in fat. Antioxidants increase the stability of food components, especially polyunsaturated lipids, and maintain nutritional value and color by preventing oxidative rancidity, degradation, and discoloration. However, it is important to note that any compound that is antioxidative under one set of conditions can become pro-oxidative under different conditions. As an example of this point, ascorbate has been found to both inhibit and accelerate lipid oxidation depending on the concentration of linoleate hydroperoxides in the system (Kanner & Mendel, 1977). In addition, antioxidants are required to be approved for the intended use. It has been suggested that an ideal antioxidant food quality should has the following characteristics:

- ✓ no harmful physiological effects
- ✓ absence of undesirable effects on color, odor, or flavor
- ✓ effective at low concentrations

✓ compatibility with the food and ease of application
✓ survive after processing and be stable in the finished product
✓ available at low cost.

The compound and its oxidation products must also be nontoxic, even at doses much larger than those that normally would be ingested in food.

Antioxidants can be classified according to the mechanism of action into two groups:

9.2.1.4.1 *Primary Antioxidants*

Primary antioxidants interfere with autoxidation by interrupting the chain propagation mechanism. Primary antioxidants are free radical acceptors (Wasowicz et al., 2004). The best-known and most effective primary antioxidant substances are polyphenols. They react with the chain-propagating radical species, which results in the formation of radical species incapable of extracting hydrogen atoms from unsaturated lipids (Coma & Kerry, 2012).

Tocopherols inhibit lipid oxidation by scavenging of aqueous and lipophilic free radicals as well as physical effects on membrane structure (Buettner, 1993; Atkinson et al., 2008). Dietary antioxidant treatments (i.e., the inclusion of antioxidants in animal feed) have been shown to stabilize lipids in membranes and reduce the extent of lipid oxidation in meat during storage, but antioxidant effects in meat can differ between muscle types (Morrissey et al., 1997; Ahn et al., 2006). Vitamin E in livestock diets has been shown to reduce lipid oxidation in meats (Morrissey et al., 1998; Álvarez et al., 2009). Batifoulier et al. (2002) reported that supplementation of turkeys with α-tocopheryl acetate increased vitamin E content of microsomal membranes and had also a protective effect on lipid oxidation.

Some endogenous enzymes also have an antioxidant effect such as superoxide dismutase, catalase or glutathione peroxidase. These enzymes inhibit lipid oxidation through the following mechanisms:

-   **Superoxide dismutase** is present in cells and extracellular fluids to remove $^-\bullet O_2$ resulting in formation of oxygen and hydrogen peroxide.
-   **Catalase**, a heme-containing enzyme reacts with H_2O_2 to form water and oxygen (Goth, 1987; Richards, 2006).
-   **Glutathione peroxidase** reduces hydrogen peroxide and lipid hydroperoxides to alcohols (Gong et al., 2010).

Therefore, the extent of these enzymes activities can be a determining factor for the different rates of lipid oxidation in meat (Pradhan et al., 2000; Min & Ahn, 2009).

In addition, some antioxidative peptides can be released from food proteins. Jurewicz and Salmonowicz (1973) reported that DL-valine, DL-methionine, OL-proline, and L-cysteine had antioxidant activity. Moreover, milk casein-derived peptides have been shown to have free radical scavenging activity to inhibit enzymatic and non-enzymatic lipid oxidation (Suetsuna et al., 2000; Rival et al., 2001a; Rival et al., 2001b).

9.2.1.4.2 Secondary Antioxidants

Secondary antioxidants, in opposite to the primary antioxidants, do not break free radical chain but are able to stop the lipid oxidation through various mechanisms (Wasowicz et al., 2004). There are some compounds who act as secondary antioxidants:

a) Reducing agents

Reductants such as ascorbic acid, which decrease the local concentration of oxygen, are also able to decrease the formation of peroxyl radicals (Ruiter & Voragen, 2002). Ascorbate is believed to scavenge tocopherol free radicals thereby regenerating tocopherol. Ascorbate can also scavenge various free radicals such as $^-\!\cdot O^2$, $\cdot OOH$, and $\cdot OH$. In addition, ascorbate reduces hypervalent to forms of heme proteins which inhibit lipid oxidation in muscle foods (Kroger-Ohlsen & Skibsted, 1997).

Although ascorbic acid is mainly recognized as antioxidant, at low concentrations ascorbic acid may act as a pro-oxidant, especially in the presence of metal-catalyzed oxidation. Ascorbic acid is able to reduce Fe^{3+} to Fe^{2+}. Nevertheless, the reduced Fe^{2+} catalyzes the breakdown of hydroperoxides to free radicals (as explained above) at a higher rate than Fe^{3+} (Ponce-Alquicira, 2006).

b) Chelating agents

As mentioned previously, the presence of metals in fats greatly accelerates the oxidation process. Inactivation of the catalysis effect of these metals can be achieved by the use of a sequestering agent (Chu & Hwang, 2002).

- **Phosphates**: Polyphosphates like sodium tri-polyphosphate are excellent metal chelators and inhibitors against lipid oxidation. However, when added to raw meat, they are ineffective due to rapid hydrolysis to monophosphate by endogenous phosphatase enzyme (Lee et al., 1998). But when this enzyme is denaturized (e.g., in cooked meat) polyphosphates inhibited lipid oxidation (Sato & Hegarty, 1971).
- **EDTA**: EDTA can inhibit lipid oxidation by forming an inactive complex with metals.
- **Citric acid**: Citrate esters improve oil solubility but at least two free carboxyl groups are needed for effective metal inactivation (Richards, 2006).
- **Desferrioxamine**: Desferrioxamine is often used as a metal chelator, but this can lead to errant results since desferrioxamine can also act as a free radical scavenger (Kanner & Harel, 1987; Richards, 2006).
- **Peptides**: Both carnosine and anserine are endogenous antioxidative dipeptides found in skeletal muscle at high concentrations (Lynch & Kerry, 2000). They are known to be the most abundant antioxidants in meats. It is capable of chelating copper, scavenging peroxyl radicals, and forming adducts with aldehydes (Decker et al., 2000). Histidine was found to inhibit non-enzymatic iron mediated lipid oxidation apparently due to formation of an inactive chelate but histidine was also found to activate enzymatic pathways of lipid oxidation (Erickson & Hulin, 1992). In addition, carnosine, anserine, histidine, lysine, albumin, and sulfur or amine containing compounds have the ability to bind aldehydes and therefore decrease rancidity in foods (Decker, 1998).

9.3 ANTIOXIDANTS

The use of molecules with antioxidant activity is the best solution for preventing oxidative processes during storage and increasing the shelf life of foods. Several molecules from different sources have been recognized possessing this ability and used as antioxidants in foods, acting through one or more of the mechanisms already described. We will comment the most relevant compounds having this property, their characteristics, and performances.

9.3.1 NATURAL ANTIOXIDANTS

Due to the damage caused by oxidations in live tissues, animals and vegetables accumulate antioxidant molecules as a mechanism of defense against
these undesirable changes. Most of these antioxidants are supplied by the
feed in the animals.

Several natural antioxidants are frequent in animal and vegetable tissues
as such or as precursors. We will shortly review the most representative:

9.3.1.1 PHENOLIC COMPOUNDS

Phenolic compounds are natural antioxidants widely distributed in vegetable
tissues. Its characteristic common chemical structure consist in a benzene
ring having an alcohol (hydroxyl) group bonded to a carbon atom (phenol).
Phenol itself has not antioxidant activity, but substitution of the hydrogen
atoms placed in the *ortho*- and *para*-positions with alkyl groups enhances its
reactivity toward free lipid radicals (Shahidi et al., 1992). Phenolics are classified as simple phenols or polyphenols, these having more than one phenol
unit in their molecules. Most of them are soluble and the smaller molecules
are usually volatiles.

Several polyphenols have antioxidant activity due to scavenging activity
on free radicals by donating a hydrogen atom or an electron to the free
radical and stabilizing it. They can also act as singlet oxygen quenchers and
also through the regulation of some concrete chelation reactions.

Briefly, three main groups of phenolic compound have a high-recognized
antioxidant activity in foods: tocopherols, flavonoids, and phenolic acids.

Tocopherols are a family of compounds naturally found in vegetable oils,
fish, nuts, and leafy green vegetables, which also have vitamin E activity.
Tocopherols derive from a common alcohol matrix named tocol (2-methyl-
2(4', 8', 12'-trimethyltridecyl)chroman-6-ol), and differ according to the
number and position of the methyl groups placed in the ring structure (chromanol ring), giving rise to different forms, called α, β, γ, and δ tocopherol.
The antioxidant activity of tocopherols increases from α to δ, while the
vitamin E activity and the reactivity with the peroxyl radicals decrease from
α to δ forms. Despite its low reactivity with the free radicals, the higher antioxidant efficiency of the γ-tocopherol when compared to the α-tocopherol
is a consequence of the high stability of the γ-tocopherol and of the difference in the products formed in both cases during the antioxidative reactions
(Belitz et al., 2009). All the tocopherol forms have a higher rate of reaction

with peroxyl radicals than BHA, due to the different nature of the radicals formed on H-abstraction.

Tocopherols are approved as food additives with different E numbers: E306 (tocopherol), E307 (α-tocopherol), E308 (γ-tocopherol), and E309 (δ-tocopherol).

Flavonoids are pigments widely distributed in vegetables where typically impart a yellow color. Chemically, they have a general structure consisting in a 15-carbon atoms skeleton integrated by two phenyl rings (named A and B) and a heterocyclic ring (named C). Such carbon structure can be abbreviated C_6-C_3-C_6 (A-C-B rings). The different classes of flavonoids differ in the degree of oxidation and pattern of substitution in the C ring, while individual compounds within a same class differ in the pattern of substitution in the A and B rings (Pietta, 2000). Flavonoids, according to their chemical structure are divided into five different classes: Anthoxanthins (which include two subgroups, flavones, and flavonols), flavonones, flavanonols, flavans (which include flavan-3-ols, flavan-4-ols, and flavan-3, 4-diols), and anthocyanidins. The capacity of flavonoids to act as antioxidants *in vitro* has been demonstrated by several studies, and important structure-antioxidant activity relationships have been established (Pietta, 2000). Flavonoids are generally primary antioxidants which act as free radical acceptors, breaking the oxidation chain. Flavonols can also chelate metal ions at the 3-hydroxy-4-keto-group, and/or the 5-hydroxy-4-keto-group (in the case in that the A ring was hydroxylated at the fifth position).

It is generally recognized that the degree of hydroxylation and the position of the hydroxyl groups determine the antioxidant activity of the flavonoids (Shahidi et al., 1992). The hydroxylation in the B ring is the major factor for antioxidant activity. The *o*-dihydroxylation in the B ring actively contributes to the antioxidant activity, and all the flavonoids with 3'-4'-dihydroxy configuration have antioxidant activity in more or less extent. Two flavones, robinetin and myricetin, have an additional hydroxyl group placed at their fifth position, which confers to these two molecules an enhanced antioxidant activity in relation to the corresponding molecules that do not possess such 5'-hydroxyl group (fisetin and quercetin). On the contrary, two other flavones, naringenin, and hesperetin, have only a hydroxyl group in the B ring, and due to this particularity they show little antioxidant activity. Besides the hydroxylation in the B ring, other structural characteristics affecting the A ring determine the antioxidant activity such as the presence of a carbonyl group at the fourth position and a free hydroxyl group at third and/or fifth positions.

Quercetin (2-(3,4-dihydroxyphenyl)-3,5,7-trihydroxy-4*H*-chromen-4-one) is the most abundant flavonoid in foods, and due to its antioxidant activity and other beneficial properties has been object of special attention in the past years (Alrawaiq & Abdullah, 2014). Although it has not been confirmed scientifically as a specific therapeutic nor approved by any regulatory agency, it is widely used as food supplement in the treatment of several health problems. The Joint FAO/WHO Expert Committee on Food Additives evaluated quercetin for use in food in 1977 (Harwood et al., 2007), but limited data on its toxicity were available at the time of the evaluation which precluded the establishing an acceptable daily intake (ADI). In Japan, quercetin is permitted as a food additive since the 1996 year (Harwood et al., 2007). Other flavonoids such as myricetin or robinetin have a recognized high antioxidant activity. All those compounds could be in the future efficient food antioxidant additives after approval by the health authorities, upon proof of their harmlessness.

Phenolic acids are substances containing a phenolic ring and a carboxylic function, therefore having a C6-C1 skeleton. They can be mono-, di- or tri-hydroxybenzoic acids depending on the number of positions hydroxylated in the phenolic ring. The antioxidant activity of the phenolic acids and their corresponding esters is determined by the number of hydroxyl groups. Some concrete phenolic acids such as cafeic, coumaric, ferulic, gallic, and protocatechuic acid are known as molecules possessing a not negligible antioxidant activity. They could be in the future successfully used for this purpose in foods after further studies on their stability and safety.

9.3.1.2   ASCORBIC ACID

Ascorbic acid $(C_6H_8O_6)$ ((5R)-[(1S)-1,2-Dihydroxyethyl]-3,4-dihydroxyfuran-2(5h)-one) is a molecule with antioxidant and vitamin (vitamin C) activities widely present in vegetables and fruits, and to a lesser extent in animal tissues. It is oxidized with successive loss of two electrons to form dehydroascorbic acid. It reacts with oxidants (reactive oxygen species), such as the hydroxyl radical. Therefore, ascorbate can terminate these chain radical reactions by electron transfer. Ascorbic acid is special because it can transfer a single electron, due to the resonance-stabilized nature of its own radical ion.

Ascorbic acid is an active antioxidant in aqueous media, because of its water-soluble character, but only at high concentrations (around 10^{-3} mol/L). A pro-oxidant activity is observed at lower concentrations (10^{-5} mol/L), especially at high oxygen tensions and when heavy metal ions are present.

This circumstance, however, seems to be irrelevant from the point of view of the food chemistry concerns.

Ascorbic acid is, of course, approved and widely used in foods, being he E300 additive.

9.3.1.3   CAROTENOIDS

Carotenoids are natural pigments which are synthesized by plants, being responsible for the bright colors of several fruits and vegetables. There are several dozen carotenoids in foods, and most of them have antioxidant activity (Paiva & Russell, 1999). Beta-carotene and lycopene; however, have been the best studied and more widely used ones.

Beta-carotene ($C_{40}H_{56}$) is a tetraterpene formed by eight isoprene units having beta-rings at the two ends of the molecule. It is well known for its provitamin A activity. With a strong lipophilic character, it acts as free radical scavenger and therefore it has antioxidant properties widely demonstrated in *in vitro* assays and in animal models. This antioxidant activity is not lost by degradation to long chain breakdown products (Mueller & Boehm, 2011). It shows, however, good antioxidant behavior only at partial pressures of oxygen lower than 150 mm Hg. At higher oxygen pressure values, β-carotene loses its antioxidant activity and, in contrary, it shows an autocatalytic, pro-oxidant effect, particularly at relatively high concentrations (higher than 5×10^{-5} mol/L). Despite of its antioxidant activity, in the food industry is more used as colorant with the number E160a(ii).

Lycopene ($C_{40}H_{56}$) is also a tetraterpene formed by eight isoprene units, but with a single aliphatic chain lacking of rings. As the β-carotene, lycopene is highly lipophilic and it has antioxidant activity (Sies & Stahl, 1998) due to their conjugated double bonds, but it lacks of provitamin A activity. Contrary to the β-carotene, it is not obtained by synthesis, and purification from natural foods, following complicated and expensive processes, is the only source of this compound. This circumstance, together with the high instability of the molecule notably limits its use as food additive, being used preferably as colorant (E160d). However, tomato powder, mainly due to its high lycopene content, was reported as an effective antioxidant in cooked pork patties (Kim et al., 2013).

Mixtures of carotenoids or associations with other antioxidants (e.g., tocopherols) can increase their scavenging free radical activity.

9.3.1.4   ESSENTIAL OILS

Due to the doubts arose on the safety of the most common synthetic antioxidants, the efforts of searching for new natural antioxidants usable in foods have been redoubled. Together with the classical natural antioxidants already described, other natural substances have been object of study and utilization for this property in the recent past years.

Essential oils (EOs) are liquid mixtures of volatile compounds obtained from plants, generally by steam distillation. Several EOs have shown a satisfactory antioxidant capacity attributed in most cases to the presence in such mixtures of molecules with antioxidant ability, mainly phenolic compounds that act as antioxidants due to their high reactivity with the peroxyl radicals. Phenolic compounds present in EOs are usually assigned to two structural families according to their hydrocarbon skeleton: (a) terpenoids, formed by an isoprene unit (hemiterpenoids) or by combination of two (monoterpenoids), three (sesquiterpenoids), four (diterpenoids) or more isoprene units, and (b) phenylpropanoids, formed by an aromatic phenyl group and the three-carbon propene tail of cinnamic acid. Some common phenolic compounds belonging to these two families (carvacrol or cymophenol, and thymol, among terpenoids, and eugenol, guaiacol, syringaldehyde, umbelliferone, and coniferyl alcohol, among phenylpropanoids) are described as principal components of several EOs.

EOs from *Allium* spp. have a chemical composition very different from most of the other EOs. They are mainly composed of sulfur-containing volatile compounds possessing antioxidant activity (Tsai et al., 2012) that has been confirmed in different model systems (Banerjee et al., 2003; Iqbal & Bhanger, 2007).

Being the EOs mixtures of various compounds, the antioxidant properties of a concrete EO should reflect the antioxidant activity of the most active or the most abundant antioxidant compounds present in it. However, it is necessary to take care with this approach, because complex interactions depending on composition and experimental conditions take place, resulting in synergistic or antagonistic behaviors among activities that notably affect the whole antioxidant properties of the EOs (Kulisic et al., 2005).

Besides the botanical source, environmental factors (e.g., soil, climate, etc.) may affect the actual composition, and therefore the antioxidant activity of each EO.

In a recent work, Amorati et al. (2013) reviewed the antioxidant activity of the EOs. For some selected EO, they summarized the data existing in literature on their main components responsible for the antioxidant activity, the tests followed for activity assessment, and the antioxidant activity in relation to reference antioxidants such as butylated hydroxytoluene (BHT), or reference EO such as those from age or bush-basil. Some EO from oregano, thyme, or clove has good antioxidant activity, comparable to that of BHT. In general, however, the antioxidant activity of EOs is medium or low; on the other hand, some concrete EOs have no antioxidant or even prooxidant activity.

EOs are promising food antioxidants when their particular aroma compatible with the organoleptic characteristics of the foods in which they are applied. EO from oregano seems to be the most successful one. Goulas et al. (2007) reported that this EO in combination with modified atmospheres and salting extend the shelf life of sea bream. Oregano EO is also able to protect the extra virgin olive oil from oxidation during storage (Asensio et al., 2012) and to protect minced meat from auto-oxidation (Fasseas et al., 2008). Very recent and encouraging applications of EOs were described in active packaging and in edible coatings as we will treat later.

9.3.1.5   PEPTIDES

Despite the fact that all the amino acids naturally present in the proteins can react with free radicals if these have high energy, the free amino acids are in the practice not generally effective in the prevention of oxidation processes in foods and biological systems (Samaranayaka & Li-Chan, 2011). Some peptides, however, possess antioxidant capacity based on their chemical structure determined by the presence of some concrete amino acidic sequences. Most of the peptides derived from food proteins having antioxidant activity show molecular weights from 0.5 to 1.8 kDa and often they have hydrophobic amino acids (as Val or Leu) in the amino-terminal position and they include the amino acids Pro, His, Tyr, Trp, Met, and Cys in their sequences.

The mechanism of action of the antioxidant peptides is generally based on the free radical scavenging; the tripeptides possessing Trp or Tyr in the carbonyl-terminal position have a strong free radical scavenging activity (Saito et al., 2003). In other cases, the action is based in the scavenging of oxygen-containing compounds. Some peptides act as antioxidants though the chelation of metal ions such as Cu or Fe. The peptides chelating Cu have the amino acid His in their sequence, being the imidazole ring of this amino acid responsible for the union with the Cu ion. Finally, it has been proven that antioxidant peptides can show synergistic affects with some other antioxidants such as phenolic compounds (Wang & González de Mejía, 2005).

Antioxidant peptides can be obtained through digestion of both vegetable and animal proteins by the action of exogenous or endogenous enzymes, by microbial fermentation, during food processing, or during gastrointestinal digestion (Samaranayaka & Li-Chan, 2011). Enzymatic hydrolysis has been widely used in the production of antioxidant peptides from food proteins. Some commercial enzymes or enzymatic preparations from microbial (such as Alcalase® from *Bacillus licheniformis*, Flavourzyme® from *Aspergillus oryzae*, and Protamex® from *Bacillus* spp.), vegetable (papain from *Carica papaya* fruits), or animal (pepsin from stomach glands, and trypsin from pancreas) sources have been used in the production of antioxidant peptides (Pihlanto, 2006; Sarmadi & Ismail, 2010; Gallegos-Tintoré et al., 2011). In foods, mainly fermented foods, the antioxidant peptides can also be produced by the action of microorganisms or indigenous proteases (Samaranayaka & Li-Chan, 2011).

In recent years, information on the obtaining of antioxidant peptides and hydrolysates from vegetables (Gallegos Tintoré et al., 2013), marine foods (Kim & Wijesekara, 2010; Di Bernardini et al., 2011), milk (Pihlanto, 2006), eggs (Dávalos et al., 2004), and meat (muscles and by-products) (Di Bernardini et al., 2011) was reviewed or directly reported, and peptidic sequences of the most active peptides were elucidated. Regarding the vegetables, conventional (proteins from soya, rice, corn, and chickpea) and non-conventional (proteins from amaranth, buckwheat, colza, and Mexican pinon) sources were assayed in the obtaining of antioxidant peptides. Marine species used as sources of peptides were also diverse (jumbo squid, oyster, blue mussel, hoki, tuna, cod, Pacific hake, capelin, scad, mackerel, Alaska pollock, conger eel, yellowfin sole, yellow stripe trevally, silver carp, grass carp, herring, and microalgae). Among dairy proteins, casein is the main source, although obtaining of peptides from β-lactoglobulin was also reported. Peptides and hydrolysates with antioxidant activity are obtained from proteins of all these sources generally via hydrolysis with the commercial enzymes or enzymatic preparations indicated in the previous paragraph.

Such peptides, and also less specific protein hydrolysates, can be used as functional ingredients in foods in order to avoid or reduce undesirable oxidation processes during storage (Samaranayaka & Li-Chan, 2011). To date, however, few products are available in markets incorporating this preservation system. Several reasons (problems of preparation of peptides and/or hydrolysates at industrial scale; lack of complete and rigorous assays confirming the activity, effectiveness, and safety; low reproducibility of the manufactured food products; modifications of the taste, color, and general organoleptic characteristics; high production costs; etc.) are responsible for the scarce actual exploitation of this potential solution (Samaranayaka & Li-Chan, 2011).

9.3.2 SYNTHETIC ANTIOXIDANTS

Natural antioxidants show some disadvantages: Low antioxidant activity (antioxidants should ideally be active at low concentrations, 0.01–0.02%), and the fact that most of them are insoluble in water. In addition, antioxidants must be stable during the food processing operations (carry through effect) in order to show all their activity in the processed foods throughout the storage process. Synthetic antioxidants comply with all these requirements and are abundantly used as food additives. We will review shortly the synthetic molecules most used in the food industry:

BHA (tertiary-butyl-4-hydroxyanisole) ($C_{11}H_{16}O_2$): It is a mix of two isomers (2-tertiary-butyl-4-hydroxyanisole and 3-tertiary-butyl-4-hydroxyanisole) obtained from 4-methoxyphenol and isobutylene. The conjugated aromatic ring from its molecule captures free radicals, sequestering them and preventing the propagation in the oxidation processes. Approved as food additive with the reference number E320 it is commercialized under various trade names. Specifications have been defined in the EU legislation in Directive 2008/128/EC and by the Joint FAO/WHO Expert Committee on Food Additives (JECFA). The purity is specified to be not less than 98.5% of tertiary-butyl-4-hydroxyanisole and not less than 85% of the 3-tertiary-butyl-4-hydroxyanisole isomer.

BHT (tertiary-butyl-4-hydroxytoluene) ($C_{15}H_{24}O$): It is a non-coloring, odorless, white solid matter. As the BHA, it is a chemical derivative of phenol with similar sequestering capacity of the free radicals due to the conjugated aromatic ring. It is approved as food additive with the reference number E321.

Tert-butylhydroquinone (TBHQ) ($C_{10}H_{14}O_2$): It is approved as food additive with the reference number E319. Addition to foods does not modify color or flavor, being very effective in the enhancing of the storage life.

Ascorbyl palmitate ($C_{22}H_{38}O_7$): It is an ester formed from ascorbic acid and palmitic acid resulting in a fat-soluble form of the ascorbic acid. Contrary to the other synthetic antioxidants, its metabolism is not suspected to generate metabolites with a potential toxic effect. Ascorbyl palmitate is known to be broken down (through the digestive process) into ascorbic acid and palmitic acid which are absorbed into the bloodstream and metabolized through the habitual routes for these two natural nutrients. It is an amphipathic molecule, meaning one end is water-soluble and the other end is fat-soluble. It is approved as food additive with the reference number E304.

Propyl, octyl, and dodecyl gallates: They are esters from the propanol, octanol or dodecanol, respectively, with the gallic acid (3,4,5-trihydroxibenzoic acid). Propyl gallate ($C_{10}H_{12}O_5$) is approved as food additive with the reference number E310. It is obtained mainly by synthesis, although it can be also obtained from a natural source (pods of the fruits of *Caesalpinia spinosa*). Octyl gallate ($C_{15}H_{22}O_5$) and dodecyl gallate ($C_{19}H_{30}O_5$) are approved as food additives with the reference numbers E311 and E312, respectively.

Ethoxyquin (6-ethoxy-1,2-dihydro-2,2,4-trimethylquinoline): It is a quinoline-based antioxidant molecule ($C_{14}H_{19}NO$) approved as food additive with the reference number E324. In contrast with the other synthetic antioxidants which are widely used, ethoxyquin is only commonly used as preservative in pet foods and in spices to prevent color loss due to oxidation of the natural carotenoid pigments. It is approved as food additive in the United States; however, it is not approved for use within the European Union nor is it permitted for use in foods in Australia, due to speculations on its responsibility on multiple pet health problems.

With the exception of ethoxyquin, whose particular situation was already commented, all these synthetic antioxidants have been proved and evaluated by both the European Food Safety Authority (EFSA) and the United States Food and Drug Administration (FDA), which in turn fixed the upper limit of addition for each concrete compound.

9.3.3 EFFICIENCY OF THE ANTIOXIDANTS

The efficiency of the antioxidants can be evaluated throughout comparative tests, by using of so-called "antioxidant factor" (AF) (Belitz et al., 2009):

$AF = I_A/I_0$, were I_A = oxidation induction period for a determined fat or oil in the presence of the considered antioxidant, and I_0 = oxidation induction period for the same fat or oil without the addition of the antioxidant. The efficiency of an antioxidant increases with the increase of the AF value.

According to the results of one of such comparative assays (Belitz et al., 2009), comparative efficiency of the most commonly used antioxidants when added at 0.02% concentrations in refined lard is (AF values): DL-γ-tocopherol (12), BHA (9.5), BHT (6), octyl gallate (6), D-α-tocopherol (5), ascorbyl palmitate (4). BHA and BHT have synergistic effects, and when added together at a given total concentration they are more effective than either antioxidant alone at the same level. In the described assay, BHA and BHT added at a concentration of 0.01% each, show an AF value of 12, similar than that of DL-γ-tocopherol. In the same way, propyl gallate increases the efficiency of BHA, but not that of BHT. Ascorbyl palmitate, which alone shows a weak antioxidant efficiency, sustains the activity of the DL-γ-tocopherol (Belitz et al., 2009).

Synergic effects can occur between antioxidants, but also between antioxidants and another molecules present in foods that enhance the antioxidant activity. These molecules called synergists are lecithin, amino acids, citric, phosphoric, citraconic, and fumaric acids, and in general molecules able to form complexes with the heavy metal ions. In this way, initiation of heavy metal-catalyzed lipid autoxidation is prevented. A synergistic effect of different nature is that exercised by phospholipids. The addition of dipalmitoylphosphatidylcholine (0.1–0.2%) to lard enhances the antioxidant activity of α-tocopherol, BHT, BHA, and propyl gallate (Belitz et al., 2009). Phosphatidylcholine, however, does not show this capacity.

9.4   MODIFIED ATMOSPHERE PACKAGING (MAP)

When not packaged, foods during storage are usually surrounded by air; the main gases in dry air at sea level are N_2 (78%, v/v), O_2 (20.99%), Argon (0.94%), and CO_2 (0.03%) (McMillin, 2008). MAP implies the presence of a barrier, normally a plastic film, that impedes the permeation for gases, water vapor included, that allows the maintaining of a constant environment of the desired gas composition during the food storage. Barrier bags are first filled with the product, previously placed in a tray, and then sealed after flushing with the selected gas mixture.

Package protects the products against deterioration, which usually include discoloration, development of off-flavor and off-odor, nutrient losses, texture changes, pathogenicity, and other measurable traits (Skibsted et al., 1994). In the case of fresh meat the main objective of packaging is the increase of shelf life which implies the maintaining the water content, color, microbial quality, lipid stability, and palatability (Renerre & Labadie, 1993; Zhao et al., 1994).

Gas composition of the modified atmosphere affects the most important meat attributes during storage which are mainly color, flavor, and microbiology.

Regarding meat color, when the choice is made consumers prefer red color above purple color, which in turn is preferred above brown color (Carpenter et al., 2001). Meat color depends on the quantity and form of the myoglobin pigment. The form of the myoglobin, and its color, depends on the state of the iron placed in the porphyrin ring in the heme group (that can be oxidized or reduced) and on the presence or absence of O_2 occupying the sixth coordination site of the iron. Purple color is due to the deoxymyoglobin that is the reduced form of myoglobin (Fe^{2+}) in the absence of O_2 (with a vacant sixth coordination site in the iron). Red bright color is due to the oxymyoglobin that is the reduced pigment (Fe^{2+}) in which O_2 occupies the ligand position. Oxymyoglobin is usually the form in contact with the air. Oxymyoglobin is formed by O_2 binding to the ferrous (Fe^{2+}) ion, which occurs at high O_2 tension values. The penetration of oxygen through the meat and therefore the oxymyoglobin layer thickness depends upon temperature, O_2 partial pressure, pH, and consumption of O_2 by other respiratory processes (Mancini & Hunt, 2005). Brown or gray color is due to the metmyoglobin form that is the oxidized state of the myoglobin (Fe^{3+}). Metmyoglobin is formed when pigment is exposed for extended times to light, heat, O_2, microbial growth, or freezing, all these factors determining the oxidation of the iron to a ferric (Fe^{3+}) state. When deoxymyoglobin is exposed to carbon monoxide, another pigment form, the carboxymyoglobin, is formed. Carboxymyoglobin formation, with a stable bright-red color, occurs when CO attaches to the vacant sixth position of deoxymyoglobin, when the environment is devoid of oxygen. In the absence of CO, the three other states of the myoglobin may coexist in varying proportions in the same meat piece depending on redox conditions. The desired bright red-bloomed color is achieved by the predominance of the oxymyoglobin pigment that is easily generated when the O_2 percentages in atmosphere are higher than 5.5%, and dominates at O_2 percentages higher than 13%. Deoxymyoglobin dominates in atmosphere conditions of less than 0.2% O_2, while metmyoglobin is the main pigment form at O_2 levels of 0.2–13% (Siegel, 2001). Metmyoglobin is easily formed in fresh meat in the range of 2.6–5.3% of O_2 (Sebranek & Houser, 2006).

Flavor attributes are, together with the tenderness, the most important factors that influence the consumers' purchase habits. Compounds determining flavor and odor are usually originated from protein and lipid components of meat (Spanier, 1992). Undesired off-flavors from proteins are normally generated through the action of microorganisms producing amines, ammonia, and other odor active compounds such as sulfur compounds, from amino acids. In this sense, MAP gas composition influences the flavor attributes through the inhibition or promotion of the growth of the different microbial groups. MAP inhibits the habitual spoiling microorganisms, but as will be discussed below favors the development of concrete species that are responsible for flavor modifications in meat. Regarding lipids, they are mainly degraded via oxidation processes. Lipid oxidation is linked to pigment oxidation and discoloration of meat, but also causes flavor deterioration via the formation of several volatile compounds from the hydroperoxides generated as products of the primary oxidation. As McMillin (2008) pointed out, although initial studies did not show enhanced lipid oxidation with increased O_2 concentrations in MAP, other more recent studies (Cayuela et al., 2004; John et al., 2005) reported higher lipid oxidation in meat packaged in high O_2 concentration atmospheres, when compared with vacuum packaged or low O_2 concentration atmospheres. Other authors, however, clearly reported this problem. Jackson et al. (1992) indicated that oxidative processes cause a real problem in meat packaged in atmospheres with more than 21% of oxygen. Unfortunately, this circumstance is the counterpoint to the beneficial effects already commented of the high O_2 concentrations on the meat color.

Microbiology and microbial growth during the storage are key factors in meat quality. Anomalous unpleasant colors and odors, and surface slime are undesirable effects of microbial growth determining the deterioration of meat. In MAP, the extended shelf life arises a new issue because pathogens have extra-time for development to reach dangerous counts (Farber, 1991). Under normal O_2 concentrations in air, aerobic microorganisms are commonly present in meat surfaces, reducing the O_2 tension, promoting discoloration, and generating surface slime due to their mobile condition. MAP notably affects the survival and growth of spoiling and pathogenic bacteria (Blakistone, 1999). Oxygen stimulates growth of aerobic microorganisms and inhibits the growth of strict anaerobic bacteria, being variable the sensitivity to the O_2 of the different anaerobic bacterial species (Church, 1994). Anoxic atmospheres favor the development of lactic acid bacteria and another facultative anaerobic microorganisms. Nitrogen has minimum effect on metabolic reactions occurring in meat, but as occurs with some gasses other than oxygen, the anoxic conditions created by the use of N_2 select for anaerobic and facultative anaerobic microorganisms (Thippareddi & Phebus, 2007). Regarding the effect of CO_2 in atmospheres, gram-negative bacteria are in general more sensitive to CO_2 than the gram-positive bacteria, because gram-positive bacteria are usually strict or facultative anaerobes (Farber, 1991). Due to the fact that CO_2 in MAP is firstly absorbed by the components of the meat, mainly water, and lipids, until an equilibrium is reached, an excess of CO_2 should be used to obtain a desired preservative effect. Concentrations of 20–60% of CO_2 are necessary for complete inhibition of aerobic spoilage microorganism, but slight or no effect is observed with CO_2 values above 50–60% (Gill & Tan, 1980). In relation to CO, the shelf life is extended by adding CO at levels above 0.5% (Clark et al., 1976). However, the effect of CO is variable on pure microbial cultures as reported by Gee and Brown (1980). Numerous works have studied the effect of concrete gas mixtures on the microbiology of MAP meat and on the growth of particular microbial groups or relevant spoiling or pathogenic species (McMillin, 2008). Viana et al. (2005) reported that counts of lactic acid bacteria increased during storage in pork loin packed in oxygen-free atmospheres, while they reached the lowest counts in 100% O_2 atmosphere after 20 days of storage; the growth of *Pseudomonas* was limited in 100% CO_2 and 1% CO + 99% CO_2 atmospheres, with the highest counts in 100% O_2. According these authors, pork treated with the 1% CO + 99% CO_2 atmosphere received the greatest acceptance by the consumers.

Regarding the gas mixtures commonly used, the high O_2 MAP can have 25–90% O_2 and 15–80% CO_2 in the headspaces (Blakistone, 1999), being 80% O_2+ 20% CO_2 the most used gas mixture (Eilert, 2005). In relation to color, the oxymyoglobin levels of minced meat were similar after four days of storage in 20, 40, 60, and 80% O_2 atmospheres, but in all the cases higher than in meat stored in air, although after seven days of storage the oxymyoglobin content decreased with the O_2 decrease. At the 10th day of storage, lipid oxidation slightly increased as the O_2 percentage was higher (O'Grady et al., 2000). Luño et al. (1998) studied the effect of a low O_2 atmosphere containing CO (24% O_2 + 50% CO_2 + 25% N_2 + 1% CO) in comparison with a high O_2 atmosphere containing CO (70% O_2 + 20% CO_2 + 9% N_2 + 1% CO) and with a high O_2 atmosphere without CO (70% O_2 + 20% CO_2 + 10% N_2) on the psychrotrophic counts and color stability throughout storage of loin steaks and ground meat. Psychrotrophic counts were greatly reduced by the low O_2 atmosphere containing CO, while the bright-red color was more stable in both atmospheres containing CO, reaching 29 days of storage without signs of oxidation, which was confirmed using sensory analysis.

9.5 ACTIVE PACKAGING

The function of food packaging has evolved from simple preservation methods to include such aspects as convenience, point of purchase marketing, material reduction, safety, tamper-proofing, and environmental issues (Han, 2014). Active packaging is a novel technology which is designed to incorporate material components in the packaging that release or absorb substances from or into the packaged food or the surrounding environment in order to extend the shelf life and maintain or improve the condition of packaged food. In contrast to traditional packaging, active, and intelligent packaging may change the composition and organoleptic characteristics of food, provided that the changes are consistent with the provisions for food.

Active packaging is receiving considerable attention as an emerging technology that can be used to improve the quality and stability of food, reducing the direct addition of chemicals, and the need for changes in formulation. The technology provides several advantages compared to direct addition, such as lower amounts of active substances required, localization of the activity to the surface, the migration from film to the food matrix (which could be used to provide antioxidant effects for longer protection), and elimination of additional steps within a standard process intended to introduce the antioxidant at the industrial processing level such as mixing, immersion, or spraying (Bolumar et al., 2011).

Active packaging is defined as a package system that deliberately incorporates components that release or absorb substances into or from the packaged food or the environment surrounding the food to extend the shelf life or to maintain or improve the condition of the packaged food (Regulation (CE) No. 450/2009 (29/05/2009)). Therefore, active packaging does something more than simply providing a barrier to external detrimental factors, as the packaging system plays an active role in food preservation and quality during the marketing process (Lopez Rubio et al., 2004; Pereira de Abreu et al., 2012). When designing an active package, issues that are of importance when designing traditional food packages, such as barrier properties to gases and moisture and the mechanical strength required for pack integrity, must still be taken into account. So, the following aspects, which are specific to the antimicrobial and antioxidant function of the packages, must principally be considered (Coma & Kerry, 2012):

- ✓ The chemical nature of the bioactive agents and their inhibition mechanism
- ✓ Physico-chemical characteristics of foods and the organoleptic property of the bioactive agents
- ✓ Packaging manufacturing processes and their influence on the efficiency of bioactive additives
- ✓ Storage environments
- ✓ Migration mechanisms of bioactive agents into foods if needed, and toxicity and regulatory issues

✓ Machinability and processability of the bioactive packaging on the packaging line materials.

A suitable selection of the antioxidant compound to be incorporated in the packaging material is crucial. The antioxidant compound and the packaging material should be compatible in order to achieve a homogeneous distribution, and the partition coefficients of the antioxidant in the different phases should favor its release to the food or headspace (Gómez-Estaca et al., 2014). Once released, the solubility characteristics of the antioxidant can determine its effectiveness, and therefore the type of antioxidant should be selected as a function of the type of food. Non-polar antioxidants would seem to be more suitable for foods with high lipid content and vice versa. Lee (2014) noticed that maximum effectiveness of antioxidant packaging systems can be achieved by fitting the antioxidant release rate with the lipid oxidation rate. Mathematical models of diffusion probed to be a valuable tool to predict the release profile of antioxidants into food systems (Piringer, 2000).

There are two main modes of action for antioxidant packages: The scavenging of undesirable compounds such as oxygen, radical oxidative species or metal ions from the headspace or from the food and the release of antioxidants to the food (Gómez-Estaca et al., 2014). Among active antioxidant packaging materials, oxygen scavengers are the ones that are being most widely produced by the extrusion technique. Oxygen scavengers are able to inhibit the growth of aerobic bacteria and molds, and stop alterations of pigments, and flavors to avoid discoloration in meat products (Vermeiren et al., 1999). According to Day (2001), oxygen scavenging adhesive labels are being used for a range of sliced cooked and cured meat products which are susceptible to deleterious color changes induced by oxygen. The most common oxygen scavengers consist of small size oxygen-permeable sachets that contain an iron-based powder along with a catalyst. These types of scavengers react with water that is produced by the packaged food and generate a hydrated metallic agent which is able to scavenge oxygen and convert it to a stable oxide. Iron and ferrous oxide fine powders, ascorbic acid, some nylons, photosensitive dyes and unsaturated hydrocarbons are being used in the manufacture of extruded films with oxygen scavenging properties, as has been reported in previous reviews (Lopez Rubio et al., 2004; Brody et al., 2008).

Another popular group of active packaging systems are moisture absorbers. Several companies manufacture moisture absorbers in the form of sachets, pads, sheets, or blankets. For packaged dried food applications, desiccants such as silica gel, calcium oxide and activated clays, and minerals are typically tear-resistant permeable plastic sachets. In addition to moisture-absorber sachets for humidity control in packaged dried foods, several companies manufacture moisture-drip absorbent pads, sheets, and blankets for liquid water control in high activity water foods such as meats, fish, fruit, and vegetables (Dobrucka & Cierpiszewski, 2014).

Regarding the antioxidant releasing packaging materials, one of the main benefits, as compared to the direct addition of antioxidants to food, is that active materials may act as a source of antioxidants that are released to the food at controlled rates, so that a predetermined concentration of the active compound is maintained in the food, compensating the continuous using up of antioxidants during storage (Mastromatteo et al., 2010). There are basically two methodologies for producing antioxidant-packaging systems (Gómez-Estaca et al., 2014):

a) **Independent devices:** An independent device such as a sachet, pad or label containing the agent separately from the food product is added to a conventional "passive" package.

b) **Antioxidant packaging materials**: Antioxidant packaging materials are used in the manufacture of the package, that is, the active agent is incorporated in the walls of the package exerting its action by absorbing undesirable compounds from the headspace or by releasing antioxidant compounds to the food or the headspace surrounding it.

Many different active packaging structures can be built with active materials prepared by extrusion processes, including coextruded, and laminated multilayers. The design is dependent on the type of agent and the type of polymer matrix, but primarily on the packaging requirements of the food product. Due to environmental motivation there is increasing interest in the use of biodegradable/compostable packaging and/or edible materials. This tendency increases when materials come from industrial waste or renewable resources (Gómez-Guillén et al., 2007). Four types of plastics materials could be used for food packaging (Mecking, 2004; Reddy et al., 2012; Peelman et al., 2013; Reddy et al., 2013):

a) **Non-bio-based non-biodegradable conventional plastic materials**, for instance: polyethylene (PE), polypropylene (PP), poly (ethyleneterephthalate) (PET), and polystyrene (PS).

b) **Bio-based biodegradable plastic materials**, for example: poly lactic acid (PLA), poly-hydroxybutyrate (PHB), and chitosan.
c) **Bio-based non-biodegradable plastic materials**, for instance: the commonly advertised as "green" PE, PP, and PET.
d) **Non-bio-based biodegradable plastic materials**, for example: aliphatic-aromatic polyesters, and polycaprolactone (PCL).

Finally, edible film and coating technology is close to active packaging technology and may also be a means of reducing oxidative spoilage in foods. The main mechanism of action is the reduction of the oxygen transmission rate, as well as the possibility of incorporating antioxidant compounds in the edible film or coating matrix; this vehicle has the advantage of close contact between coating and food. An edible film or coating does not act as a package itself, but it may reduce the barrier requirements of the package.

Research on active packaging for muscle foods has focused predominantly on the use of antimicrobial agents, while the development of antioxidant applications is growing. Different antioxidant agents, such as tocopherol, ascorbic acid, and different plant extracts, and EOs from herbs such as rosemary (*Rosmarinus officinalis* L.), oregano (*Origanum vulgare* L.), and green tea (*Camellia sinensis* L.) may be successfully included in bio-based films, to decrease oxidative reactions in meat products (Table 9.1). Rosemary extract is one of the plant extracts that has already been incorporated into food packaging. It is composed by flavones (apigenin, genkwanin, hesperetin, and cirsimaritin), phenolic diterpenes (carnosic acid, carnosol, rosmadial, epirrosmanol, rosmanol, carnosic acid o-quinone), and phenolic acids (caffeic acid and rosmarinic acid). Carnosic acid is one of the most important compounds responsible for antioxidant capacity (Sanches-Silva et al., 2014). Nerin et al. (2006) observed that the active film containing rosemary extract efficiently enhanced the stability of both myoglobin and fresh meat against oxidation processes, thus being a promising way to extend the shelf life of fresh meat. So, active packaging was able to extend the shelf life of beef packaged in modified atmosphere and displayed under conventional illumination by ~2 days, that is, 17%. In addition, Camo et al. (2008) noticed that the use of rosemary and oregano active films resulted in enhanced oxidative stability of lamb steaks packed in modified atmosphere. Active films with oregano were significantly more efficient than those with rosemary, exerting an effect similar to that of direct addition of the rosemary extract.

TABLE 9.1   Meat and Meat Products Packed with Antioxidant Films.

Antioxidant source	Meat product	Reference
Oregano and pimento essential oils	Fresh beef meat	Oussalah et al. (2004)
Rosemary extract	Fresh beef meat	Nerín et al. (2006)
Rosemary and oregano EOs	Fresh lamb meat	Camo et al. (2008)
Thyme and oregano EOs	Fresh lamb meat	Karabagias et al. (2011)
Oregano	Fresh beef meat	Camo et al. (2011)
Rosemary extract	Chicken meat patties	Bolumar et al. (2011)
Thymol, carvacrol, and eugenol	Fresh beef meat	Park et al. (2012)
Green tea	Pork sausages	Siripatrawan and Noipha (2012)
Brewery and rosemary	Fresh beef meat	Barbosa-Pereira et al. (2014)
Oregano and green tea	Fresh foal meat	Lorenzo et al. (2014)
Citrus extract	Cooked turkey meat	Contini et al. (2014)
Carvacrol, eugenol, and thymol	Fresh beef meat	Tornuk et al. (2015)
Nisin, chitosan, potassium sorbate, and zeolite	Fresh chicken meat	Soysal et al. (2015)

The main constituents of oregano EO are carvacrol, thymol, p-cymene, and linalool (Karabagias et al., 2011). Oussalah et al. (2004) evaluated the ability of milk protein-based edible film containing 1% EOs of oregano, pimento or both applied on beef muscle slices, and showed that oregano-based films stabilized lipid oxidation in beef muscle samples, whereas pimento-based films presented the highest antioxidant activity. Camo et al. (2011) reported that a concentration of at least 1% oregano extract was needed for obtaining a significant increase of beef meat display from 14 to 23 days, while a concentration of 4% in the package gave rise to an unacceptable oregano smell.

On the other hand, the use of active packaging system (packed with active film containing a 2% of an oregano EO and 1% of a green tea extract) showed a preservative effect in foal meat, stored under refrigeration, exerting a protective effect against microbial spoilage, lipid and protein oxidation, and consequently preserving color and sensorial properties (Lorenzo et al., 2014). Green tea is a good source of polyphenolic compounds like catechin, theaflavins, and thearubigins, which have the ability to scavenge reactive oxygen and nitrogen species (Siripatrawan & Harte, 2010). In line with this, the incorporation of green tea extract into chitosan film could enhance the antioxidant and antimicrobial properties of the film and thus maintained the qualities and prolonged the shelf life of the sausages (Siripatrawan & Noipha, 2012).

According to Park et al. (2012), the antioxidant films improved the efficacy of vacuum packaging and stabilized raw beef patties against oxidation. The antioxidant property of prepared film containing 3% eugenol was similar to that of the film containing 0.3% eugenol, which suggests that eugenol is a strong antioxidant and 0.3% concentration was sufficient to completely retard lipid oxidation in beef patties. Contini et al. (2014) also showed that antioxidant active packaging with citrus extract is effective in reducing the lipid oxidation of cooked turkey meat during storage and in maintaining its sensory characteristics, particularly tenderness and overall acceptability. In line with this, lowest lipid oxidation was observed for samples packed with chitosan-incorporated bags, which might be helpful to limit lipid oxidation of the drumsticks during the storage period (Soysal et al., 2015). Barbosa-Pereira et al. (2014) observed a reduction in lipid oxidation by up to 80% in beef wrapped in films coated with natural extracts obtained from a brewery residual waste, while Tornuk et al. (2015) suggested that the active nanocomposite packaging had potential to extend shelf life of fresh beef by reducing *E. coli* O157:H7 numbers and retarding the meat discoloration.

Finally, active packaging can be used in combination with other processing treatments to extend food shelf life. High pressure processing has shown to be an effective technology to improve safety of meat and meat products (Marcos et al., 2008; Realini et al., 2011). However, high pressure processing may accelerate meat oxidation (Ma et al., 2007; McArdle et al., 2011, 2013). In line with this, Bolumar et al. (2011) observed that the antioxidant compounds from the antioxidant active package (rosemary extract) suppress the oxidation in both the surface part and the inner part of chicken meat patties and prevent formation of secondary lipid oxidation products. So, the antioxidant active packaging was able to delay the oxidation induced by high pressure processing and consequently extend the shelf life of chicken patties.

9.6   CONCLUSION

The active antioxidant packaging is a promising emerging technology to avoid the undesirable effects of lipid oxidation during meat storage. This solution provides several advantages compared to direct addition of antioxidants, such as lower amounts of active substances required, localization of the antioxidant activity to the surface, progressive migration from film to the food matrix that provides long-term antioxidant effects, and elimination of additional steps in the manufacturing process for introduction of antioxidants in foods.

Positive and encouraging results were obtained in recent works. More studies, however, are needed to find new materials (both plastic polymers and edible coatings) with higher performances, and to select the most appropriate antioxidant agents for each concrete meat product in each group of storage conditions.

KEYWORDS

- active packaging
- natural antioxidant
- lipid oxidation
- modified atmosphere packaging

REFERENCES

Ahn, D. U.; Lee, E. J.; Mendonca, A. Meat Decontamination by Irradiation. In *Advanced Technologies for Meat Processing;* Nollet, L. M. L., Toldra, F., Eds.; CRC Press: London, 2006; pp 155–191.

Alfaia, C. M.; Alves, S. P.; Lopes, A. F.; Fernandes, M. J.; Costa, A. S.; Fontes, C. M.; Casto, M. L.; Bessa, R. J.; Prates, J. A. Effect of Cooking Methods on Fatty Acids, Conjugated Isomers of Linoleic Acid and Nutritional Quality of Beef Intramuscular Fat. *Meat Sci.* **2010**, *84,* 769–777.

Alrawaiq, N. S.; Abdullah, A. A Review of Flavonoid Quercetin: Metabolism, Bioactivity and Antioxidant Properties. *Int. J. Pharm. Tech. Res.* **2014**, *6,* 933–941.

Álvarez, I.; De La Fuente, J.; Cañeque, V.; Lauzurica, S.; Pérez, C.; Díaz, M. T. Changes in the Fatty Acid Composition of M. *Longissimus dorsi* of Lamb during Storage in a High-Oxygen Modified Atmosphere at Different Levels of Dietary Vitamin E Supplementation. *J. Agr. Food Chem.* **2009**, *57,* 140–146.

Amorati, R.; Foti, M. C.; Valgimigli, L. Antioxidant Activity of Essential Oils. *J. Agr. Food Chem.* **2013**, *61,* 10835–10847.

Ang, C. Y. W.; Lyon, B. G. Evaluations of Warmed-Over Flavor during Chilled Storage of Cooked Broiler Breast, Thigh and Skin by Chemical, Instrumental, and Sensory Methods. *J. Food Sci.* **1990**, *55,* 644–648.

Argyri, A. A.; Panagou, E. Z.; Nychas, G. J. E. Advances in Vacuum and Modified Atmosphere Packaging of Poultry Products. In *Advances in Meat, Poultry and Seafood Packaging;* Kerry, J. P., Ed.; Woodhead Publishing: Philadelphia, 2012; pp 205–247.

Asensio, C. M.; Nepote, V.; Grosso, N. R. Sensory Attribute Preservation in Extra Virgin Olive Oil with Addition of Oregano Essential Oil as Natural Antioxidant. *J. Food Sci.* **2012,** *77,* S294–S301.

Asghar, A.; Gray, J. I.; Buckley, D. J.; Pearson, A. M.; Booren, A. M. Perspectives on Warmed-Over Flavour. *Food Technol.* **1988,** *42,* 102–108.

Atkinson, J.; Epand, R. F.; Epand, R. M. Tocopherols and Tocotrienols in Membranes: A Critical Review. *Free Radical Bio. Med.* **2008,** *44,* 739–764.

Barbosa-Pereira, L.; Aurrekoetxea, G. P.; Angulo, I.; Paseiro-Losada, P.; Cruz, J. M. Development of New Active Packaging Films Coated with Natural Phenolic Compounds to Improve the Oxidative Stability of Beef. *Meat Sci.* **2014,** *97,* 249–254.

Batifoulier, F.; Mercier, Y.; Gatellier, P.; Renerre, M. Influence of Vitamin E on Lipid and Protein Oxidation Induced by H_2O_2 -Activated MetMb in Microsomal Membranes from Turkey Muscle. *Meat Sci.* **2002,** *61,* 389–395.

Belitz, H. D.; Grosch, W.; Schieberle, P. Coffee, Tea, Cocoa. In *Food Chemistry;* Belitz, H. D., Grosch, W., Schieberle, P., Eds.; Springer: Leipzig, Saxony, 2009; pp 938–951.

Banerjee, S. K.; Mukherjee, P. K.; Maulik, S. K. Garlic as an Antioxidant: The Good, the Bad and the Ugly. *Phytother. Res.* **2003,** *17,* 97–106.

Blakistone, B. A. Meats and Poultry. In *Principles and Applications of Modified Atmosphere Packaging of Foods;* Blakistone B. A., Ed.; Aspen Publishers: Gaithersburg, ML, 1999; pp 240–290.

Bolumar, T.; Andersen, M. L.; Orlien, V. Antioxidant Active Packaging for Chicken Meat Processed by High Pressure Treatment. *Food Chem.* **2011,** *129,* 1406–1412.

Brody, A. L.; Bugusu, B.; Han, J. H.,; Sand, C. K.; McHugh, T. H. Innovative Food Packaging Solutions. *J. Food Sci.* **2008,** *73,* R107–R116.

Broncano, J. M.; Petrón, M. J.; Parra, V.; Timón, M. L. Effect of Different Cooking Methods on Lipid Oxidation and Formation of Free Cholesterol Oxidation Products (COPs) in *Latissimus dorsi* Muscle of Iberian Pigs. *Meat Sci.* **2009,** *83,* 431–437.

Buckley, D. J.; Morrissey, P. A.; Gray, J. I. Influence of Dietary Vitamin E on the Oxidative Stability and Quality of Pig Meat. *J. Anim. Sci.* **1995,** *73,* 3122–3130.

Buettner, G. R. The Pecking Order of Free Radicals and Antioxidants: Lipid Peroxidation, α-Tocopherol, and Ascorbate. *Arch. Biochem. Biophys.* **1993,** *300,* 535–543.

Byrne, D. V.; Bredie, W. L. P.; Mottram, D. S.; Martens, M. Sensory and Chemical Investigations on the Effect of Oven Cooking on Warmed-Over Flavour Development in Chicken Meat. *Meat Sci.* **2002,** *61,* 127–139.

Camo, J.; Beltrán, J. A.; Roncalés, P. Extension of the Display Life of Lamb with an Antioxidant Active Packaging. *Meat Sci.* **2008,** *80,* 1086–1091.

Camo, J.; Lorés, A.; Djenane, D.; Beltrán, J. A.; Roncalés, P. Display Life of Beef Packaged with an Antioxidant Active Film as a Function of the Concentration of Oregano Extract. *Meat Sci.* **2011,** *88,* 174–178.

Carpenter, C. E.; Cornforth, D. P.; Whittier, D. Consumer Preferences for Beef Color and Packaging did not Affect Eating Satisfaction. *Meat Sci.* **2001,** *57,* 359–363.

Cava, R.; Estévez, M.; Ruiz, J.; Morcuende, D. Physicochemical Characteristics of Three Muscles from Free-Range Reared Iberian Pigs Slaughtered at 90 kg Live Weight. *Meat Sci.* **2003,** *63,* 533–541.

Cayuela, J. M.; Gil, M. D.; Bañón, S.; Garrido, M. D. Effect of Vacuum and Modified Atmosphere Packaging on the Quality of Pork Loin. *Eur. Food Res. Technol.* **2004,** *219,* 316–320.

Cheah, P. B.; Ledward, D. A. High Pressure Effects on Lipid Oxidation in Minced Pork. *Meat Sci.* **1996,** *43,* 123–134.

Cheah, P. B.; Ledward, D. A. Catalytic Mechanism of Lipid Oxidation Following High Pressure Treatment in Pork Fat and Meat. *J. Food Sci.* **1997,** *62,* 1135–1138.

Cheftel, J. C.; Culioli, J. Effects of High Pressure on Meat: A Review. *Meat Sci.* **1997,** *46,* 211–236.

Chu, Y. H.; Hwang, L. S. Food Lipids. In *Chemical and Functional Properties of Food Components;* Sikorski Z. E., Ed.; CRC Press: London, 2002; pp 115–132.

Church, N. Developments in Modified-Atmosphere Packaging and Related Technologies. *Trends Food Sci. Technol.* **1994,** *5,* 345–352.

Clark, D. S.; Lentz, C. P.; Roth, L. A. Use of Carbon Monoxide for Extending Shelf-Life of Prepackaged Fresh Beef. *Can. Inst. Food Sci. Technol. J.* **1976,** *9,* 114–117.

Coma, V.; Kerry, J. P. Antimicrobial and Antioxidant Active Packaging for Meat and Poultry. In *Advances in Meat, Poultry and Seafood Packaging;* Kerry, J. P. Ed.; Woodhead Publishing: Oxford, 2012; pp 477–503.

Contini, C.; Álvarez, R.; O'Sullivan, M.; Dowling, D. P.; Gargan, S. Ó.; Monahan, F. J. Effect of an Active Packaging with Citrus Extract on Lipid Oxidation and Sensory Quality of Cooked Turkey Meat. *Meat Sci.* **2014,** *96,* 1171–1176.

Cross, H. R.; Leu, R.; Miller, M. F. Scope of Warmed-Over Flavour and its Importance to the Meat Industry. In *Warmed-over Flavour of Meat;* Angelo, A. J. S. T., Bailey M. E., Eds.; Academic Press: New York, 1987; pp 1–18.

Dávalos, A.; Miguel, M.; Bartolomé, B.; López-Fandiño, R. Antioxidant Activity of Peptides Derived from Egg White Proteins by Enzymatic Hydrolysis. *J. Food Protect.* **2004,** 1939–1944.

Day, B. P. F. Active Packaging – a Fresh Approach. *J. Brand Technol.* **2001,** *1,* 32–41.

Decker, E. A.; Welch, B. Role of Ferritin as a Lipid Oxidation Catalyst in Muscle Foods. *J. Agr. Food Chem.* **1990,** *38,* 674–677.

Decker, E. A.; Antioxidant Mechanisms. In *Food Lipids. Chemistry, Nutrition, and Biotechnology;* Akoh, C. C., Min D. B., Eds.; Marcel Dekker: New York, 1998; pp 397–421.

Decker, E. A.; Livisay, S. A.; Zhou, S. A Re-Evaluation of the Antioxidant Activity of Purified Carnosine. *Biochemistry(Moscow).* **2000,** *65,* 766–770.

Di Bernardini, R.; Harnedy, P.; Bolton, D.; Kerry, J.; O'Neill, E.; Mullen, A.M.; Hayes, C. Antioxidant and Antimicrobial Peptidic Hydrolysates from Muscle Protein Sources and by-Products. *Food Chem.* **2011,** *124,* 1296–1307.

Dobrucka, R.; Cierpiszewski, R. Active and Intelligent Packaging Food–Research and Development–A Review. *Pol. J. Food Nutr. Sci.* **2014,** *64,* 7–15.

Domínguez, R.; Gómez, M.; Fonseca, S.; Lorenzo, J. M. Effect of Different Cooking Methods on Lipid Oxidation and Formation of Volatile Compounds in Foal Meat. *Meat Sci.* **2014a,** *97,* 223–230.

Domínguez, R.; Gómez, M.; Fonseca, S.; Lorenzo, J. M. Influence of Thermal Treatment on Formation of Volatile Compounds, Cooking Loss and Lipid Oxidation in Foal Meat. *LWT-Food Sci. Technol.* **2014b,** *58,* 439–445.

Domínguez, R.; Martínez, S.; Gómez, M.; Carballo, J.; Franco, I. Fatty Acids, Retinol and Cholesterol Composition in Various Fatty Tissues of Celta Pig Breed: Effect of the Use of Chestnuts in the Finishing Diet. *J. Food Compos. Anal.* **2015,** *37,* 104–111.

Eilert, S. J. New Packaging Technologies for the 21st Century. *Meat Sci.* **2005,** *71,* 122–127.

Erickson, M.C. Lipid Oxidation of Muscle Foods. In *Food Lipids. Chemistry, Nutrition, and Biotechnology;* Akoh, C. C.; Min D. B. Eds.; Marcel Dekker: New York, 2022; pp 321-364.

Erickson, M. C.; Hultin, H. O. Influence of Histidine on Lipid Peroxidation in Sarcoplasmic Reticulum. *Arch. Biochem. Biophys.* **1992,** *292,* 427–432.

Farber, J. M. Microbiological Aspects of Modified-Atmosphere Packaging Technology – A Review. *J. Food Protect.* **1991,** *54,* 58–70.

Fasseas, M. K.; Mountzouris, K. C.; Tarantilis, P. A.; Polissiou, M.; Zervas, G. Antioxidant Activity in Meat Treated with Oregano a Sage Essential Oils. *Food Chem.* **2008,** *106,* 1188–1194.

Frankel, E. N. Hydroperoxide Decomposition. In *Lipid Oxidation;* Frankel, E. N., Ed.; The Oily Press: Scotland, 1998; pp 55–77.

Gallegos Tintoré, S.; Chel Guerrero, L.; Corzo Rios, L. J.; Matínez Ayala, A. L. Péptidos con Actividad Antioxidante de Proteínas Vegetales. En *Bioactividad de Péptidos Derivados de Proteínas Alimentarias;* Segura Campos, M., Chel Guerrero, L., Betancur Ancona, D., Eds.; OmniaScience: Barcelona, Spain, 2013; pp 111–122.

Gallegos-Tintoré, S.; Torres-Fuentes, C.; Martínez-Ayala, A. L.; Solorza-Feria, J.; Alaiz, M.; Girón-Calle, J.; Vioque, J. Antioxidant and Chelating Activity of *Jatropha curcas* L. Protein Hydrolysates. *J. Sci. Food Agr.* **2011,** *91,* 1618–1624.

Gee, D. L.; Brown, W. D. The Effect of Carbon Monoxide on Bacterial Growth. *Meat Sci.* **1980,** *5,* 215–222.

Gill, C. O.; Tan, K. H. Effect of Carbon Dioxide on Growth of Meat Spoilage Bacteria. *Appl. Environ. Microbiol.* **1980,** *39,* 317–319.

Gómez-Estaca, J.; López-de-Dicastillo, C.; Hernández-Muñoz, P.; Catalá, R.; Gavara, R. Advances in Antioxidant Active Food Packaging. *Trends Food Sci. Tech.* **2014,** *35,* 42–51.

Gómez-Guillén, M. C.; Ihl, M.; Bifani, V.; Silva, A.; Montero, P. Edible Films Made from Tuna-Fish Gelatin with Antioxidant Extracts of Two Different Murta Ecotypes Leaves (*Ugni molinae* Turcz). *Food Hydrocoll.* **2007,** *21,* 1133–1143.

Gong, Y.; Parker, R. S.; Richards, M. P. Factors Affecting Lipid Oxidation in Breast and Thigh Muscle from Chicken, Turkey and Duck. *J. Food Biochem.* **2010,** *34,* 869–885.

Goth, L. Heat and pH Dependence of Catalase. A Comparative Study. *Acta Biol. Hung.* **1987,** *38,* 279–285.

Goulas, A. E.; Kontominas, M. G. Combined Effect of Light Salting, Modified Atmosphere Packagingand Oregano Essential Oil on the Shelf-Life of Sea Bream (*Sparus aurata*): Biochemical and Sensory Attributes. *Food Chem.* **2007,** *100,* 287–296.

Gray, J. L.; Pearson, A. M. Lipid-Derived off-Flavours in Meat Formation and Inhibition. In *Flavour of Meat and Meat Products;* Shahidi, F., Ed.; Blackie Academic: London, 1994; pp 116–143.

Gutteridge, J. M. C.; Halliwell, B. The Measurement and Mechanism of Lipid Peroxidation in Biological Systems. *Trends Biochem. Sci.* **1990,** *15,* 129–135.

Han, J. H. A Review of Food Packaging Technologies and Innovations. In *Innovations in Food Packaging;* Han, J. H., Ed.; Academic Press: San Diego, 2014; pp 3–12.

Harwood, M.; Danielewska-Nikiel, B.; Borzelleca, J. F.; Flamm, G. W.; Williams, G. M.; Lines, T. C. A Critical Review of the Data Related to the Safety of Quercetin and Lack of Evidence of *in vivo* Toxicity, Including Lack of Genotoxic/Carcinogenic Properties. *Food Chem. Toxicol.* **2007,** *45,* 2179–2205.

Hendrickx, M.; Ludykhuyze, L.; Van Den Broeck, I.; Weemaes, C. Effects of High Pressure on Enzymes Related to Food Quality. *Trends Food Sci. Tech.* **1998,** *9,* 197–203.

Hogan, S. A.; Kerry, J. P. Smart Packaging of Meat and Poultry Products. In *Smart Packaging Technologies for Fast Moving Consumer Goods;* Kerry J. P., Butler P., Eds.; John Wiley & Sons: West Sussex, UK, 2008; pp 33–54.

Iqbal, S.; Bhanger, M. I. Stabilization of Sunflower Oil by Garlic Extract during Accelerated Storage. *Food Chem.* **2007,** *100,* 246–254.

Jackson, T. C.; Acuff, G. R.; Vanderzant, C.; Sharp, T. R.; Savell, J. W. Identification and Evaluation of Volatile Compounds of Vacuum and Modified Atmosphere Packaged Beef Strip Loins. *Meat Sci.* **1992**, *31*, 175–190.

Jakobsen, M.; Jespersen, L.; Juncher, D.; Becker, E. M.; Risbo, J. Oxygenand Light-Barrier Properties of Thermoformed Packaging Materials used for Modified Atmosphere Packaging. Evaluation of Performance under Realistic Storage Conditions. *Packag. Technol. Sci.* **2005**, *18*, 265–272.

Jayasingh, P.; Cornforth, D. P.; Brennand, C. P.; Carpenter, C. E.; Whittier, D. R. Sensory Evaluation of Ground Beef Stored in High - Oxygen Modified Atmosphere Packaging. *J. Food Sci.* **2002**, *67*, 3493–3496.

John, L.; Cornforth, D.; Carpenter, C. E.; Sorheim, O.; Pettee, B. C.; Whittier, D. R. Color and Thiobarbituric Acid Values of Cooked Top Sirloin Steaks Packaged in Modified Atmospheres of 80% Oxygen, or 0.4% Carbon Monoxide, or Vacuum. *Meat Sci.* **2005**, *69*, 441–449.

Jurewicz, I.; Salmonowicz, J. Pro-and Antioxidant Effects of Some Amino Acids upon Fish Oil. *Zeszyty. Probl. Postepow. Nauk. Roln.* 1973.

Kanatt, S. W.; Chander, R.; Sharma, A. Effect of Radiation Processing on the Quality of Chilled Meat Products, *Meat Sci.* **2005**, *69*, 269–275.

Kanner, J.; Mendel, H. Prooxidant and Antioxidant Effect of Ascorbic Acid and Metal Salts in Beta Carotenelinoleate Model System. *J. Food Sci.* **1977**, *42*, 60–64.

Kanner, J.; Harel, S. Desferrioxamine as an Electron Donor. Inhibition of Membranal Lipid Peroxidation Initiated by H_2O_2-Activated Metmyoglobin and other Peroxidizing Systems. *Free Rad. Res. Comms.* **1987**, *3*, 1–5.

Kanner, J. Oxidative Processes in Meat and Meat Products: Quality Implications. *Meat Sci.* **1994**, *36*, 169–189.

Karabagias, I.; Badeka, A.; Kontominas, M. G. Shelf Life Extension of Lamb Meat using Thyme or Oregano Essential Oils and Modified Atmosphere Packaging. *Meat Sci.* **2011**, *88*, 109–116.

Katusin-Razem, B.; Mihaljevic, K. W.; Razem, D. Time-Dependent Post Irradiation Oxidative Chemical Changes in Dehydrated Egg Products. *J. Agr. Food Chem.* **1992**, *40*, 1948–1952.

Kerry, J. P.; O'grady, M. N.; Hogan, S. A. Past Current and Potential Utilization of Active and Intelligent Packaging Systems for Meat and Muscle-Based Products: A Review. *Meat Sci.* **2006**, *74*, 113–130.

Kim, I. S.; Jin, S. K.; Yang, M. R.; Chu, G. M.; Park, J. H.; Rashid, R. H. I.; Kim, J. Y.; Kang, S. N. Efficacy of Tomato Powder as Antioxidant in Cooked Pork Patties. *Asian-Australas. J. Anim. Sci.* **2013**, *26*, 1339–1346.

Kim, S. K.; Wijesekara, I. Development and Biological Activities of Marine-Derived Bioactive Peptides: A Review. *J. Funct. Foods.* **2010**, *2*, 1–9.

Kroger-Ohlsen, M, LH Skibsted. Kinetics and Mechanism of Reduction of Ferrylmyoglobin by Ascorbate and D-Isoascorbate. *J. Agr. Food Chem.* **1997**, *45*, 668–676.

Krotcha, J. M. Introduction to Frozen Food Packaging. In *Handbook of Frozen Food Processing and Packaging;* Sun D. W., Ed.; CRC Press: London, 2006; pp 615–640.

Kulisic, T.; Radonic, A.; Milos, M. Inhibition of Lard Oxidation by Fractions of Different Essential Oils. *Grasas Aceites.* **2005**, *56*, 284–291.

Ladikos, D.; Lougovois, V. Lipid Oxidation in Muscle Foods: A Review. *Food Chem.* **1990**, *35*, 295–314.

Lee, B. J.; Hendricks, D. G.; Cornforth, D. P. Effect of Sodium Phytate, Sodium Pyrophosphate and Sodium Tropolyphosphate on Physico-Chemical Characteristics of Restructured Beef. *Meat Sci.* **1998**, *50*, 273–283.

Lee, D. S. Antioxidant Packaging System. In *Innovations in Food Packaging;* Han, J. H., Ed.; Academic Press: San Diego, 2014; pp 111–131.

Lopez-Rubio, A.; Almenar, E.; Hernandez-Muñoz, P.; Lagarón, J. M.; Catalá, R.; Gavara, R. Overview of Active Polymer-Based Packaging Technologies for Food Applications. *Food Rev. Int.* **2004,** *20,* 357–387.

Lorenzo, J. M.; Batlle, R.; Gómez, M. Extension of the Shelf-Life of Foal Meat with Two Antioxidant Active Packaging Systems. *LWT-Food Sci. Technol.* **2014,** *59,* 181–188.

Luño, M.; Beltrán, J. A.; Roncalés, P. Shelf-Life Extension and Colour Stabilization of Beef Packaged in a Low O_2 Atmosphere Containing CO: Loin Steaks and Ground Meat. *Meat Sci.* **1998,** *48,* 75–84.

Lynch, P. B.; Kerry, J. P. *Utilizing* Diet to Incorporate Bioactive Compounds and Improve the Nutritional Quality of Muscle Foods. In *Antioxidants in Muscle Foods;* Decker, E., Faustman, C., Lopez-Bote C. J., Eds.; Wiley & Sons: New York, 2000; pp 455–480.

Ma, H. J.; Ledward, D. A.; Zamri, A. I.; Frazier, R. A.; Zhou, G. H. Effects of High Pressure/ Thermal Treatment on Lipid Oxidation in Beef and Chicken Muscle. *Food Chem.* **2007,** *104,* 1575–1579.

Mancini, R. A.; Hunt, M. C. Current Research in Meat Color. *Meat Sci.* **2005,** *71,* 100–121.

Marcos, B.; Aymerich, T.; Monfort, J. M.; Garriga, M. High-Pressure Processing and Anti-microbial Biodegradable Packaging to Control *Listeria monocytogenes* during Storage of Cooked Ham. *Food Microbiol.* **2008,** *25,* 177–182.

Mastromatteo, M.; Conte, A.; Del Nobile, M. A. Advances in Controlled Release Devices for Food Packaging Applications. *Trends Food Sci. Tech.* **2010,** *21,* 591–598.

McArdle, R. A.; Marcos, B.; Kerry, J. P.; Mullen, A. M. Influence of HPP Conditions On Selected Beef Quality Attributes and their Stability during Chilled Storage. *Meat Sci.* **2011,** *87,* 274–281.

McArdle, R. A.; Marcos, B.; Mullen, A. M.; Kerry, J. P. Influence of HPP Conditions On Selected Lamb Quality Attributes and their Stability during Chilled Storage. *Innov. Food Sci. Emerg. Technol.* **2013,** *19,* 66–72.

McMillin, K. W. Where is MAP going? A Review and Future Potential of Modified Atmo-sphere Packaging for Meat. *Meat Sci.* **2008,** *80,* 43–65.

Mecking, S. Nature or Petrochemistry? – Biologically Degradable Materials. *Angewandte Chemie. Int. Ed.* **2004,** *43,* 1078–1085.

Min, B.; Ahn, D. U. Mechanism of Lipid Peroxidation in Meat and Meat Products-A Review. *Food Sci. Biotechnol.* **2005,** *14,* 152–163.

Min, B.; Nam, K. C.; Cordray, J.; Ahn, D. U. Endogenous Factors Affecting Oxidative Stability of Beef Loin, Pork Loin, and Chicken Breast and Thigh Meats. *J. Food Sci.* **2008,** *73,* C439–C446.

Min, B.; Ahn, D. U. Factors in Various Fractions of Meat Homogenates that Affect the Oxida-tive Stability of Raw Chicken Breast and Beef Loin. *J. Food Sci.* **2009,** *74,* C41–C48.

Min, B.; Ahn, D. U. Sensory Properties of Packaged Fresh and Processed Poultry Meat. In *Advances in Meat, Poultry and Seafood Packaging;* Kerry, J. P. Ed.; Woodhead Publishing: Oxford, 2012; pp 112–153.

Morehouse, K. M. Food Irradiation: US Regulatory Considerations. *Radiat. Phys. Chem.* **2002,** *63,* 281–284.

Morrissey, P. A.; Brandon, S.; Buckley, D. J.; Sheehy, P. J. A.; Frigg, J. Tissue Content of a-Tocopherol and Oxidative Stability of Broilers Receiving Dietary-Tocopheryl Acetate Supplementation for Various Periods Pre-Slaughter. *Brit. Poultry Sci.* **1997,** *38,* 84–88.

Morrissey, P. A.; Sheehy, P. J. A.; Galvin, K. J.; Kerry, P.; Buckley, D. J. Lipid Stability in Meat and Meat Products. *Meat Sci.* **1998,** *49,* S73–S86.

Mottram, D. S. Meat. In *Volatile Compounds in Foods and Beverages;* Maarse H. Ed.; M. Decker: New York, 1991; pp 107–177.

Mueller, L.; Boehm, V. Antioxidant Activity of β-Carotene Compounds in Different *in vitro* Assays. *Molecules.* **2011,** *16,* 1055–1069.

Nabrzyski, M. *Mineral Components.* In *Chemical and Functional Properties of Food Components;* Sikorski, Z. E. Ed.; CRC Press: London, 2002; pp 51–79.

Nam, K. C.; Ahn, D. U. Combination of Aerobic and Vacuum Packaging to Control Lipid Oxidation and Off-Odor Volatiles of Irradiated Raw Turkey Breast. *Meat Sci.* **2003a,** *63,* 389–395.

Nam, K. C.; Ahn, D. U. Use of Double Packaging and Antioxidant Combinations to Improve Color, Lipid Oxidation and Volatiles of Irradiated Raw and Cooked Turkey Breast Patties. *Poultry Sci.* **2003b,** *82,* 850–857.

Nerín, C.; Tovar, L.; Djenane, D.; Camo, J.; Salafranca, J.; Beltrán, J. A.; Roncalés, P. Stabilization of Beef Meat by a New Active Packaging Containing Natural Antioxidants. *J. Agr. Food Chem.* **2006,** *54,* 7840–7846.

O'Grady, M. N.; Kerry, J. P. Smart Packaging Technologies and their Application in Conventional Meat Packaging Systems. In *Meat Biotechnology;* Toldrá F. Ed.; Springer: New York, 2008; pp 425–451.

O'Grady, M. N.; Monahan, F. J.; Burke, R. M.; Allen, P. The Effect of Oxygen Level and Exogenous A-Tocopherol on the Oxidative Stability of Minced Beef in Modified Atmosphere Packs. *Meat Sci.* **2000,** *55,* 39–45.

O'Sullivan, M. G.; Kerry, J. P. Sensory and Quality Properties of Packaged Fresh and Processed Meats. In *Advances in Meat, Poultry and Seafood Packaging;* Kerry, J. P., Ed.; Wiley-Blackwell: Oxford, 2012; pp 86–111.

Orlien, V.; Hansen, E. Lipid Oxidation in High Pressure Processed Chicken Breast Muscle during Chill Storage: Critical Working Pressure in Relation to Oxidation Mechanism. *Eur. Food Res. Technol.* **2000,** *211,* 99–104.

Oussalah, M.; Caillet, S.; Salmieri, S.; Saucier, L.; Lacroix, M. Antimicrobial and Antioxidant Effects of Milk Protein-Based Film Containing Essential Oils for the Preservation of Whole Beef Muscle. *J. Agr. Food Chem.* **2004,** *52,* 5598–5605.

Paiva, S. A; Russell, R. M. Beta-Carotene and Other Carotenoids as Antioxidants. *J. Am. Coll. Nutr.* **1999,** *18,* 426–433.

Park, H. Y.; Kim, S. J.; Kim, K. M.; You, Y. S.; Kim, S. Y.; Han, J. Development of Antioxidant Packaging Material by Applying Corn-Zein to LLDPE Film in Combination with Phenolic Compounds. *J. Food Sci.* **2012,** *77,* E273–E279.

Peelman, N.; Ragaert, P.; De Meulenaer, B.; Adons, D.; Peeters, R.; Cardon, L.; Van Impe, F.; Devlieghere, F. Application of Bioplastics for Food Packaging. *Trends Food Sci. Tech.* **2013,** *32,* 128–141.

Pereira de Abreu, D. A.; Cruz, J. M., Paseiro Losada, P. Active and Intelligent Packaging for the Food Industry. *Food Rev. Int.* **2012,** *28,* 146–187.

Pietta, P. G. Flavonoids as Antioxidants. *J. Nat. Prod.* **2000,** *63,* 1035–1042.

Pihlanto, A. Antioxidative Peptides Derived from Milk Proteins. *Int. Dairy J.* **2006,** *16,* 1306–1314.

Piringer, O. Transport Equations and their Solutions. In *Plastic Packaging Materials for Foods;* Piringer, O., Baner A. L., Eds.; Wiley-VCN: Weinhem, Germany, 2000; pp 183–219.

Ponce-Alquicira, E. Flavor of Frozen Foods. In *Handbook of Food Science, Technology, and Engineering;* Hui Y. H., Ed.; CRC Press: London, 2006; pp 60.1–60.7.

Pradhan, A. A.; Rhee, K. S.; Hernández, P. Stability of Catalase and its Potential Role in Lipid Oxidation in Meat. *Meat Sci.* **2000,** *54,* 385–390.

Realini, C. E.; Guardia, M. D.; Garriga, M.; Perez-Juan, M.; Arnau, J. High Pressure and Freezing Temperature Effect on Quality and Microbial Inactivation of Cured Pork Carpaccio. *Meat Sci.* **2011,** *88,* 542–547.

Reddy, M. M.; Misra, M.; Mohanty, A. K. Bio-Based Materials in the New Bio-Economy. *Chem. Eng. Prog.* **2012,** *108,* 37–42.

Reddy, M. M.; Vivekanandhan, S.; Misra, M.; Bhatia, S. K.; Mohanty, A. K. Biobased Plastics and Bionanocomposites: Current Status and Future Opportunities. *Prog. Polym. Sci.* **2013,** *38,* 1653–1689.

Renerre, M.; Labadie, J.In *Fresh Red Meat Packaging and Meat Quality,* Proceedings 39th International Congress of Meat Science and Technology, Calgary, Alberta, Canada, Aug 1–6, 1993; pp 361–387.

Richards, M. P. Lipid Chemistry and Biochemistry. In *Handbook of Food Science, Technology, and Engineering;* Hui Y. H., Ed.; CRC Press: London, **2006;** pp 8.1–8.21.

Rival, S. G.; Boeriu, C. G.; Wichers, H. J. Caseins and Casein Hydrolysates: 2. Antioxidative Properties and Relevance to Lipoxygenase Inhibition. *J. Agr. Food Chem.* **2001a,** *49,* 295–302.

Rival, S. G.; Fornaroli, S.; Boeriu, C. G.; Wichers, H. J. Caseins and Casein Hydrolysates: 1. Lipoxygenase Inhibitory Properties. *J. Agr. Food Chem.* **2001b,** *49,* 287–294.

Ruiter, A.; Voragen, A. G. J. Major Food. In *Chemical and Functional Properties of Food Components;* Sikorski, Z. E., Ed.; CRC Press: London, 2002; pp 273–89.

Saito, K.; Hao, J. D.; Ogawa, T.; Muramoto, K.; Hatakeyama, E.; Yasuhara, T.; Nokihara, K. Antioxidative Properties of Tripeptide Libraries Prepared by Combinatorial Chemistry. *J. Agr. Food Chem.* **2003,** *51,* 3668–3674.

Sanches-Silva, A.; Costa, D.; Albuquerque, T. G.; Buonocore, G. G.; Ramos, F.; Castilho, M. C.; Costa, H. S. Trends in the use of Natural Antioxidants in Active Food Packaging: A Review. *Food Addit. Contam. Part. A.* **2014,** *31,* 374–395.

Samaranayaka, A. G. P.; Li-Chan, E. Food-Derived Peptidic Antioxidants: A Review of their Production, Assessment, and Potential Applications. *J. Funct. Foods* **2011,** *3,* 229–254.

Sarmadi, B. H.; Ismail, A. Antioxidative Peptides from Food Proteins: A Review. *Peptides.* **2010,** *31,* 1949–1956.

Sato, K.; Hegarty, G. R. Warmed-Over Flavor in Cooked Meats. *J. Food Sci.* **1971,** *36,* 1098–1102.

Sebranek, J. G.; Houser, T. A. Use of CO for Red Meats: Current Research and Recent Regulatory Approvals. In *Modified Atmosphere Processing and Packaging of Fish;* Otwell W. S., Kristinsson H. G., Balaban M. O., Eds.; Blackwell Publishing: Ames, Iowa, 2006; pp 87–101.

Shahidi, F. Lipid-Derived Flavors in Meat Products. In *Meat Processing: Improving Quality;* Kerry, J., Kerry, J., Ledward, D., Eds.; CRC Press: London, 2002; pp 105–121.

Shahidi, F.; Janitha, P. K.; Wanasundara, P. D. Phenolic Antioxidants. *Crit. Rew. Food Sci. Nutr.* **1992,** *32,* 67–103.

Siegel, D. G. Case Ready Concepts- Packaging Technologies. In *Western Science Research Update Conference on Technologies for Improving the Quality and Safety of Case- Ready Products,* Annual Meeting of National Meat Association; Las Vegas, Nevada, Feb 22, 2001; <http://www.meatscience.org/ meetings/WSC/2001/default.htm> (Accessed 26.02.08.).

Sies, H.; Stahl, W. Lycopene: Antioxidant and Biological Effects and its Bioavailability in the Human. *Proc. Soc. Exp. Biol. Med.* **1998**, *218*, 121–124.

Siripatrawan, U., Harte, B. R. Physical Properties and Antioxidant Activity of an Active Film from Chitosan Incorporated with Green Tea Extract. *Food Hydrocoll.* **2010**, *24*, 770–775.

Siripatrawan, U.; Noipha, S. Active Film from Chitosan Incorporating Green Tea Extract for Shelf Life Extension of Pork Sausages. *Food Hydrocoll.* **2012**, *27*, 102–108.

Skibsted, L. H.; Bertelsen, G.; Qvist, S.In *Quality Changes during Storage of Meat and Slightly Preserved Meat Products,* Proceedings 40th International Congress of Meat Science and Technology; The Hague, Netherlands, Aug 28–Sept 2, 1994; S-II.MP1, pp 1–10.

Smiddy, M.; Papkovskaia, N.; Papkovsky, D. B.; Kerry, J. P. Use of Oxygen Sensors for the Non-Destructive Measurement of the Oxygen Content in Modified Atmosphere and Vacuum Packs of Cooked Chicken Patties: Impact of Oxygen Content on Lipid Oxidation. *Food Res. Int.* **2002**, *35*, 577–584.

Soysal, Ç.; Bozkurt, H.; Dirican, E.; Güçlü, M.; Bozhüyük, E. D.; Uslu, A. E.; Kaya, S. Effect of Antimicrobial Packaging on Physicochemical and Microbial Quality of Chicken Drumsticks. *Food Control.* **2015**, *54*, 294–299.

Spanier, A. M. Current Approaches to the Study of Meat Flavor Quality. *Develop. Food Sci.* **1992**, *29*, 695–709.

Suetsuna, K.; Ukeda, H.; Ochi, H. Isolation and Characterization of Free Radical Scavenging Activities Peptides Derived from Casein. *J. Nutr. Biochem.* **2000**, *11*, 128–131.

Takama, K. Changes in the Flesh Lipids of Fish during Frozen Storage. V. Accelerative Substances of Lipid Oxidation in the Muscle of Rainbow Trout. *Bull. Faculty Fisher.* **1974**, *25*, 256–263.

Thippareddi, H.; Phebus, R. K. *Modified Atmosphere Packaging (MAP): Microbial Control and Quality;* Pork Information Gateway Factsheet, National Pork Board/American Meat Science Association Fact Sheet: US, 2007; 12–05–07, pp 1-5.

Tornuk, F.; Hancer, M.; Sagdic, O.; Yetim, H. LLDPE Based Food Packaging Incorporated with Nanoclays Grafted with Bioactive Compounds to Extend Shelf Life of Some Meat Products. *LWT- Food Sci. Technol.* **2015**, *64*, 540–546.

Tsai, C. W.; Chen, H. W.; Sheen, L. Y.; Lii, C. K. Garlic: Health Benefits and Actions. *Biomedicine.* **2012**, *2*, 17–29.

Vermeiren, L.; Devlieghere, F.; van Beest, M.; de Kruijf, N.; Debevere, J. Developments in the Active Packaging of Foods. *Trends Food Sci. Tech.* **1999**, *10*, 77–86.

Viana, E. S.; Gomide, L. A. M.; Vanetti, M. C. D. Effect of Modified Atmospheres on Microbiological, Color and Sensory Properties of Refrigerated Pork. *Meat Sci.* **2005**, *71*, 696–705.

Walsh, H. M.; Kerry, J. P. Meat Packaging. In *Meat Processing: Improving Quality;* Kerry, J., Kerry, J., Ledward, D., Eds.; CRC Press: London, 2002; pp 417–452.

Wang, W.; Gonzalez de Mejia, E. A New Frontier in Soy Bioactive Peptides that May Prevent Age-Related Chronic Diseases. *Compr. Rev. Food Sci. Food Safety.* **2005**, *4*, 63–76.

Wasowicz, E.; Gramza, A.; Hes, M.; Jelen, H. H.; Korczak, J.; Malecka, M. Oxidation of Lipids in Food. *Pol. J. Food. Nutr. Sci.* **2004**, *13*, 87–100.

Zhao, Y.; Wells, J. H.; McMillin, K. W. Applications of Dynamic Modified Atmosphere Packaging Systems for Fresh Red Meats: A Review. *J. Muscle Foods* **1994**, *5*, 299–328.

INDEX

Printed and bound by CPI Group (UK) Ltd, Croydon, CR0 4YY

23/10/2024

01777705-0007